Geology of the c
Newcastle upon
Gateshead and C

This memoir is the first comprehensive geological account of the area covered by Sheet 20 of the Geological Map of England and Wales, which includes part of the Northumberland and Durham Coalfield. Newcastle upon Tyne, Gateshead, Chester-le-Street and Consett are the main population centres. The account is based primarily on the Geological Survey's work in the region, but also brings together much additional information previously scattered through the geological literature. The book is intended to provide basic information for geologists, civil and mining engineers, land-use planners, environmentalists and those involved in education and research in earth science topics.

A description of the Upper Carboniferous rocks, more particularly of the Coal Measures, makes up the bulk of the book, and the Quaternary drift cover is also described in detail. Shorter chapters are devoted to the concealed Lower Carboniferous formations, the Lower Palaeozoic basement and its granitic pluton, and the tectonic structure of the district. Economic geology, which has been such a powerful social and industrial factor in the development of the region, is also considered.

Cover photograph

One of the more characteristic features of Newcastle upon Tyne is the spectacular set of road and rail bridges which link the city to Gateshead on the south bank of the River Tyne. There have been major crossing-points of the river in this general area since at least Roman times. The present course of the river, which is here somewhat incised into glacial deposits, dates from late or postglacial times. The relatively steep valley sides, and the need until recently to allow ships upstream access to transport coal, has dictated the high level of the bridges above the river. The bedrock hereabouts, Middle Coal Measures, is below OD.

Plate 1 Hown's Gill, Consett [095 491], a glacial meltwater channel cut through the Derwent–Browney interfluve. L1746.

BRITISH GEOLOGICAL SURVEY

D A C MILLS and
D W HOLLIDAY

Geology of the district around Newcastle upon Tyne, Gateshead and Consett

Memoir for 1:50 000 Geological Sheet 20
(England and Wales)

London: The Stationery Office 1998

First published 1998

The grid used on the figures is the National Grid taken from the Ordnance Survey maps. Figure 2 is based on material from Ordnance Survey 1:50 000 scale maps, numbers 87 and 88.

ISBN 0 11 884538 1

Bibliographical reference

MILLS, D A C, and HOLLIDAY, D W. 1998. Geology of the district around Newcastle upon Tyne, Gateshead and Consett. *Memoir of the British Geological Survey*, Sheet 20 (England and Wales).

Authors

D A C Mills, BSc, FGS, CGeol
formerly *British Geological Survey*

D W Holliday, MA, PhD, FGS, CGeol
British Geological Survey, Keyworth

Printed in the UK for The Stationery Office
J57636 C6 10/98

Other publications of the Survey dealing with this district and adjoining districts

BOOKS

British Regional Geology
Northern England, 4th Edition, 1971

Memoirs
Geology of the country around Bellingham (Sheet 13), 1980
Geology of the Tynemouth district (Sheet 15), 1974
Geology of the country around Sunderland (Sheet 21), 1994
Geology of the country bewtween Durham and West Hartlepool (Sheet 27), 1967
Geology of the Northern Pennine Orefield, Volume 1, Tyne to Stainmore (2nd Edition), 1990
The structure and evolution of the Northumberland–Solway Basin and adjacent areas, 1995

Mineral Assessment Reports
The sand and gravel resources of the country around Hexham (sheets NY86 and 96). Report 65, 1981
The sand and gravel resources of the country around Blaydon, Tyne and Wear (sheets NZ06 and 16). Report 74, 1981

Open-file reports
Edmundbyers and Hedley on the Hill. Geological notes and local details for sheets NZ 05 NW, NE, SW and SE, 1982
Carboniferous geology of Corbridge and Prudhoe. Geological notes and local details for sheets NZ 06 and eastern parts of NY 96 NE and SE, 1990
Chopwell, Rowlands Gill, Consett and Stanley. Geological notes and local details for sheets NZ 15 NW, NE, SW and SE, 1982
Kibblesworth, Birtley, Craghead and Chester-le-Street. Geological notes and local details for sheets NZ 25 NW, NE, SW and SE, 1983
Newcastle upon Tyne and Gateshead. Geological notes and local details for sheets NZ 26 NW, NE, SW and SE, 1983

MAPS

1:625 000
Solid geology (north sheet), 1979
Quaternary geology (north sheet), 1977
Bouguer gravity anomaly (north sheet), 1981
Aeromagnetic (north sheet), 1972

1:250 000 Solid Geology
Borders (55°N-04°W), 1986
Farne (55°N-02°W), 1988
Lake District (54°N-04°W), 1980
Tyne- Tees (54°N-02°W), 1981
Bouguer gravity and aeromagnetic anomaly maps are also available

1:50 00 or 1:63 360
Sheet 13 Bellingham (Solid), 1980
Sheet 13 Bellingham (Drift), 1980
Sheet 14 Morpeth (Solid), reprinted 1970
Sheet 14 Morpeth (Drift), reprinted 1977
Sheet 15 Tynemouth (Solid), reprinted 1975
Sheet 15 Tynemouth (Drift), 1968
Sheet 19 Hexham (Solid), reprinted 1975
Sheet 20 Newcastle upon Tyne (Solid), 1989
Sheet 20 Newcastle upon Tyne (Drift), 1992
Sheet 21 Sunderland (Solid and Drift), 1978
Sheet 25 Alston (Solid and Drift), reprinted 1973
Sheet 26 Wolsingham (Solid), 1977
Sheet 26 Wolsingham (Drift), 1977
Sheet 27 Durham and West Hartlepool (Solid), 1965
Sheet 27 Durham and West Hartlepool (Drift), 1965

CONTENTS

FIGURES

TABLES

PLATES

PREFACE

The geological mapping of an area is not a one-off exercise, for new information becomes available and new geological concepts evolve. Urban areas, especially, require periodic revision if the best possible maps are to be available in support of resource exploitation and environmental protection. The Newcastle upon Tyne district provides an example of this need.

The Primary Survey maps of the Newcastle upon Tyne district were completed long before the peak of coal mining in the early part of the 20th century, and they soon came to be regarded as inadequate for their purpose. Resurvey of the district began after the First World War, but, because attention was preferentially given to other parts of northern England, partly in the hope of finding new coal resources, the work proceeded slowly and was not completed until after the Second World War. This coincided with the nationalisation of the coal industry and a time of rapid expansion of exploration and exploitation for opencast coal. The staff of the Survey's Newcastle Office were heavily committed at that time to borehole examination on behalf of the National Coal Board. This led to a flood of new geological information, often in areas previously poorly known and with little or no former mining, so that significant areas of the resurvey became outdated.

New data continued to accumulate in the succeeding three decades, not only from the coal industry but also from the heavy capital input into the infrastructure of the region. This was a time of great advances in the understanding of the regional geology. The drilling of the Survey's borehole in 1963–64 at Throckley, just within the adjacent Morpeth district, significantly altered previous views of Namurian stratigraphy in the north-west of the district. Desk revision of the six-inch maps, and reconstitution of certain 'County' maps on to National Grid sheets during this period, made useful contributions to revising the solid geology, including the highlighting of several areas of opencast mine potential. However, the largely unrevised depiction of the Quaternary cover was not deemed adequate for modern purposes.

To meet the urgent need for modern geological maps and to aid planning in several parts of the district, a new survey of these areas began in 1968. Remapping, or revision with partial resurvey, was extended to the remaining parts of the district and was completed in 1983. The publication of the 1:50 000 maps in 1989 (Solid edition) and 1992 (Drift edition), and of this memoir, bring this latest phase of resurvey to a close, though as pointed out earlier there will undoubtedly be a need for further improvements in the future.

Although this memoir is much concerned with resources, and in particular with coal and its associated strata, it also contains a wealth of geological information most of which has not been published previously, or is only available in open file reports. These new data have considerable bearing on regional studies of Carboniferous rocks, and make a substantial additional contribution to the understanding of their stratigraphy. The study of the glacial and postglacial landforms and deposits allows, for the first time, an adequate appreciation of the Devensian and Holocene history of the district. This also has wider implications, not only for environmental geology but also for our understanding of past climatic change.

The city of Newcastle upon Tyne and its environs are virtually synonymous with coal, and feature prominently in any history of coal mining in Britain. Although this was a major centre of human activity long before mining became a major industry, coal extraction, and the parallel growth of heavy industry dependent on this energy source, have been crucial to the history and development of the city and its region. The completion of the 1:50 000 maps, and of this memoir, following closely on the closure of the last deep mine in the district, provide a timely epitaph to a once-great industry. Their publication will help to ensure that the great wealth of geological information gained from the mining of coal will continue to be of use in the future development of the district, and will contribute to its future prosperity.

David A Falvey, PhD
Director

British Geological Survey
Kingsley Dunham Centre
Keyworth
Nottingham
NG12 5GG

HISTORY OF SURVEY OF THE NEWCASTLE UPON TYNE SHEET

The original geological survey of the district was carried out by H H Howell and D Burns between 1865 and 1870, the maps being published on the scale of one inch to one mile as Geological Sheet 105 NW (Old Series England and Wales) in separate Solid and Drift editions in 1870 and 1884 respectively. Further editions of these maps, incorporating minor amendments, were issued in 1889 and 1892. Over broadly the same period, Six-Inch 'County' sheets for Northumberland and County Durham were published covering the whole district. No memoir was published, and virtually no notes survive from that period.

The district was resurveyed between 1926 and 1946 on the scale of six inches to one mile by W Anderson, G Burnett, R G Carruthers, J Maden and T Robertson. Much underground information was obtained for the first time during this period. Six-Inch New Meridian 'County' sheets were published for most, but not all, of the district in the late 1940s and 1950s. However, no one-inch maps or memoir based on this work were published.

Desk revision of the solid geology of certain 'County' maps, and their reconstitution on to National Grid sheets, was undertaken by W Anderson (1959–60) and P M Allen, D W Holliday, A J Reedman and G Richardson (1966–67). These maps were not published, but remain in manuscript.

The present 1:50 000 scale geological maps (sheet 20) derive from revision and resurvey work carried out between 1968 and 1983 by F C Cox, J R Davies, E A Edmonds, D V Frost, D W Holliday, D A C Mills, G Richardson and J G O Smart. The solid edition was published in 1989, and the drift edition in 1992.

Geological National Grid six-inch scale maps included wholly or partially in sheet 20 are listed below with the initials of the surveyors and the dates of survey. The surveyors were F C Cox, J R Davies, E A Edmonds, D V Frost, D W Holliday, D A C Mills, G Richardson and J G O Smart. Several of the maps include areas surveyed during the revision of adjacent districts. The additional surveying officers were E A Francis, D H Land and D B Smith, and also C R Thewlis of the (then) NCB Opencast Executive. All the maps are available from BGS as dyeline prints or litho-printed maps, with the exception of a small area on the western margin of the district. Manuscript copies of maps of this area are, however, available for consultation.

Sheet	Name	Surveyors	Dates
NY 94 NE	Edmundbyers Common	JRD, DACM	1973; 1980–1982
NY 95 NE	Reaston Burn and Slaley Hall	EAE	1981
NY 95 SE	Derwent Reservoir (west)	JRD, DACM	1980–1982
NY 96 NE	Stagshaw and Sandhoe	DWH, DHL	1968–1971
NY 96 SE	Corbridge and Riding Mill Burn	DWH	1977–1978
NZ 04 NW	Muggleswick Park	DACM	1973
NZ 04 NE	Castleside and Waskerley	DACM	1974
NZ 05 NW	Healey Hill and Minsteracres	EAE	1981
NZ 05 NE	Hedley on the Hill	EAE	1981
NZ 05 SW	Edmundbyers and Derwent Reservoir (east)	JRD	1979–1980
NZ 05 SE	Shotley Bridge	JRD, DACM	1977–1980
NZ 06 NW	Aydon Castle and Newton Hall	DWH	1976–1977
NZ 06 NE	Harlow Hill and Horsley	DWH	1976–1977
NZ 06 SW	Riding Mill	DWH, DHL	1968; 1976–1977
NZ 06 SE	Ovington and Prudhoe	DWH	1977
NZ 14 NW	Delves	GR	1970–1971
NZ 14 NE	Lanchester and Burnhope	GR, CRT	1969
NZ 15 NW	Chopwell and Ebchester	DACM	1978–1980
NZ 15 NE	Rowlands Gill and Burnopfield	DACM	1977–1979
NZ 15 SW	Consett and Leadgate	DACM	1976–1978
NZ 15 SE	Annfield Plain and Stanley	DACM	1976–1977
NZ 16 NW	Heddon-on-the-Wall	GR	1977–1980
NZ 16 NE	Throckley and Westerhope	DVF, GR	1977–1980
NZ 16 SW	West Wylam and Crawcrook	GR	1976–1980
NZ 16 SE	Ryton, Blaydon and Lemington	GR	1977–1981
NZ 24 NW	Edmondsley and Sacriston	GR	1970
NZ 24 NE	Great Lumley and Plawsworth	EAF	1961
		GR	1970
		DBS	1975
		FCC, DACM	1981–1983
NZ 25 NW	Sunniside and Kibblesworth	FCC	1981–1982
NZ 25 NE	Wrekenton and Birtley	FCC, DBS	1966–1967; 1981
NZ 25 SW	Beamish and Craghead	FCC	1981
NZ 25 SE	Urpeth and Chester-le-Street	FCC, DBS	1966–1975; 1979–1980
NZ 26 NW	Gosforth and Newcastle (NW)	DACM, GR, JGOS	1981–1982
NZ 26 NE	Longbenton and Newcastle (NE)	DHL, DBS, GR	1959–1982
NZ 26 SW	Newcastle (W and central), Whickham and Gateshead (W)	GR	1975
NZ 26 SE	Newcastle (SE) and Gateshead (central and E)	GR, DBS	1967–1974; 1980–1982

ACKNOWLEDGEMENTS

This memoir has been written largely by D A C Mills; D W Holliday wrote Chapter 2 and has contributed substantially to the others. Following the resurvey, reports describing the 'Geology and Local Details' were published for some, but not all, of the National Grid sheets included in the district (Edmonds and Davies, 1982; Mills, 1982; Cox, 1983; Richardson, 1983; Holliday and Pattison, 1990). The authors have made extensive use of these, and of palaeontological reports by J Pattison, M A Calver, B Owens and N J Riley. Data acquired during the Mineral Assessment Surveys by J R A Giles and J H Lovell have also been incorporated. The resurvey has benefitted greatly from access to information obtained or held by British Coal (and its precursors), local authorities, and many commercial organisations. Information in Appendix 3 is reproduced by kind permission of Northumbrian Water Ltd. and the Babtie Group. These contributions are gratefully acknowledged. The memoir has been edited by D W Holliday, J I Chisholm and A A Jackson.

NOTES

Throughout the memoir the word 'district' refers to the area covered by the 1:50 000 Newcastle upon Tyne (20) sheet.

National Grid references are given in square brackets; unless otherwise stated all lie within the 100 km square NZ.

Details of boreholes and shafts marked* are given in Appendix 2. Location of boreholes and shafts are shown in Figure 30.

ONE

Introduction

This memoir describes the geology of the area covered by the Newcastle upon Tyne sheet (20) of the 1:50 000 scale geological map of England and Wales. The location of this district, with reference to northern England, is shown in Figure 1, and the main geographical elements are indicated in Figure 2. The north-western and western parts of the district lie within Northumberland, and large areas in the south and south-east are within County Durham. An area in the centre and north-east formed part of the former Metropolitan County of Tyne and Wear, but, under more recent legislation, the administration of that area has become the responsibility of several smaller authorities.

The major historical, industrial and social factors which have fashioned the north-east of England, including the city of Newcastle and this district, are well documented (Middlebrook, 1950; Miller, 1968; Manders, 1973; Rogers, 1974; McCord and Rowe, 1977; McCord, 1979; Atkinson, 1979, 1989). Newcastle upon Tyne, with a population of about 280 000, is the main administrative and commercial centre for north-east England. The city has its origins in Roman times, in a fortification close to the (then) eastern end of Hadrian's Wall. Although the city was well established by the Middle Ages, it was not until the 18th and 19th centuries that development took place to any extent in the remainder of the district.

The greatest single natural resource of the district was coal. The expression 'taking coals to Newcastle' is still widely used to describe the gift of a commodity with which the recipient is already over-endowed. The earliest workings date back to the late 12th and early 13th centuries, when it was probably dug at outcrop along the banks of the River Tyne, and also at higher levels in Gateshead and Newcastle. Subsequent workings probably extended both up the Tyne Valley and progressively farther away from the river.

The zenith of the coal industry in this district was reached in the closing years of the last century and the first half of the present century. The depletion of reserves, combined with other economic and social factors, have seen the contraction of the industry at a steadily accelerating pace over the last 25 years. All deep mines in this district are now closed, and production is limited to a few opencast operations.

The town of Consett developed in another direction. The occurrence of ironstone bands and nodules in the Coal Measures of the area, together with abundant local coal and limestone, led to the establishment of the Consett Iron Company (Anon, 1858, 1954), iron and steel becoming a major factor in the local economy for over 140 years until the closure of the plant in 1979.

Other industries based on ground resources have played a significant, though subsidiary, role in the district. These are quarrying, mainly for sandstone but also some limestone; brick manufacture from fireclay, mudstone and clay; and aggregate production from sand and gravel. Recently, the availability of minerals within part of the district has been the subject of a study by the former Tyne and Wear County (Anon, 1984a, 1984b).

The advent of the modern economy has brought a wide range of new industries into the district, and though few of these have any relation to the local ground resources, support for the infrastructure is partly provided by continuing brick manufacture and aggregate production. The leisure and tourist industries have also shown significant growth in the last 20 years. A side effect

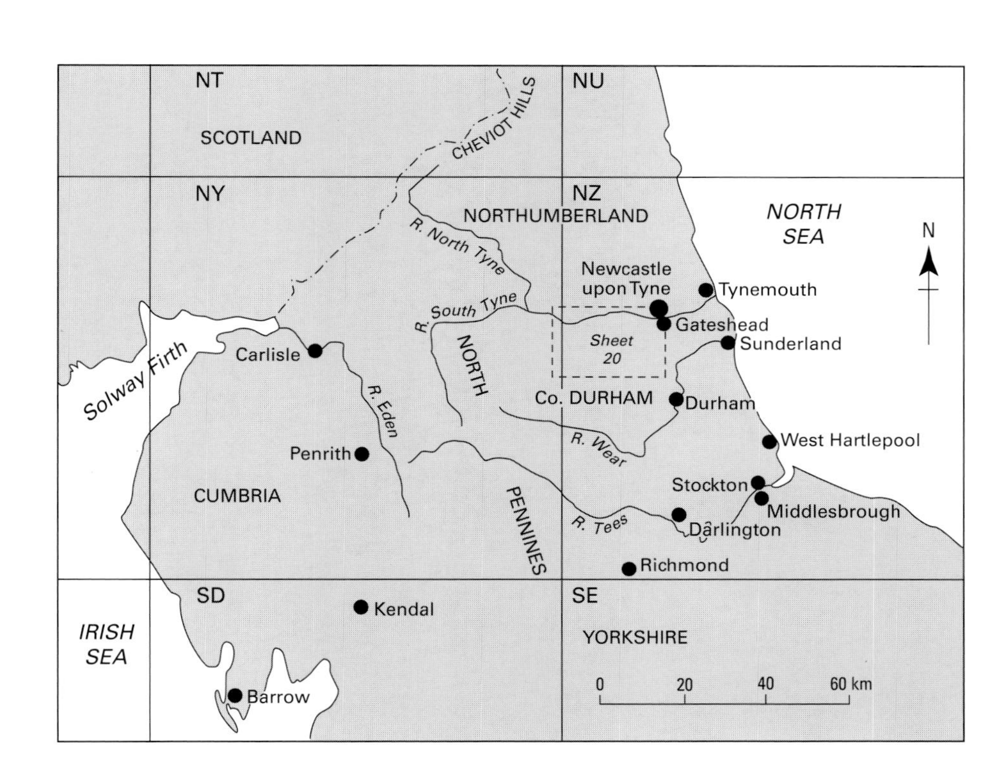

Figure 1 Sketch map showing the location of the district.

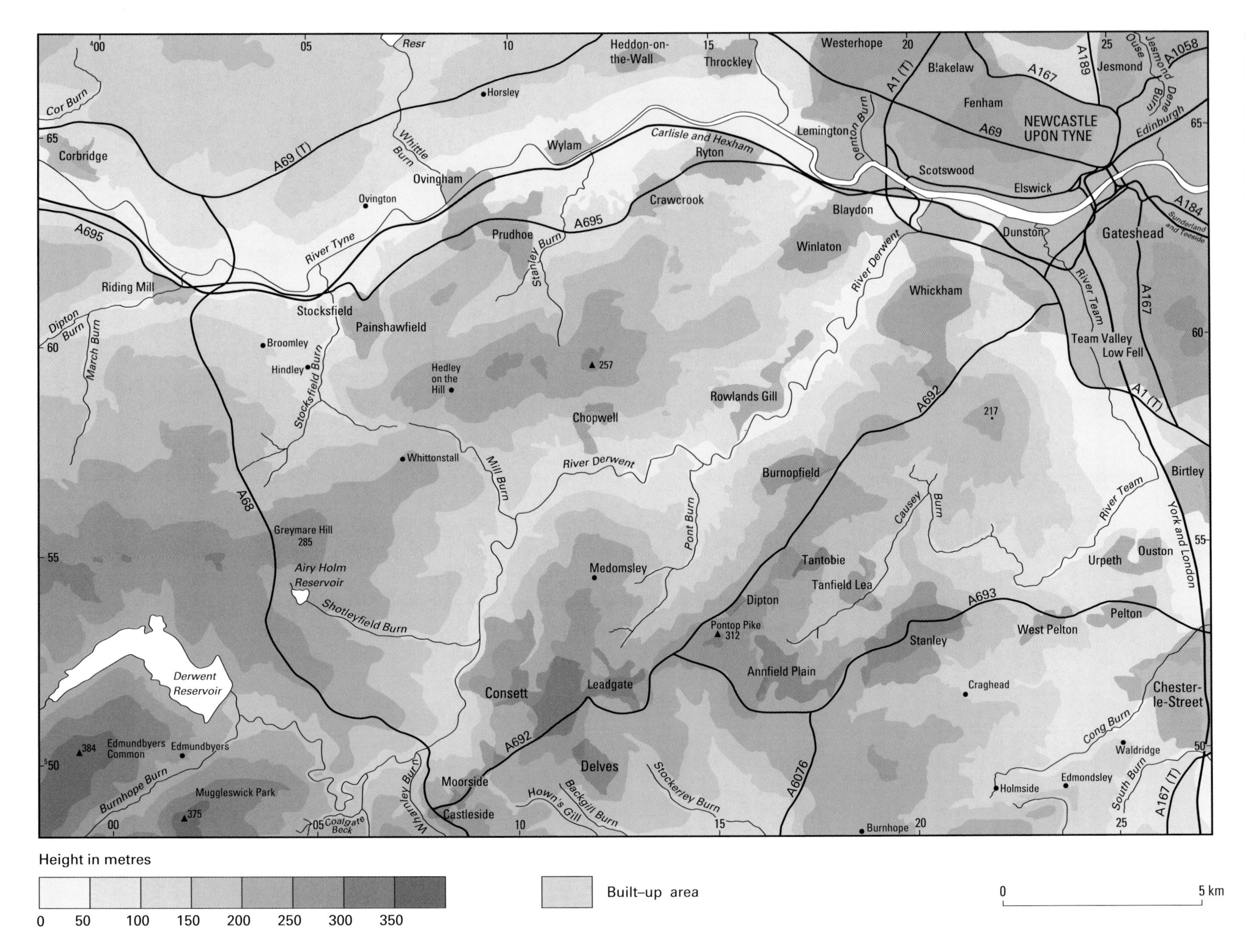

Figure 2 Main geographical elements of the district.

to the rundown of mining has been the extensive reclamation of industrial and mining landscapes, so that a visitor could be forgiven for not knowing that much of this district was until comparatively recently part of a major coalfield.

Most of the district lies within the drainage system of the River Tyne, but a small area in the south-east includes streams that drain into the River Wear. The valley of the Tyne dominates the northern part of the district (Figure 2), extending from Corbridge in the west to Newcastle and Gateshead in the east; below Wylam the river is tidal. Throughout most of its length, the river is characterized by broad alluvial and terrace flats locally over 1 km in width. However, below Dunston, to the west of Gateshead, the valley becomes narrower.

A major tributary of the Tyne, the River Derwent, runs in a broadly south-west to north-east direction and enters the Tyne below Blaydon. Across the district the Derwent has a relatively rapid fall of some 230 m, and locally it is quite sharply incised into the surrounding countryside, as, for example, some 4 km south-west of Consett, where the river flows through a deep gorge-like meander. North of Edmundbyers, the river has been dammed for water supply.

Near the eastern margin of the district a broad but pronounced valley extends north-north-westwards from the Wear Valley near Birtley and Chester-le-Street to join the Tyne Valley west of Gateshead. The bottom of the valley lies at less than 30 m above OD throughout its length, and from Birtley northwards carries the River Team, barely more than a large stream.

The area as a whole forms part of a plateau dissected by the Tyne, Derwent and Team river systems and falling very gently to the east. The highest ground (384 and 375 m above OD) is in the extreme south-west.

The north-eastern part of the district is dominated by the major conurbations of Newcastle upon Tyne and Gateshead. Considerable urban development is also present west and south of this area, notably along the Tyne and Team valleys. More isolated centres of population somewhat farther south-west, including Consett, Annfield Plain, Stanley and Burnopfield, largely owe their origins to the growth of coal mining and other heavy industry. Elsewhere in the district the population is limited to a number of villages of which Corbridge on the western margin of the district and Rowlands Gill in the Derwent Valley are the largest; elsewhere the population is widely scattered.

There are many publications which refer to the geology of this district, but the descriptions are either of a generalised nature, or consider only some specific aspect or local detail. The geology of the wider region has been summarised by Carruthers et al. (1931), Hickling (1970), Taylor et al. (1971) and Robson (1980). This memoir is the first fully comprehensive account of the geology of the district. A summary of previous research is included at the start of each chapter. Further information is published in reports describing the 'Geology and local details' for some, but not all, of the National Grid sheets covered during the present survey. The available reports cover Edmundbyers and Hedley on the Hill (Edmonds and Davies, 1982); Corbridge and Prudhoe (Holliday and Pattison, 1990); Chopwell, Rowlands Gill, Consett and Stanley (Mills, 1982); Kibblesworth, Birtley, Craghead and Chester-le-Street (Cox, 1983); and Newcastle upon Tyne and Gateshead (Richardson, 1983).

The geological succession in the district is shown inside the front cover. The area is underlain entirely by Carboniferous rocks (Figure 3), those at outcrop belonging mainly to the Lower and Middle Coal Measures. Rocks of the Stainmore Group crop out in the north-west and south-west, the lowest bed to crop to surface being a sandstone underlying the Crag Limestone. Strata below this level are proved only in boreholes, or are inferred from geophysical data or by analogy with adjacent districts. A dolerite dyke of Tertiary age is at or near the surface for a distance of some 2 km near Walbottle. Over much of the district the Carboniferous rocks are masked by Quaternary deposits of varying thickness and character. As a result, exposure of bedrock is generally not good and the sections are discontinuous. Drift-free areas are largely, but not entirely, restricted to higher ground, steeper slopes, stream sections and quarries.

The stratigraphy of the Carboniferous rocks, particularly that of the Coal Measures, has been worked out principally from shaft, borehole and other underground data. The more important boreholes and shafts are listed in Appendix 1, and their locations are shown in Figure 30.

Little is known about the pre-Carboniferous and lowest Carboniferous rocks of this district. Indications from adjacent districts suggest that the pre-Carboniferous basement is made of Lower Palaeozoic strata, probably of Ordovician age, which were strongly folded and faulted, and intruded by the Weardale Granite, during the late Caledonian (Acadian) Orogeny in early Devonian times. Extensive erosion and peneplanation of these rocks took place prior to the deposition of the earliest Carboniferous strata. The erosion was probably sufficiently deep locally to reveal the upper surface of the granite. These basement rocks are masked with great unconformity by the overlying Carboniferous strata.

During Carboniferous times, the north-western part of the district lay along a major syndepositional hingeline, the Ninety Fathom–Stublick Fault System, which separated the relatively stable Alston Block, to the south, from the more rapidly subsiding Northumberland Trough in the north. As a result of the differential subsidence, strata laid down at the southern margin of the trough are significantly thicker than those immediately adjacent on the block. Major thickness changes between the two areas, resulting from this syndepositional faulting, were especially marked in Lower Carboniferous (Dinantian) times, but became progressively less significant during the Namurian and into the Westphalian.

The broad pattern of sedimentation during the Carboniferous shows an alternation of fluviodeltaic deposits laid down by rivers flowing from the north-east, and marine rocks deposited when the sea invaded the district from the south and west. In Upper Carboniferous times, particularly the Westphalian, the marine influence was

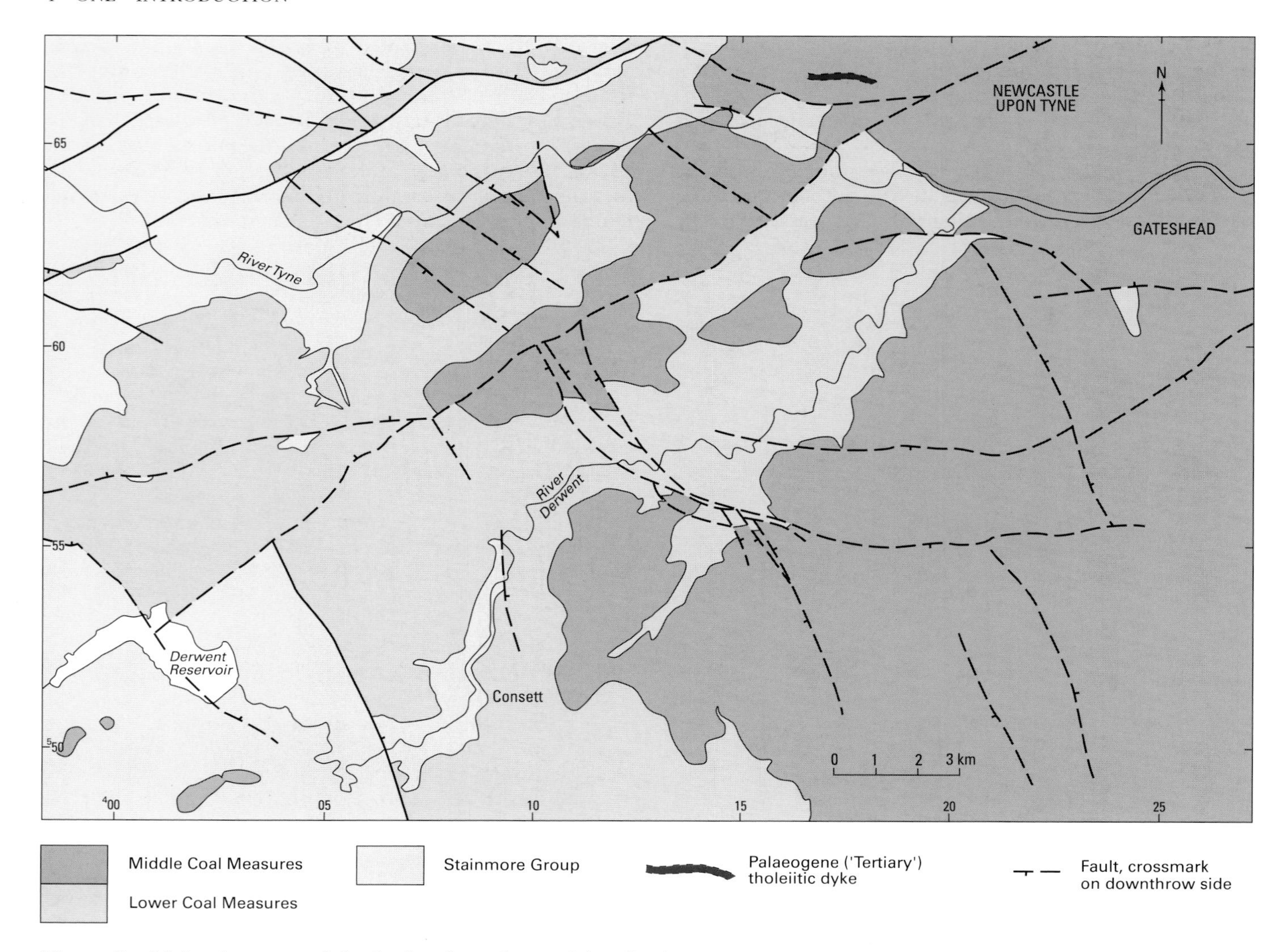

Figure 3 Main elements of the bedrock geology of the district.

much reduced. All the Carboniferous strata show a broadly rhythmic, or cyclothemic, pattern of sedimentation. This was produced by the alternation of marine and nonmarine environments, or by sedimentary processes within the deltaic and fluvial areas. The most complete and laterally persistent cyclothems have limestone and marine shale at the base, passing up into deltaic shales and sandstones commonly capped by coal. Such cyclothems are typical of the Alston and Liddesdale groups and the lower part of the Stainmore Group. Coal Measures cyclothems generally lack marine strata, but may nevertheless exhibit considerable lateral persistence.

Folding and faulting, associated with basin inversion and regional uplift, took place towards the end of Carboniferous times and continued into the early Permian. The Whin Sill was also emplaced at this time. This period, one of transpressional deformation, formed part of the Variscan Orogeny. The faulting seen in the Carboniferous rocks at or near the surface in the district today largely dates from this time, although considerable reactivation of previously established lines of weakness also took place. No rocks younger than the Middle Coal Measures have been preserved in the district, but it is probable that a significant thickness of later Westphalian rocks was deposited, prior to the Variscan uplift and erosion.

Following the Variscan events, Permian and Mesozoic rocks were also probably laid down in the district. These were eroded away during one or more periods of Tertiary (Neogene and Palaeogene) basin inversion and regional uplift, when old fault lines were reactivated, some new structures were established, and a tholeiitic dolerite dyke was intruded. Perhaps the most important effect of the Tertiary earth movements was the establishment of an overall easterly tilt to the rocks of the district, a result of the uplift of the Pennines and the continued subsidence of the North Sea area. This episode marked the start of an erosional cycle in which the present drainage system evolved.

Substantial modification of the landscape took place with the advent of the Quaternary Period, and the district became covered by a variable thickness and range of glacial and postglacial deposits. The district probably underwent several glaciations, but all the known deposits are Late Devensian or younger in age. Modification of the drainage system established in the late Tertiary also took place, whereby pre-existing valleys, some of which had been cut to sea level or below, became choked with glacial deposits to produce buried valleys. The processes of modification and change continue to the present day, with rivers depositing sands, silts and clays, and locally cutting through earlier deposits. The present landscape has also undergone considerable local modification by human activities — mining, quarrying, urban development and post-industrial land reclamation.

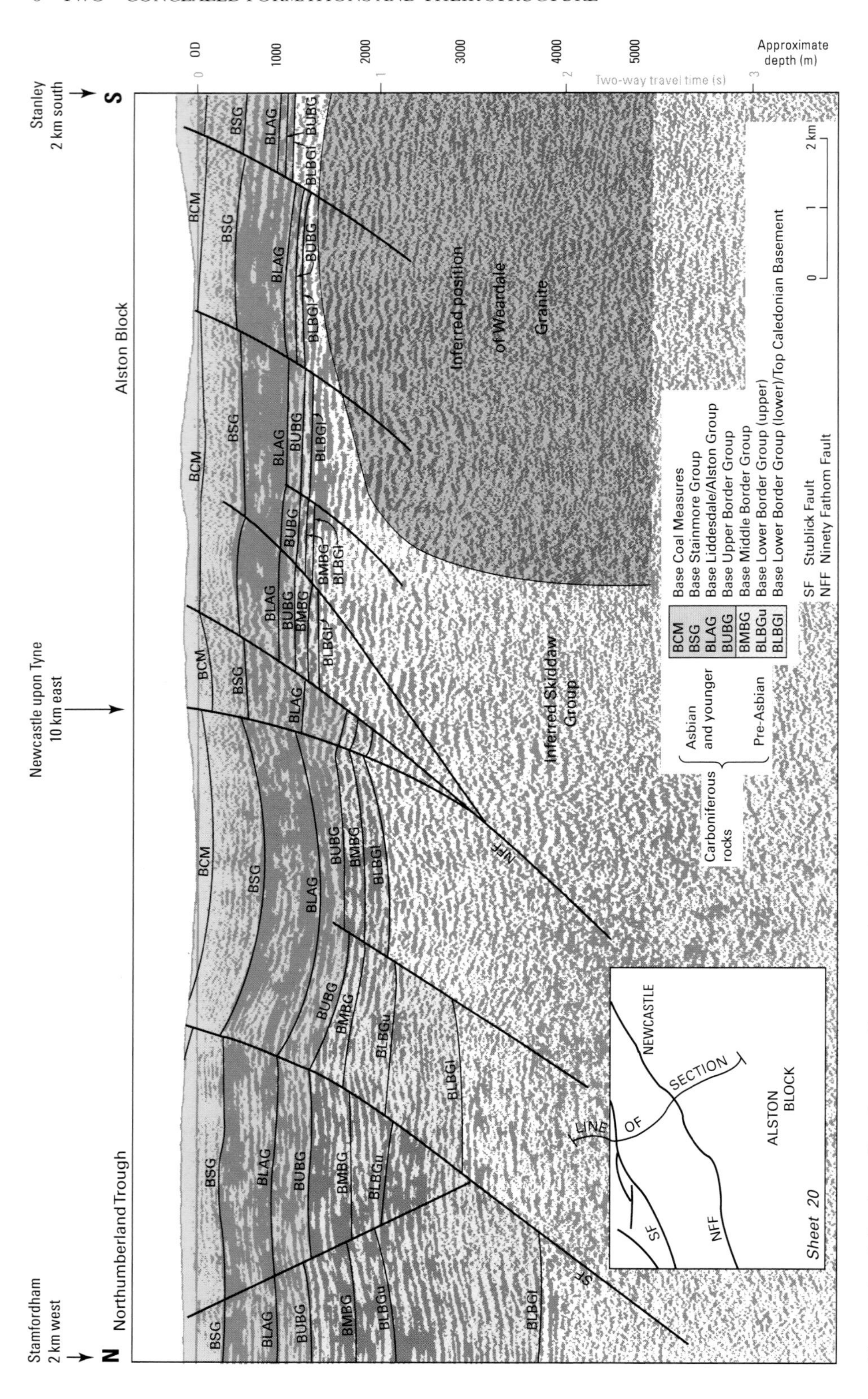

Figure 4 Seismic reflection profile across the Stublick and Ninety Fathom faults showing the approximate position of the Weardale Granite as inferred from gravity data. Modified from Kimbell et al. (1989) and Chadwick et al. (1995).

TWO

Concealed formations and their structure

The oldest rocks to crop out within the district occur north of Corbridge, near to the north-west corner of the sheet. They are of Pendleian age, just below the level of the Crag (Oakwood) Limestone. Somewhat older strata, of late Dinantian age, were proved in the Chopwell and Throckley boreholes, but otherwise there is little direct evidence from within the district relating to the older concealed rocks. However, geophysical data, taken in conjunction with information from adjacent districts, provide a number of insights into the nature, thickness and structure of the concealed Carboniferous rocks, and of the Lower Palaeozoic basement (with Devonian intrusions) on which they unconformably rest. Seismic reflection surveys, supplemented by gravity, magnetic and magnetotelluric data, have proved particularly useful in this respect (Bott and Masson Smith, 1957; Bott, 1967; Beamish, 1986; Beamish and Smythe, 1986; Evans et al., 1988; Kimbell et al., 1989; Chadwick and Holliday, 1991; Chadwick et al., 1993a, 1995).

The Lower Palaeozoic and early Devonian rocks of northern England and southern Scotland provide evidence of the closure of a major ocean (Iapetus) as a result of the convergence and subsequent collision of the palaeo-North American continent (Laurentia) and the palaeo-European continent (Avalonia). The line of junction between the two continents, the Iapetus Suture, is believed to lie between the Silurian rocks of the Southern Uplands of Scotland, which were deposited on the southern margin of Laurentia, and the mainly Ordovician rocks of the northern Lake District, which were laid down on the northern margin of Avalonia (McKerrow and Soper, 1989; Fortey et al., 1989; Soper et al., 1992). The Iapetus Suture extends north-east beneath the Northumberland Trough, and is located at depth not far north of the Newcastle district. The periods of deformation associated with this continental collision collectively form the Caledonian Orogeny, the latest phase of which (Acadian), in late Silurian to early Devonian times, was largely responsible for the folding, faulting and cleavage of the Lower Palaeozoic rocks of the Lake District (Soper et al., 1987; Cooper et al., 1993). Caledonian deformation in the Southern Uplands was diachronous and mainly completed in Silurian times (Stone et al., 1987).

In much of the area of collision and attendant terrane accretion, called the Iapetus Convergence Zone, the Lower Palaeozoic and early Devonian rocks are now concealed by a cover of Carboniferous and younger strata. The deposition of the Carboniferous cover rocks, and their thickness and facies, were strongly influenced by extensional reactivation of basement structures (Johnson, 1967, 1980; Leeder, 1982; Dunham, 1990; Chadwick et al., 1993a, 1995). Much of the district forms part of the Alston Block, a structurally high, granite-centred area, where the strata are relatively thin and rocks deposited in the earlier part of Lower Carboniferous times are largely absent. This is separated by the Ninety Fathom–Stublick Fault System from a major extensional basin to the north, the Northumberland Trough, in which the rocks are markedly thicker and span a much greater part of Lower Carboniferous time (Figures 4 and 5).

The concealed Carboniferous rocks of the district are intruded by a quartz-dolerite sill, the Whin Sill. These igneous rocks are considered in Chapter 5.

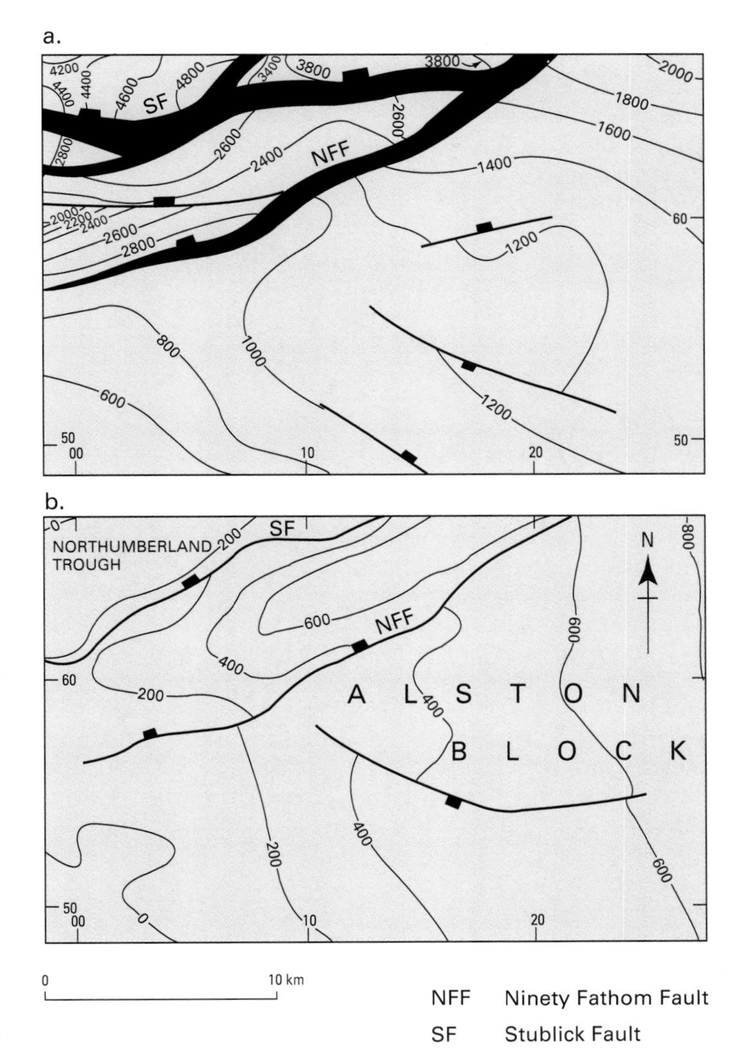

Figure 5 Structure contours for the district in metres below OD. a. Base Carboniferous; b. Base Stainmore Group. From Chadwick et al. (1995).

ALSTON BLOCK

Lower Palaeozoic rocks

There are no provings of pre-Carboniferous strata within the district, so that the nature of the basement rocks can only be inferred from evidence in adjacent areas. Boreholes at Allenheads, Rookhope and Roddymoor, to the south-west and south of the district (Woolacott, 1923; Dunham, 1990; Chadwick et al., 1995), and outcrop evidence from the Lower Palaeozoic rocks of the Cross Fell and Teesdale inliers, at the western edge of the Alston Block (Burgess and Holliday, 1979; Arthurton and Wadge, 1981), suggest that the pre-Carboniferous basement of the Alston Block largely comprises Ordovician mudstones, siltstones and sandstones like those of the Skiddaw Group of the Lake District. These rocks exhibit Acadian cleavage and low-grade metamorphism, and were intruded in early Devonian times by the Weardale Granite.

The rocks of the Skiddaw Group in the Cross Fell and Teesdale inliers are similar in lithology and structure to those of the Northern Fells and Central Fells belts of the Lake District (Cooper et al., 1993; Hughes et al., 1993 and references therein). It is probable, therefore, that the structure of the Alston Block broadly resembles that of the Lake District, and that rocks comparable to the Northern and Central Fells belts underlie the district. There is evidence from seismic reflection profiles that the Causey Pike Fault, a major northward-dipping Acadian thrust and strike-slip fault which separates these belts in the Lake District, is present in the pre-Carboniferous rocks of the present district, close to the northern margin of the Weardale Granite (Chadwick and Holliday, 1991).

In the Northern Fells Belt, the Skiddaw Group is mostly of late Tremadoc to early Llanvirn age. It comprises at least 5000 m of uniform grey siltstone and mudstone, locally interlaminated with fine-grained sandstone. Two coarse-grained sandstone formations are present. The rocks are largely turbiditic in origin and deposited in mid and distal fan environments (Cooper et al., 1993). The 3000 m thick Skiddaw Group in the Central Fells Belt comprises an early Llanvirn olistostrome deposit containing slumped and reworked material of Tremadoc and late Arenig age, overlain by Llanvirn mudstones with subordinate volcaniclastic sandstone turbidites and metabentonites (Cooper et al., 1993). The rocks of the Skiddaw Group have a complex deformation history (Hughes et al., 1993). Syndepositional deformation led to soft-sediment folding and slumping. Late Silurian to early Devonian (Acadian) deformation produced a north-easterly to easterly trending regional cleavage, which is axial-planar to large-scale folds and was followed, particularly in the Northern Fells Belt, by southward-directed thrusting with associated minor folds and crenulation cleavage. The Skiddaw Group also shows regional metamorphism resulting from deep burial and tectonic thickening (Fortey et al., 1993). Contact metamorphism adjacent to intrusive rocks is an important feature of exposed Skiddaw Group rocks, and is likely to be particularly well developed within the district, resulting from proximity to the Weardale Granite.

By analogy with the Lake District, and the Cross Fell and Teesdale inliers, mid-Ordovician volcanic rocks, correlatives of the Borrowdale and Eycott Volcanic groups of the Northern and Central Fells belts, may also be locally present in the district, but beds of the late Ordovician to Silurian Windermere Group, more typical of the Southern Fells Belt, are probably absent.

Weardale Granite

Detailed descriptions of the petrology and geochemistry of the Weardale Granite, proved in the Rookhope Borehole in the Alston district, have been given by Dunham et al. (1965), Holland (1967, 1980), Holland and Lambert (1970) and Dunham (1990). Radiometric dating of samples from the borehole suggests that the granite was intruded at around 400 Ma (early Devonian) (Holland, 1980). It is not known how representative the borehole material is of the intrusion elsewhere.

The form of the Weardale Granite has been investigated using gravity data (Bott, 1967; Evans et al., 1988). This work has suggested that a subsidiary cupola of the main granite batholith underlies the present district, centred beneath Rowlands Gill, where its top probably lies at or close to the sub-Carboniferous unconformity (Figures 4 and 6). Modelling based on the gravity data suggests that the northern flank of the granite is steep, closely parallel to the Ninety Fathom Fault, and its position is probably controlled by the location of the continuation of the Causey Pike Fault. This would imply that the granite is probably intruded into Skiddaw Group rocks forming a continuation of the Central Fells Belt of the Lake District (see above).

The Weardale Granite is associated with a high heat flow anomaly, caused by its relatively high content of uranium and thorium (Gebski et al., 1987; Rollin, 1987; Evans et al., 1988). Measurements in the Rowlands Gill Borehole gave an estimated heat flow value of 99.5 mW per square metre. This concealed heat source has had a profound effect on the surrounding rocks; as discussed in Chapter 7, it has led to the formation of relatively high rank coals and hence contributed significantly to the economic potential of the region.

Dinantian rocks

The seismic reflection data show that the sub-Carboniferous surface on the Alston Block is largely planar, with little relief, indicating that there was a long period of erosion and peneplanation between the time of intrusion of the Weardale Granite and the beginning of Carboniferous sedimentation (Figure 4). The concealed Dinantian rocks on the Alston Block in the district are up to 600 m thick, and mainly of late Asbian to Brigantian age, but thin developments of older (Chadian to early Asbian) strata are probably also present (Chadwick et al., 1995). Table 1 shows the current nomenclature of Dinantian strata on the Alston Block, and how these

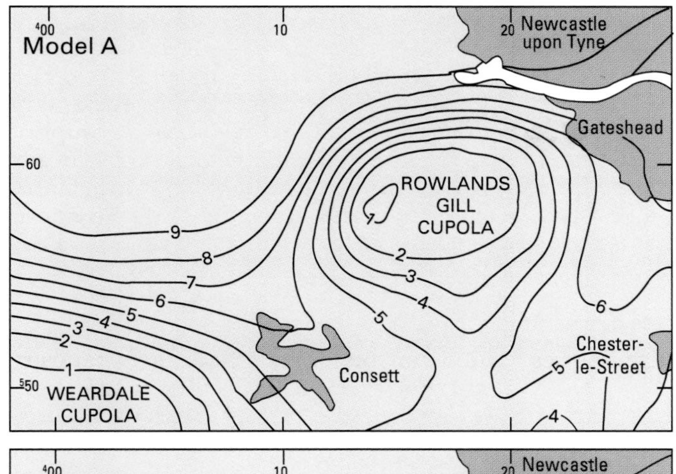

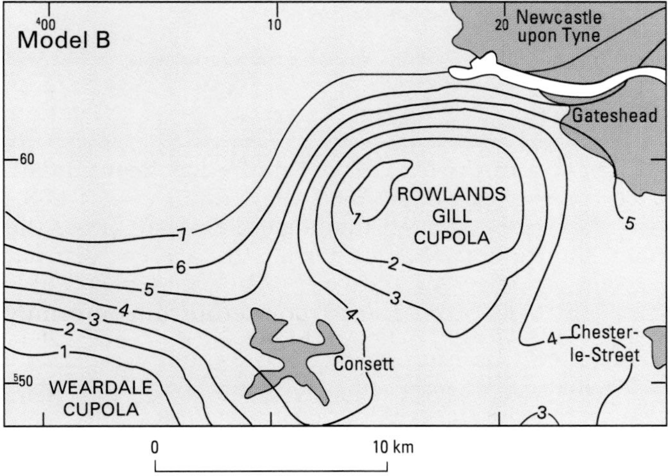

Figure 6 Three-dimensional models of the Weardale Granite in the district, showing alternative interpretations of the gravity data. Contours to top of granite in km below OD. Model A, base of granite at 10 km. Model B, base of granite at 8 km at the perimeter and 11.5 km at its centre [located at NY 95 40, in adjacent Alston district]. From Evans et al. (1988).

names relate to those used in the Northumberland Trough.

The nature of the pre-late Asbian rocks of the district is uncertain, and they may be laterally variable, both in thickness and lithology. In the Northern Pennines, these beds, belonging to the **Orton Group**, are up to a few tens of metres thick and comprise limestones and shales with basal sandstones and conglomerates (Dunham, 1990). However, in the Harton Borehole, east of the present district, they are over 200 m thick and contain a relatively high proportion of sandstone (Ridd et al., 1970; Smith, 1994).

The **Alston Group**, of late Asbian and Brigantian age, crops out extensively in the Northern Pennines to the west and south-west of the district (Dunham, 1990), and has also been proved to the east in the Harton Borehole (Ridd et al., 1970; Smith, 1994). The group comprises a cyclothemic sequence containing limestone, shale and sandstone, with sporadic thin coal seams, formed by alternating marine and deltaic conditions. It is around 500 m thick on the block within the district.

In the Chopwell Borehole* (Simpson, 1904) in the Derwent Valley, 2 km west-south-west of Rowlands Gill, nearly 52 m of strata were proved underlying the Great Limestone. The strata recorded are referred to the Alston Group and appear to consist largely of sandstone. They are similar to correlative strata elsewhere on the Alston Block (cf. Dunham, 1990). A thin sandstone, shale and ganister are present underlying the Great Limestone. The Four Fathom Limestone, 6.15 m thick and described merely as 'blue,' is present 41.2 m below the Great Limestone. The Iron Post Limestone, always very thin (Dunham, 1990), was not recorded.

THE NINETY FATHOM FAULT

The long-held view that the Ninety Fathom Fault defines the boundary between the Alston Block and the Northumberland Trough has been confirmed by seismic reflection surveys (Kimbell et al., 1989; Chadwick and Holliday, 1991; Chadwick et al., 1993a, 1995). These provide evidence of syndepositional normal displacement on the Ninety Fathom Fault and on its *en-échelon* continuation, the Stublick Fault, particularly in early Dinantian times (Figure 4, Lower and Middle Border groups). During Asbian to Westphalian times (Figure 4, Upper Border, Liddesdale/Alston and Stainmore groups, Coal Measures), syndepositional faulting continued, but on a much reduced scale, and regional subsidence became the main mechanism for basin subsidence (Figure 5).

The early Carboniferous down-to-the-north displacements of the Ninety Fathom–Stublick Fault System are believed to have resulted from extensional reactivation of an older line of weakness in the basement rocks, forming in the hanging-wall block of the Causey Pike Fault (Chadwick and Holliday, 1991). Regional investigations have shown that the fault system has a long and complex history of repeated reactivation, although only part of this history can be demonstrated within the district. In other areas, where significant amounts of post-Carboniferous rocks are preserved, seismic reflection data (Chadwick et al., 1993b, 1995) reveal that periods of extensional normal faulting related to basin subsidence (in the Carboniferous and again during the Mesozoic era), alternated with periods of reversed displacement associated with basin margin inversion and regional uplift (during the late Carboniferous and in Tertiary times).

NORTHUMBERLAND TROUGH

Lower Palaeozoic rocks

The Iapetus Suture is a deep-seated structure concealed beneath the Carboniferous rocks of the Northumberland Trough. The nature and precise position of the suture is unknown, but it is probably a northward dipping thrust zone located to the north of the Ninety Fathom-Stublick

Table 1 Nomenclature and classification of Lower Carboniferous rocks in the region.

Alston block	Northumberland Trough	Stage	Series	Subsystem
Alston Group	Upper Liddesdale Group	Brigantian	Viséan	Lower Carboniferous (Dinantian)
	Low Tipalt Limestone			
Melmerby Scar Limestone	Lower Liddesdale Group	Asbian		
hiatus	Redesdale Limestone			
	Upper Border Group			
Orton Group	Middle Border Group	Holkerian ? Arundian		
Basement Group	Lower Border Group	? Chadian		
hiatus	Courceyan	Tournaisian		

Fault System (Soper et al., 1992). The lack of a related fault system in the Dinantian cover (Chadwick et al., 1993a, 1995) suggests that the suture was not significantly reactivated during Carboniferous sedimentation and later times, and that it is probably a complex and ill-defined feature.

Along the northern margin of the Northumberland Trough, to the north of the suture and of the district, the Carboniferous rocks rest unconformably on Silurian (Wenlock) greywackes, sandstones, siltstones and mudstones (Lumsden et al., 1967; Stone et al., 1987). Within the district, the basement rocks of the southern margin of the trough are probably similar to those of the Skiddaw Group of the northern part of the Alston Block, but towards the north these may be inter-thrust with Silurian rocks like those of the Southern Uplands.

Dinantian rocks

The Northumberland Trough is markedly asymmetrical, with between 4000 and 4500 m of Dinantian rocks at its southern margin, in the north-western part of the district (Kimbell et al., 1989). Classification of the Dinantian rocks is shown in Table 1. Much of this thickness is made up of rocks of the Lower and Middle Border groups (Figure 4), which crop out extensively to the west, along the strike of the basin, in the Langholm and Bewcastle districts (Lumsden et al., 1967; Day, 1970), and also to the north in the Rothbury district (Fowler, 1936). The **Lower Border Group** comprises marine limestones and shales, indicative of marine incursions from the west, alternating with deltaic and fluvial sandstones and shales derived from the north-east (Leeder et al., 1989; Chadwick et al., 1995). The Alston Block was emergent during much of early Dinantian times (Figure 4), and the concealed correlatives of the Lower Border Group beneath the north-western part of the district, up to around 3000 m thick, probably contain much detritus derived from this source. Any relief on the Alston Block that remained during deposition of the **Middle Border Group** was probably much reduced (Chadwick et al., 1995), and the correlative rocks of the district, around 400 m thick, are probably sandstones and shales of fluvial, lacustrine and deltaic origin deposited by the north-easterly sourced river. Thin marine limestone and shale intercalations, similar to those proved in the Stonehaugh Borehole of the Bellingham district (Frost and Holliday, 1980; Smith and Holliday, 1991), are probably also present.

Younger Dinantian rocks in the north-west of the district, including the Upper Border and Liddesdale groups, are also believed to be similar to the limestones, shales, sandstones and thin coals, of alternating marine and deltaic origin, exposed in the Bellingham district (Johnson, 1959; Frost and Holliday, 1980). The correlatives of the **Upper Border Group** vary from 0 to 200 m on the block and reach around 500 m in the basin (Chadwick et al., 1995) (Figure 4). The broadly equivalent **Liddesdale** and **Alston** groups record similar styles of sedimentation in both block and basin areas, with only relatively modest northward increase of thickness from around 500 to 700 m (Figure 4).

The Throckley Borehole*, 6 km to the north of the Ninety Fathom Fault and just north of the district, proved 14.3 m of strata of the Upper Liddesdale Group, underlying the Great Limestone. An alternating sequence of siltstone, mudstone and sandstone was recorded, in which the sandy fraction predominates. Three thin coals are also present, the middle one being overlain by a calcareous sandstone and mudstone containing a fauna of ribbed brachiopods, bivalves, gastropods and abundant crinoid debris, probably at the horizon of the Iron Post Limestone.

STRUCTURE OF CONCEALED DINANTIAN ROCKS

The structure of the concealed Lower Carboniferous rocks is illustrated by structure contour maps (Figure 5). These have been derived from the study of relatively widely spaced commercial seismic reflection profiles (Chadwick et al., 1995), and present a somewhat generalised view of the structure. A more detailed review of the structure of near-surface Upper Carboniferous rocks is given and illustrated in Chapter 4.

A prominent feature of both structure contour maps is the dominant east-north-easterly tilt of the strata on the Alston Block. Two factors have contributed to this tilt; one is the general easterly thickening of the Carboniferous sequence, the other is the uplift of the Pennines and subsidence of the North Sea during Variscan and, more particularly, Tertiary tectonic episodes. The tilt is less evident in the part of the district that lies within the Northumberland Trough. The Base Carboniferous map (Figure 5a) reveals the large cumulative northward downthrow of the Ninety Fathom-Stublick Fault System of around 3000 m, resulting largely from early Carboniferous syndepositional faulting and extensional basin formation (Figure 4). The throw of the faults at Base Stainmore Group (Namurian) level is considerably less, up to 200 m (Figure 5b). The structure revealed at this level is broadly similar to that in the overlying strata (Chapter 4), and shows the relative decline in importance of syndepositional extensional faulting, and corresponding increase in the role of regional subsidence, as basin-forming mechanisms in the district, from late Dinantian times onwards. Several east-north-easterly trending folds can be recognised in the Base Stainmore Group map (Figure 5b), closely related to the Ninety Fathom–Stublick faults (see Figure 4). These provide evidence of late Carboniferous (Variscan) basin margin inversion, and dextral transpressive displacement (Chapter 4).

THREE

Upper Carboniferous

Carboniferous rocks occupy the entire district (Figure 3); those that crop out at the surface or beneath Quaternary cover belong to the Upper Carboniferous (Silesian). The classification of the Upper Carboniferous rocks adopted in this memoir, and some others previously used in this and adjacent districts, are shown in Table 2. At least two thirds of the district is underlain by Westphalian-age rocks of the Lower and Middle Coal Measures 480 m thick, whereas older strata of the Namurian-age Stainmore Group, 470 m thick, are present at the surface only in the north-west and south-west. The lower part of the Stainmore Group does not crop out within the district but is known from boreholes and from exposures in adjacent areas.

DEPOSITIONAL SETTING

Detailed consideration of the sedimentary processes involved in the deposition of the Upper Carboniferous rocks of the district is beyond the scope of this memoir. However, in recent years studies have been carried out which have added substantially to understanding of the controls affecting the depositional environments of the Stainmore Group and Coal Measures in this and adjacent districts (e.g. Elliott, 1974, 1975, 1976; Percival, 1983, 1986; Collier, 1989; Fielding, 1984a, 1984b, 1984c, 1986; Fielding and Johnson, 1986; Haszeldine, 1983a, 1983b, 1984a, 1984b; Haszeldine and Anderton, 1980).

As described in Chapter 2, during Carboniferous times the district lay across the northern margin of the Alston Block and the southern fringes of the Northumberland Trough, the dividing line between the two being the Ninety Fathom — Stublick Fault System. The block area was relatively buoyant and stable in comparison to the subsiding trough, with the result that strata laid down on the block are thinner than their time equivalents in the trough. These effects were especially marked in Lower Carboniferous times, but during the Upper Carboniferous the pattern of fault-related subsidence characteristic of the Lower Carboniferous gave way to more widespread regional subsidence, and any thickness differences between block and trough resulted more from differential compaction. By mid Westphalian (Duckman-

Table 2 Nomenclature and classification of Upper Carboniferous rocks in the district.

MB = Marine Band

Former classifications			Present classification		
Lebour (1876 and 1886)	Fowler (1936); One inch to one mile sheet 14 (Morpeth)				
The Coal Measures	Coal Measures	Upper Coal Group	Middle	Bolsovian (Westphalian C)	Upper Carboniferous (Silesian)
			Ryhope MB	Aegiranum MB	
		High Main Coal	Coal Measures	Duckmantian (Westphalian B)	
		Middle Coal Group	Harvey MB	Vanderbeckei MB	
Brockwell (or Main) Coal		Brockwell Coal	Lower Coal Measures	Langsettian (Westphalian A)	
The Gannister & Millstone Grit Series		Lower Coal Group (Ganister Clay Coal)			
	Millstone Grit		Quarterburn MB	Subcrenatum MB	
Felltop Limestone	Carboniferous Limestone Series (part)	Upper Limestone Group	Stainmore Group	Namurian	
Bernician		Middle Limestone Group (part)	Great Limestone		

tian) times, there was little differential subsidence between the two areas.

Deltaic sedimentation was dominant during Upper Carboniferous times. Repeated marine invasion of the deltaic coastline, followed by progradational rebuilding of the delta, led to the deposition of the cyclic (cyclothemic) sequence characteristic of Carboniferous rocks in north-east England. Through most of Namurian times, the district is thought to have formed a lower delta plain environment close to sea level. The frequency of the marine and semimarine incursions broadly decreased with time, and during late Namurian and early Westphalian times the deposition of a complex sequence of erosive-based sandstones points to a substantial increase in the supply of fluvial sediments from the north and north-east. Later in Westphalian times, the depositional environment is thought to have changed to that of an upper deltaic plain, still near to sea level but somewhat further from the coastline. At this time, the district formed a small part of the Pennine province of cyclical Coal Measures deposition (Calver, 1969). Forest swamps were widely developed, and thick deposits of peat accumulated from time to time on this surface in an environment of impeded drainage. Continued subsidence led to the repeated drowning of these swamps, the formation of lakes and the burial of the peat by mud, silt and sand. Plants then recolonised the new surfaces. Rare marine incursions resulted in the deposition of thin bands of black mudstone containing the brachiopod *Lingula*. Although the broad pattern of deposition is clear, the detailed sedimentary geometry is complex, with a wide lateral variation in thickness, completeness and extent of individual sedimentary cycles (Fielding, 1984a). Despite these variations in detail, the thicknesses of both Stainmore Group and Coal Measures as a whole are remarkably uniform across the district (Figures 7 to 22).

LITHOLOGY

The following account of the main rock types found in the district is based largely on descriptions by Land (1974), Lawrence and Jackson (1990) and Holliday and Pattison (1990).

Limestone

At outcrop, limestone is found only in the Stainmore Group (Figures 7 to 12), although it is a common rock type in the concealed Dinantian strata. The Stainmore Group limestones are typically grey and dark grey, fine- to very fine-grained in texture, and predominantly biomicritic. Pale grey limestone with a sparry cement occurs in the upper part of the Thornbrough (Upper Felltop) Limestone. The thinner limestones contain varying proportions of sand, silt and mud, and commonly contain laminae, ribs and thin layers of calcareous mudstone, locally shelly. Carbonaceous laminae, and blebs and sheets of bituminous material, may also be present. Many silty and sandy limestones are ferruginous and bioturbated. On weathering, with solution of calcium carbonate, such limestones commonly form an orange-yellow rottenstone known locally in the Northern Pennines as 'famp'.

With the exception of the Great Limestone, and of the thin but persistent and uniform Whitehouse Limestone, all the limestones exhibit considerable lateral variation in thickness and bedding characteristics. Variations in lithology and purity are particularly notable, with lateral transitions into calcareous and shelly mudstone and/or sandstone commonly recorded (Figures 8 to 11). To some extent, such variation may reflect differences in the logging or interpretation of these often complex and difficult lithologies in borehole and shaft sections. However, adequate exposures exist in numbers large enough to confirm the broad extent of the lithological variation in the marine intervals shown in Figures 8 to 11.

Mudstone and siltstone

Mudstones and siltstones form a significant proportion of the sequence. Calcareous fossiliferous mudstones commonly occur in the strata immediately overlying limestone; they generally weather to sharp blocky fragments and locally are cut by numerous fine carbonate veinlets. Calcareous mudstones are not generally present in rocks of the Coal Measures, the marine phases here being marked generally by very dark grey to black mudstone, containing *Lingula*, pectinoid bivalves and, less commonly, productoid brachiopods. Mudstones in the Coal Measures normally contain a brackish-water or lacustrine fauna of nonmarine bivalves, ostracods and fish debris, or contain only plant fragments. Most mudstones are finely micaceous and carbonaceous, and the dominant clay minerals are illite and kaolinite. Ironstone ribs and nodules are commonly present, and are characteristic of certain intervals, such as the Coalcleugh Shell Bed. They tend to be more common in the higher parts of the Westphalian sequence and occur particularly above certain coals.

In an upward sequence, the mudstones become progressively more silty, commonly passing through gradational lithologies to siltstone or very fine-grained sandstone. Commonly, siltstone and fine-grained sandstone in equal proportions are interlaminated and interbedded, forming the aptly named 'striped beds'. A wide range of sedimentary structures is found, including ripple marks, small-scale cross-bedding, planar lamination, load casts, slumps and micro-faulted layers. The upper part of such a striped bed sequence is commonly overlain by a thin-bedded and flaggy sandstone with small-scale cross-bedding.

Sandstone

A considerable range of sandstone lithologies and types is present. In general they vary from greyish-white to grey, but near to the surface, or along joints, they are commonly greyish brown, rusty brown or yellowish white. The sandstones are generally subarkosic, with dominant quartz and subordinate feldspar, muscovite and biotite.

The principal cements are kaolinite, quartz, chlorite and more rarely biotite and calcite. In the coarser pebbly sandstones, the larger grains and pebbles are fragments of quartzite and more locally derived, reworked intraclast rock types. The tops of some sandstones contain roots, recording the development of plant growth. Such seatearth-like rocks commonly underlie coals. In some cases, the sandstones are strongly leached and silicified to yield ganisters (Percival, 1983, 1986).

The sandstones of the Coal Measures have two main geometrical forms, 'sheets' and 'channels', both of which may show considerable lateral and vertical variation. Some sandstones bodies are composite, showing characteristics of both types.

Sheet sandstones are generally less than 5 m thick and are of wider lateral extent than channel sandstones. They are commonly fine to medium grained, more rarely coarse grained, and are often cross-bedded; thin siltstone and mudstone ribs occur, more particularly at the top and base. In general, sheet sandstones have gradational bases with the underlying siltstones and striped beds, and form the upper members of coarsening-upwards intervals. Sheet sandstones with erosive bases, forming the lower members of fining-upwards intervals, are less common and generally comprise coarser-grained rocks.

Channel sandstones, up to 30 m thick, are medium to coarse or very coarse grained, and may be pebbly. They generally show well marked large-scale cross-bedding. A sharp, erosive base is a general feature of these sandstones, many of which occupy deep excavations ('washouts') cut into the underlying strata, from which key marker beds and a significant part of the succession locally may have been removed. Such sandstones generally form fining-upwards units and commonly contain basal channel-lag conglomeratic deposits incorporating intraclast fragments of siltstone, mudstone, ironstone, coalified plant debris and more rarely impressions of tree trunks. Although rarely more than a few kilometres wide, some channel sandstones have been proved from mine workings to have linear or gently curving lengths of 20 to 30 km (Fielding, 1984a, fig. 4).

Sandstones in the upper part of the Stainmore Group and the lower part of the Coal Measures are commonly coarse grained and erosive based, and some may be channelised, but generally they appear to be more sheet-like in distribution. They locally weather and decompose to a coarse kaolinitic rock which is soft and friable, not only at the surface but also to considerable depth. Sandstone of this type, in the upper part of the Stainmore Group, has been worked as a source of moulding sand in the Wolsingham district, and similar 'rotted grit' was recorded at a lower stratigraphical level when the Letch House–Airy Holm–Derwent Tunnel (Appendix 3) was driven.

Coal, seatearth and fragmental clayrock

Land plants normally colonised emergent surfaces resulting from Carboniferous deltaic and fluvial sedimentation. Evidence of this is provided by the presence of fossilised plants, and of numerous coals and seatearths, in the succession. The majority of coals are believed to have formed by the preservation of in-situ accumulations of plant debris, and the seatearths are the remains of the soils in which the plants grew.

Seatearths (or fireclays) are common, especially in the Coal Measures, and consist generally of grey, brown or green non-bedded mudstone with abundant rootlets and sporadic irregular ironstone nodules. Seatearths have an irregular fracture with curved, polished ('listric') surfaces. Locally seatearths are silty or sandy, grading into seatearth-sandstones or ganisters. Typically, seatearths are associated with an overlying coal, but this is not always so. Although the thickness of a seatearth is probably related to the length of time of emergence, and of plant growth, there is no general correlation between the thickness of a seatearth and that of the overlying coal. Seatearths are commonly more persistent laterally than their related coals.

Coals in the district are generally of a high-rank bituminous character, with a high carbon content and low to moderate volatile matter. The rank generally increases towards the south, and also downwards in the sequence. An increase in rank is evident in the Medomsley, Burnopfield, Rowlands Gill and Stanley areas, which overlie a cupola of the concealed Weardale Granite batholith (Chapters 2 and 7). The majority of coals hereabouts have a carbon content of 82 to 89 per cent, and volatiles of 20 to 30 per cent. The coals are bright and vitrinite-dominant. Cleat is strongly developed, becoming closely spaced in the high-rank coals of the district and causing the seams to become soft (Jones and Magraw, 1980).

Many coal seams contain persistent thin partings, locally termed 'band', of seatearth, carbonaceous mudstone and, less commonly, sandstone. In some instances, these partings thicken into significant splits of several metres thickness, over both limited or wider areas (Fielding 1984a, 1984c). Locally, where the intervening strata die out, two or more seams may unite. Coal seams are also subject to washouts, generally by the downcutting of channels infilled by medium- to coarse-grained sandstone (see above).

Thin cannel coals occur separately, or are associated with normal coals. They are typically found in the immediate roof of a coal seam. They commonly contain fish debris and the crustacean *Estheria*. Cannel coals are thought to comprise transported plant material, and have no direct relationship with seatearths.

Fragmental clayrocks (FCRs) in coal-bearing sequences in Scotland and north-east England have been described by Richardson and Francis (1971). They are generally 5–10 cm thick, and are characterised by brecciation and flow structures, a high proportion of kaolinite, an association with coal, and limited lateral extent. The most significant occurrence of such rocks within the district is as thin beds overlying the Brockwell and Harvey coals. They are interpreted as deposits which, having originated in shallow lakes within the coal swamps, have undergone intense diagenetic change.

CYCLOTHEMS

The cyclical, or cyclothemic, nature of Carboniferous strata in northern England has been widely described and discussed (e.g. Dunham, 1950, 1990; Armstrong and Price, 1954; Smith and Francis, 1967; Johnson, 1967, 1970, 1980; Land, 1974; Frost and Holliday, 1980; Leeder and Strudwick, 1987; Holliday and Pattison, 1990; Maynard and Leeder, 1992; Smith, 1994). A full discussion of the origin of these sedimentary cycles is beyond the scope of this memoir. The causative factors include eustatic changes of sea level, compaction, subsidence, tectonic history, and sedimentary processes inherent to deltaic and fluvial environments.

The idealised 'complete' cyclothem, listed below, comprises a basal marine member, overlain by a coarsening-upward prograding siliciclastic deltaic and fluvial interval, capped by seatearth and coal.

Coal
Seatearth
Sandstone, fine- to very coarse-grained, massive or cross-bedded
Sandstone, very fine- to fine-grained, flaggy, with small-scale cross-bedding
Siltstone, sandy; sandstone laminae and ribs
Siltstone, muddy
Mudstone, silty
Mudstone; ironstone nodules
Mudstone; calcareous, often fossiliferous
Limestone, muddy limestone, shelly mudstone or calcareous shelly siltstone and sandstone.

In detail, the cyclothems vary widely in thickness, extent, and the relative proportions of the different items listed above. Some cycles are incomplete, and some may pass into thinner cycles within a major unit. Where the siliciclastic beds form a sharply based fining-upwards unit, some or all of the lower members may be absent. The thickest cycles tend to be the most complete and extensive, and are bounded by the thickest and most persistent limestones or coals. In the lower and middle part of the Stainmore Group, marine strata including relatively thick limestones and shelly beds are prominent, but higher in the sequence marine beds are relatively rare. Coals are generally thin and discontinuous in the Stainmore Group, but persistent and relatively thick coals become more prominent higher in the succession, and are characteristic of all but the lower part of the Coal Measures.

Although the thicknesses of individual cyclothems may show wide local variations, particularly in the Coal Measures, the strata overall are remarkably uniform in thickness (Figures 8 to 22). Thus where a cyclothem is locally thin, adjacent cycles above or below tend to be thicker than normal, and a group of three or more cyclothems generally maintains a broadly constant thickness over a wide area. It is unlikely that such thickness changes result from widely fluctuating, but highly localised, rates of crustal subsidence. They are more probably due to the effects of differential compaction between the underlying sands, silts, muds and peats, and are particularly related to the geometry of the deltaic and fluvial sand bodies. Cyclothems tend to be at their thickest where channel sandstones occur in the sequence, because of the low degree of compaction of sand relative to that of the adjacent argillaceous sediments. For the same reason, the immediately succeeding sediments are likely to be thin over the channel and thicken away from it. There are exceptions to this, and where two or more channel sandstones overlie each other, some local tectonic control on sedimentation, and on the location of channel sandstones, is suggested (Fielding, 1984a, 1984b).

STAINMORE GROUP

The Stainmore Group is defined as extending from the base of the Great Limestone up to the base of the Quarterburn Marine Band, the assumed equivalent of the Subcrenatum Marine Band at the base of the Coal Measures (Mills and Hull, 1976; Ramsbottom et al., 1978; Burgess and Holliday, 1979; Frost and Holliday, 1980; Dunham, 1990). The group therefore spans most of the Namurian Series (Johnson et al., 1962; Hull, 1968), although an unknown but probably small thickness of strata below the Great Limestone is also probably of Namurian age (cf. Dunham and Wilson, 1985). The adoption of this classification (Figure 7) has led to the abandonment of the terms Upper Limestone Group and Millstone Grit of former classifications (Table 2). The term 'Millstone Grit Series', used as a synonym for 'Namurian Series' on the published 1:50 000 scale map, is also discontinued. As a result of these changes only two of the three 'grits' (sandstones) mapped during the primary survey now form part of the Stainmore Group, the higher 'Third Grit' being included in the lower part of the Coal Measures. The maximum thickness of the group within the district is 470 m, the lowest 110 m having been proved only in boreholes. In the district, and in north-east England generally, the rocks of the group are generally much thinner than their equivalents to the north, in Scotland, and to the south, in Yorkshire (Hull, 1968; Ramsbottom et al., 1978).

The classification, nomenclature and generalised vertical succession of the Stainmore Group in the district are shown in Figure 7. The age-diagnostic ammonoid (goniatite) fossils, on which the zonation of Namurian strata in the Pennine region is mainly based, are generally scarce. To a great extent, therefore, correlation within the group has been based on the recognition and mapping of key lithological and faunal marker beds, supported by other faunal and palynological data. Much confusion has arisen in the past from the miscorrelation of these markers, and from the use of local or inconsistent stratigraphical nomenclature. However, sufficient evidence has now accumulated to be reasonably certain that most of the Namurian stages are present, although some are attenuated and represented by only a few metres of strata (Hull, 1968; Ramsbottom et al., 1978). The greater part of the sequence falls within the

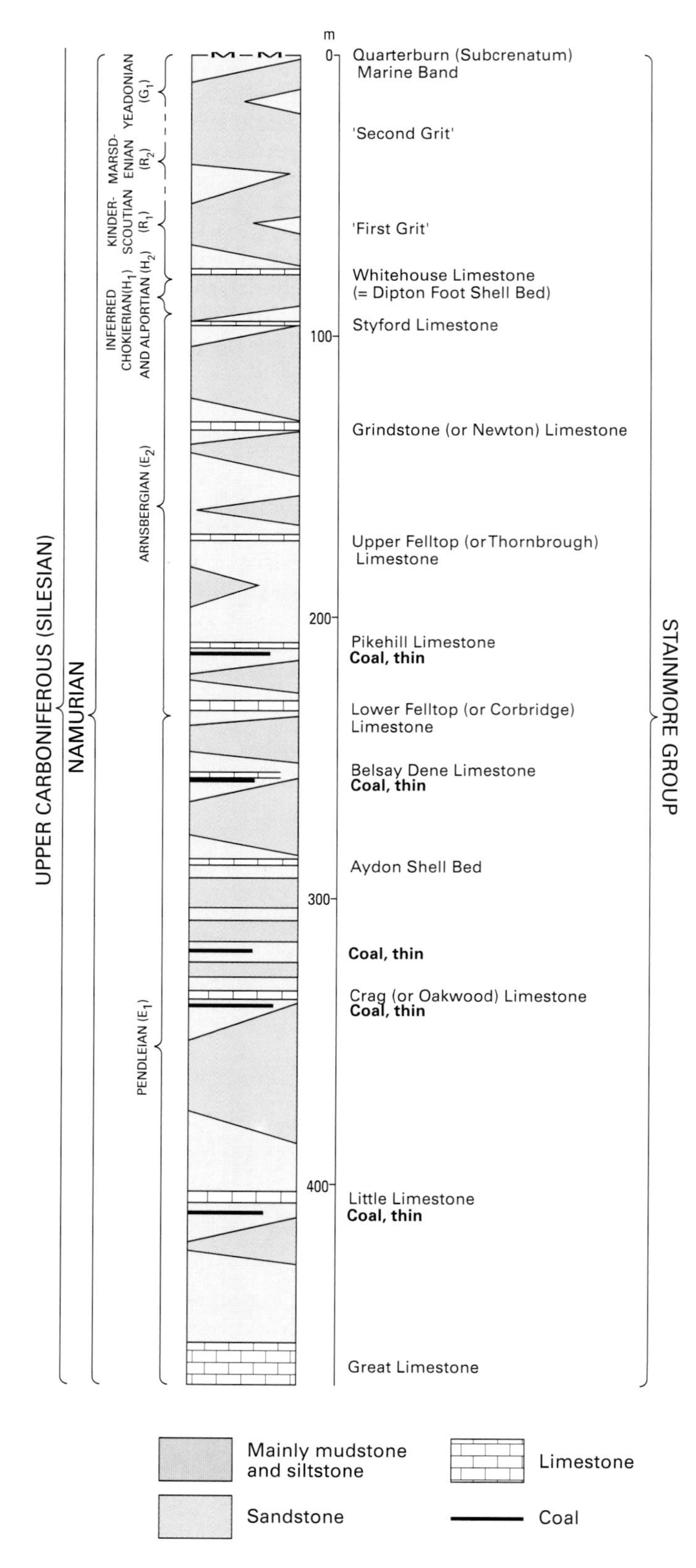

Figure 7 Stainmore Group: generalised succession, nomenclature and classification.

Pendleian (E_1) and Arnsbergian (E_2) stages. There is only little evidence for the presence of the Chokierian (H_1) and Alportian (H_2) stages, but as there is no indication of a regional unconformity in the sequence, beds of this age are probably present but thin. There is sufficient evidence, from rare ammonoid finds and from palynology, to confirm the presence of the remaining Namurian stages, the Kinderscoutian (R_1), Marsdenian (R_2) and Yeadonian (G_1). The base of the Quarterburn Marine Band (Mills and Hull, 1976) is thought to equate with the Namurian–Westphalian boundary, but although a fragment of *?Gastrioceras* has been recorded at one locality, the key fossil *G. subcrenatum* has not been found.

Exposure of rocks of the Stainmore Group is generally poor, being confined largely to discontinuous sections along some streams, to sporadic quarries, and to scarp features exposing sandstones and limestones. In the south-western part of the district, especially on the south side of the Derwent Valley, thick sandstones at the top of the group form well-developed features on the flanks of the Northern Pennines. North-eastwards these features become partly concealed by drift and are less easy to trace.

Lebour (1875, 1885) published some details of the stratigraphy, particularly of the area around Corbridge, and faunal lists for the limestones were published by Smith (1912). Additional faunal data were given by Hedley and Waite (1929) in an account of the stratigraphy of the lower part of the group. Hedley (1931) discussed the age of these rocks including part of the Coal Measures. The coal and mineral deposits of the north-western part of the district were described by Smith (1912, 1923). Significant and important advances in the understanding of the Millstone Grit and Lower Coal Measures of the Tyne Valley and adjoining areas were made in unpublished PhD theses by Green (1954) and Whittaker (1963). A borehole drilled by the Geological Survey in 1964–5 near Throckley, just within the adjacent Morpeth district, provided a complete stratigraphical sequence of the Stainmore Group. This borehole (Richardson, 1965, 1966; Hull, 1968; Ramsbottom et al., 1978) was of immense value in the resurvey of the western part of the sheet and provided a standard section to which other sections and boreholes could be referred. Pattison (1980) in an unpublished report has provided a comprehensive list of the Namurian macrofossils of the district and Owens (1972), also in an unpublished report, has described the palynology of the Throckley Borehole. Supporting palaeontological studies (unpublished) with particular reference to the upper part of the Stainmore Group include those by Owens (1978a, 1978b) and Riley (1982).

Stratigraphy

The lowest strata that crop to surface in the district are sandstones underlying the Crag Limestone, in the extreme north-west, but neither they nor the limestone are exposed, due to drift cover. For the underlying strata, the Chopwell Borehole* has provided an outline section of the sequence between the Great and Crag limestones. Where necessary to implement the description of the sequence, more especially in this lower part, data have been taken from records and sections in the nearby parts of the adjacent districts.

In this account, the lithostratigraphical nomenclature adopted is that generally used for the Alston Block (Figure 7; see also Dunham, 1990); local names, mainly those used in the Northumberland Trough, and alternative names, are quoted as they arise. Much of the information is taken from the investigations of Edmonds and Davies (1982) and Holliday and Pattison (1990), supplemented by work undertaken by the senior author during the period 1978–83. Supporting palaeontological studies include those by Owens (1972, 1978a, 1978b), Pattison (1980) and Riley (1982).

The nearest surface exposures of the **Great Limestone** are in the south-eastern part of the Bellingham district (Johnson, 1959; Frost and Holliday, 1980). There, the limestone is up to 15 m thick, and is characterised by the continuity of individual limestone beds and mudstone partings (Fairbairn, 1980). Three major divisions have been recognised in these outcrops, the highest of them, the 'Tumbler Beds', containing numerous mudstone beds. In the present district, the limestone (Figure 8) has only been proved in the Chopwell Borehole*, which recorded 17.4 m of 'blue limestone'. Near the top, a dark grey shale bed, 4.11 m thick, provided evidence for the presence of the 'Tumbler Beds'. In the Throckley Borehole* just to the north of the district, the limestone was only 13.8 m thick and most of the limestone is metamorphosed and partially recrystallised due to the proximity of the Whin Sill. Crinoid debris was common throughout, and abundant shell debris was recorded at a number of levels, including ribbed brachiopods, gastropods and bivalves near the top. Sporadic corals were present in an apparently massive 6 m-thick bed 1 m above the base. This is probably equivalent to the Main Posts, the middle division of the Great Limestone of the Bellingham district (cf. Frost and Holliday, 1980). Mudstone

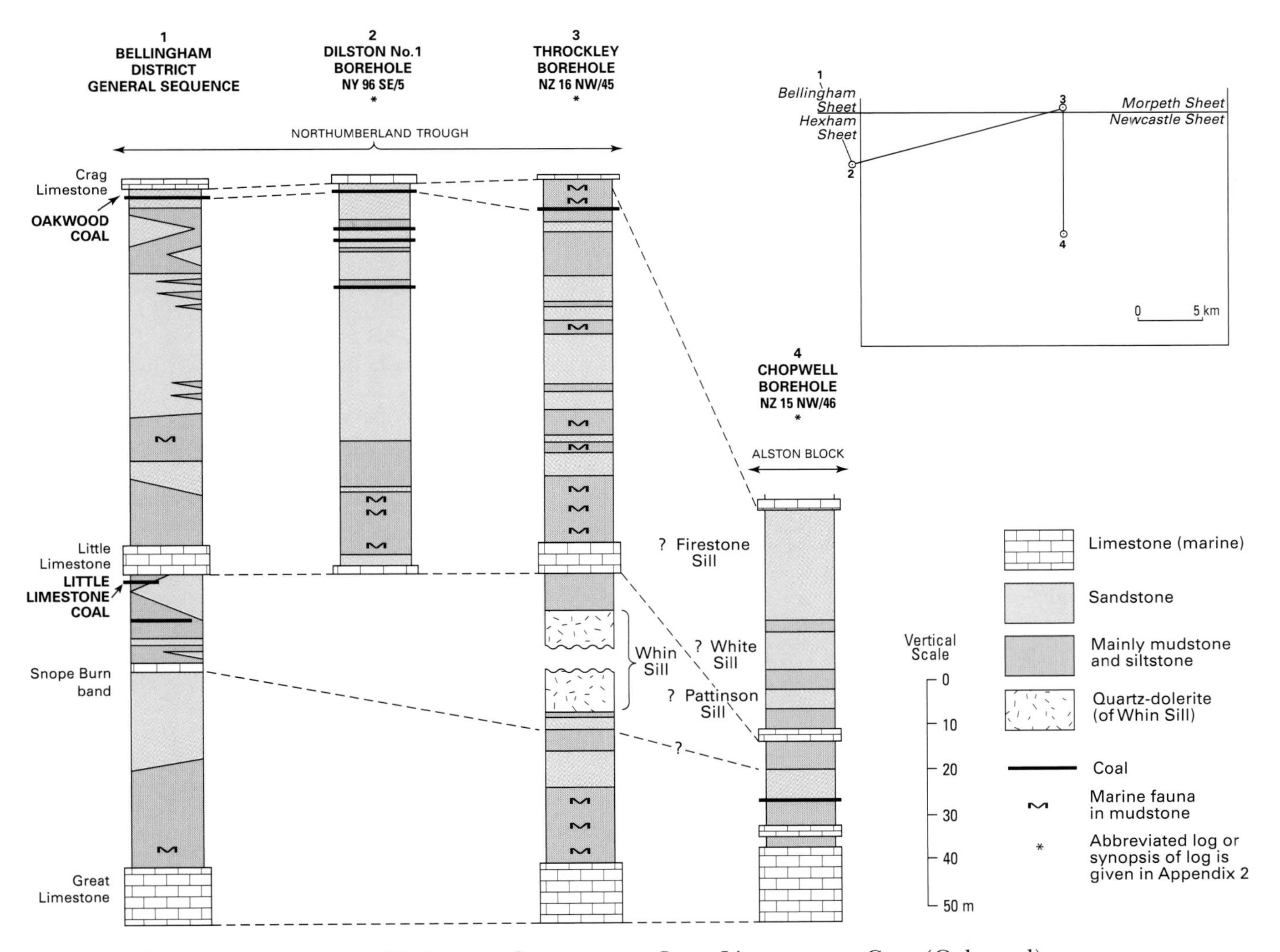

Figure 8 Comparative sections of Stainmore Group strata: Great Limestone to Crag (Oakwood) Limestone.

beds in the uppermost 2.5 m suggest the presence of the 'Tumbler Beds'.

The strata between the Great and Little limestones (Figure 8) range up to about 65 m thick and, as in the Bellingham district, can be divided into two broad divisions, although there is much local variation (see also Elliott, 1974, 1975). The lower division is a single broadly coarsening-upward sequence. Above, a more variable marine and deltaic upper division comprises several thin cyclothems and is characterised by the presence of thick coals, collectively known as the **Little Limestone Coal;** some of these seams were worked in the past in adjacent districts. A sandy marine interval, the **Snope Burn Band** (Trotter and Hollingworth, 1927; Hedley and Waite, 1929; Johnson, 1959) forms the lower bed of the upper division. Broadly similar sequences occur in the Northern Pennines, to the west and south of the district (Mills and Hull, 1976; Burgess and Holliday, 1979; Dunham, 1990; Hodge and Dunham, 1991). The only proving of this part of the sequence in the district is again provided by the Chopwell Borehole*, where the beds are only 20.44 m thick and the poor record is not easy to interpret (Figure 8).

The Little Limestone (Figure 8) is only 3.7 m thick in the Chopwell Borehole*, somewhat thinner than recorded in the Bellingham and Morpeth districts, where it ranges between 4 and 7.5 m in thickness and is in part argillaceous with mudstone partings. The overlying beds, up to the Crag Limestone (Figure 8), range between about 47 m thick in the Chopwell Bore to over 85 m thick in the Bellingham and Morpeth districts, where they were described by Lebour (1875), Hedley and Waite (1929) and Frost and Holliday (1980). Broadly, the sequence tends to be predominantly argillaceous near the base, with progressively thicker sandstones occurring in the middle and higher parts, and with coals and seatearths commonly present towards the top. Although the unit is often regarded as a major cyclothem, in detail the interval comprises two or more subcycles (Frost and Holliday, 1980). In the Chopwell Borehole*, three main sandstones are present. The details of the log are inadequate to provide a firm correlation, but it seems possible that the sandstones are broadly equivalent to the **Pattinson Sill**, **White Sill** and **Firestone Sill** of the Alston Block (Dunham, 1990), and also to three sandstones recognised in the Bellingham district (Frost and Holliday, 1980). The **Crag (Oakwood) Coal**, though not recorded in the district, occurs in adjacent areas and has been worked locally. The coal is normally separated from the base of the Crag Limestone by a lithologically variable interval ranging up to 5 m thick.

The **Crag (Oakwood) Limestone** (Figures 8 and 9), together with the underlying Crag (Oakwood) Coal, forms a readily recognised marker bed in boreholes but is not exposed in the district. The limestone ranges between 0.61 and 2.18 m in thickness. It is generally grey and argillaceous, with calcareous mudstone bands, and is commonly sandy towards the base. Comminuted fossil debris is generally common. Whereas in the Bellingham and Morpeth districts it is generally between 0.6 and 1 m thick, in the Chopwell* and DW/C boreholes* the limestone was 2.18 m and 2 m in thickness respectively.

The strata between the Crag and Lower Felltop limestones (Figure 9) range between 62 and 110 m thick within the district, forming a complex rhythmic unit in which several minor cycles are generally present. This interval contains several coarse-grained sandstones, forming complex bodies made up of overlapping and coalesced lobes which only rarely unite to form a continuous arenaceous sequence. There is little evidence of any deep channelling associated with these sandstones within the district. The sandstones are in part, equivalent to the Grit Sills and Slate Sills of the Alston Block (Dunham, 1990). The sandstones are generally interbedded with argillaceous measures which contain several thin marine bands and coals. Three marine bands, in particular, are persistent and widely recognised (Figure 9). The lowest of these, comprising shelly calcareous mudstone and siltstone, is the probable equivalent of the **Plankey Shell Bed** of the Hexham district and possibly of the Knucton Shell Beds of the Alston Block. Typically the marine band is characterised by abundant *Lingula* and other brachiopods. The second marine band, the **Aydon Shell Bed,** is lithologically more variable, locally comprising limestone, shelly sandstone, mudstone, or combinations of these. Faunal collections from this horizon, made from the Ouston and Throckley* boreholes in the Morpeth district, have yielded a characteristic Pendleian (E_1) shelly fauna (Pattison, 1980; Holliday and Pattison, 1990). The uppermost and best developed marine band in this part of the sequence generally forms a thin limestone, the probable correlative of the **Belsay Dene Limestone** of the Morpeth district and the Rookhope Shell Beds of the Alston Block. The limestone is characteristically sandy, with argillaceous partings, and is up to 2.5 m thick. Several thin coals and associated seatearths are generally present between the Crag and Lower Felltop limestones. Of these the **Crowhall Coals**, between the Crag Limestone and Aydon Shell Bed, and the **Corbridge Coal** below the Belsay Dene Limestone, are the only named seams. Near Corbridge, a seatearth mudstone underlying the Lower Felltop Limestone was formerly worked for industrial pipe manufacture.

The **Lower Felltop (Corbridge) Limestone** (Figure 10) marks a new phase of sedimentation during which, in the beds up to and including the Grindstone (Newton) Limestone, a series of relatively thick and persistent limestones was deposited in the southern part of the Northumberland Trough. The limestone ranges up to 6 m in thickness around Corbridge but is usually somewhat thinner than this; argillaceous partings are present more especially near the top. The fauna collected from the limestone (Smith, 1910; Hedley and Waite, 1929; Pattison, 1980) includes bryozoa, brachiopods and chaetetids (organisms of uncertain coralline affinities).

The thickness of the strata between the Lower Felltop and Upper Felltop limestones ranges between 31 and 65 m, being significantly less on the block than in the trough area (Figure 10). The sequence is characterised by marked lateral variation. The lower part tends to be dominated by more arenaceous beds, but this is not everywhere the case; the upper part is predominantly argillaceous but with several thin sandstones. The two

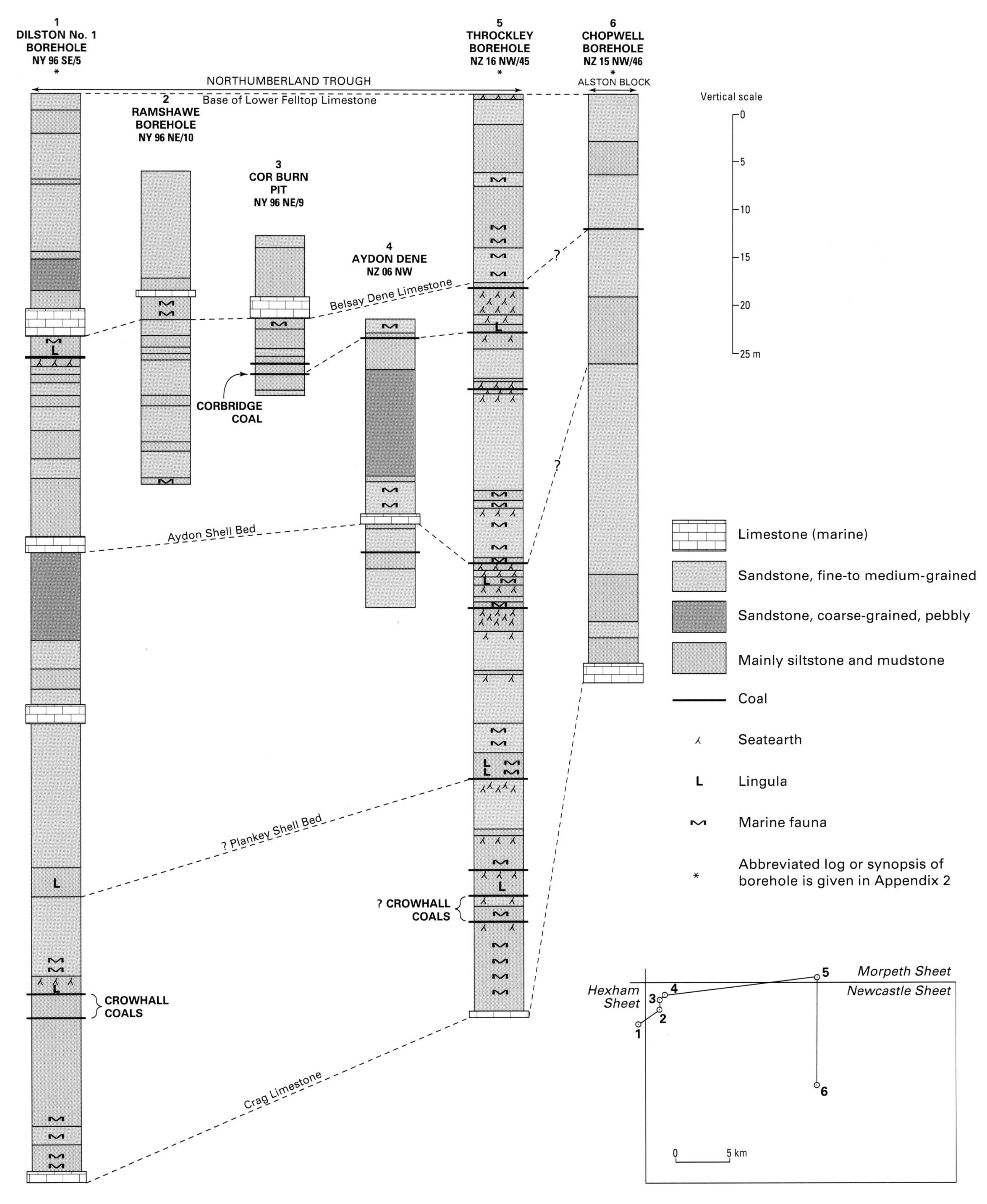

Figure 9 Comparative sections of Stainmore Group strata: Crag (Oakwood) Limestone to Lower Felltop (Corbridge) Limestone.

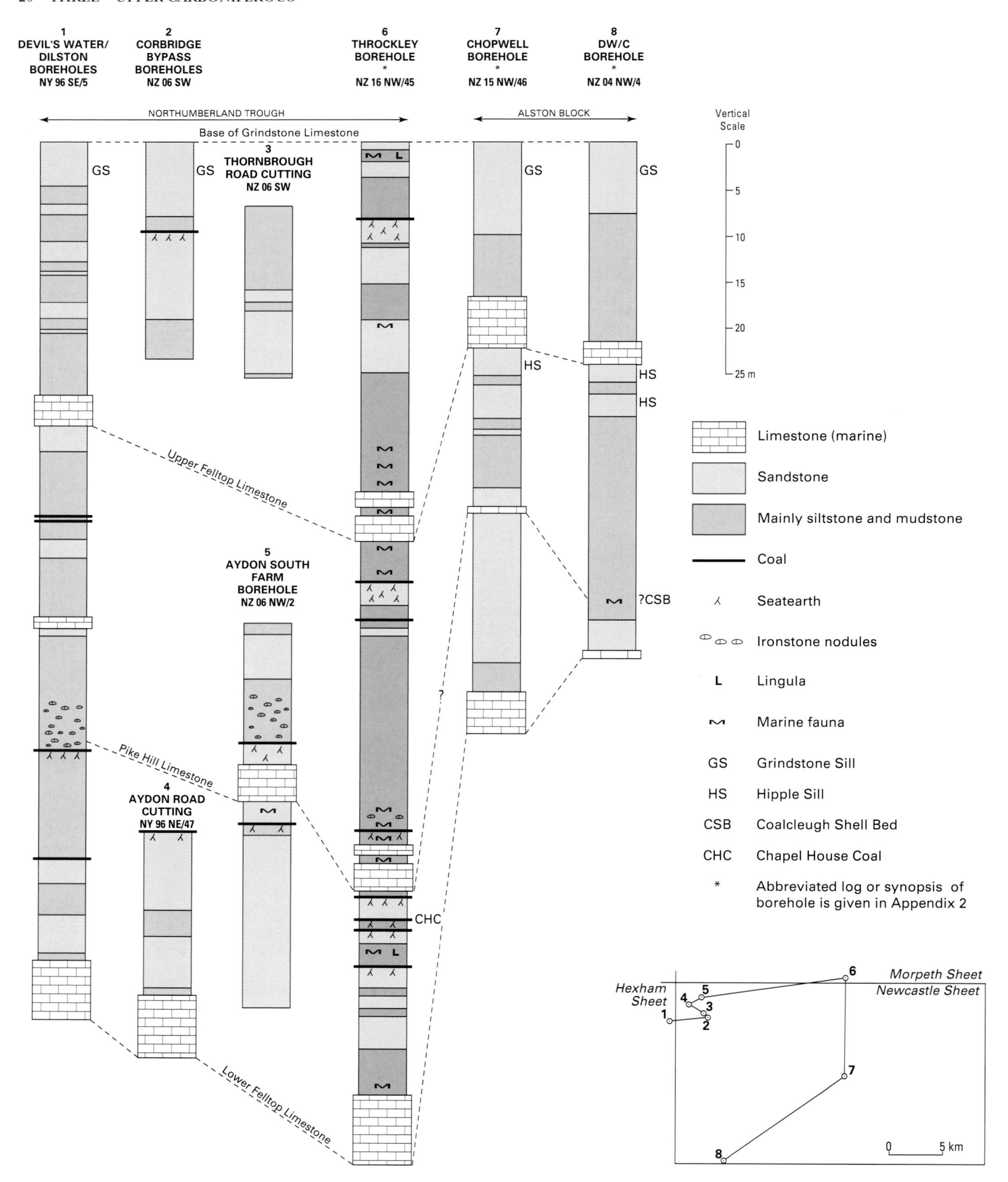

Figure 10 Comparative sections of Stainmore Group strata: Lower Felltop (Corbridge) Limestone to Grindstone (Newton) Limestone.

parts of the sequence are separated by the **Pike Hill Limestone**, named from Pike Hill in the Morpeth district, and the inferred equivalent of the **Coalcleugh Shell Bed** of the Alston Block. The Pike Hill Limestone is usually an argillaceous limestone, or a shelly calcareous mudstone with ironstone nodules, and ranges between 3.5 and 5 m in thickness. Faunal collections made from this limestone in the Ouston and Throckley* boreholes in the Morpeth district include clisiophyllid corals, bryozoa, brachiopods and a few bivalves. Thin subordinate coals and seatearths also occur in this sequence. One coal, the **Chapel House Coal**, which occurs in the strata below the Pike Hill Limestone, has been worked in the Stamfordham area of the Morpeth district.

The **Upper Felltop (Thornbrough) Limestone** (Figure 10) is one of the thickest and most consistent limestones in this part of the Namurian sequence. The limestone ranges between 2 and 7.3 m in thickness and is generally pale grey to grey, impure, medium to coarse grained, and crinoidal. It locally contains argillaceous mudstone partings, and in some places exhibits palaeokarstic features. The limestone contains one of the richest Namurian faunas in the district, with rugose corals, abundant bryozoa, brachiopods and rare bivalves recorded. The faunal content, and particularly the occurrence of the coral *Aulophyllum fungites*, suggests a direct correlation with the Lower Foxton Limestone of the Alnwick district of northern Northumberland.

The strata between the Upper Felltop and Grindstone limestones (Figure 10) range between about 17 and 40 m in thickness, and broadly comprise an alternating sequence of sandstone, siltstone and mudstone. Up to five sandstones, none exceeding 9 m thickness, are locally present and tend to predominate in the upper part of the interval. A well-developed sandstone, the **Grindstone Sill** of the Alston Block, is generally present under the Grindstone Limestone, and is commonly split into two parts in the south-west of the district. A thin coal has been recorded locally in the upper part of the sequence. Marine faunas, including brachiopods, were recorded in some boreholes from roof measures of the Upper Felltop Limestone, and also in siltstone and mudstone underlying the Grindstone Sill. The latter horizon has yielded the youngest known British graptolite, at a locality just south of the district (see p.28).

The **Grindstone (Newton) Limestone** (Figure 11) generally ranges between 2 and 10 m thick and forms a grey, fine-grained, bioclastic bed which is often argillaceous and nodular; thick mudstone interbeds have been recorded at some localities. In the south-western part of the district the limestone is locally replaced by hard calcareous mudstone and sandstone. In the type area of the Newton Limestone between Corbridge and Harlow Hill (Smith, 1917; Smith and Yu, 1943) the limestone contains a distinctive fauna of coral colonies, including *Aulina* and *Lithostrotion*, and the brachiopod *Latiproductus priscus*. Although this particular assemblage has not been recorded in any other Stainmore Group limestone of the Northumberland Trough or Alston Block, similar faunas are present in the equivalent Botany Limestone of the Stainmore Trough (Ramsbottom et al., 1978; Burgess and Holliday, 1979). The diagnostic E_{2b} nautiloid *Tylonautilus nodiferus* has also been recorded from this limestone.

The thickness of the strata between the Grindstone and Whitehouse limestones (Figure 11) ranges between 14 and 55 m. The lower part of the sequence consists largely of alternating thin sandstones, siltstones and mudstones; some of the argillaceous beds contain marine faunas, including *Lingula*. Two thin unnamed limestones and two thin coals have also been recorded. A thick channel sandstone is present near the base of this interval, in this and adjacent districts, and at one locality [NY 970 627] just to the west of the district, in the Devil's Water near Dilston, it appears to cut out the Grindstone Limestone. An important phase of channel sandstone development in the Midland Valley of Scotland, at or about the same horizon, has been reported by Read (1981). The **Styford Limestone** occurs broadly midway between the Grindstone and Whitehouse limestones. The limestone is imperfectly exposed, but has been recorded in several boreholes. At the type locality [0285 6215] in the banks of the River Tyne east of Styford Park, it consists of between 2 and 3 m of argillaceous limestone and calcareous shelly mudstone. Borehole evidence in the district supports the view that the limestone shows much variation and may be represented by shelly mudstone, or by up to 5 m or more of limestone, commonly lenticular, or by alternations of mudstone and limestone, locally with shelly sandstone. The Styford Limestone has no named correlative on the Alston Block, although calcareous limy bands or thin limestones have been recorded at a similar level in several boreholes, including that at Woodland (Mills and Hull, 1968) in the Barnard Castle district. A fauna of abundant brachiopods has been collected from the type locality at Styford. The varied brachiopod fauna, the highest in the Stainmore Group to contain plentiful productoids other than *Productus carbonarius*, suggests an age not later than Arnsbergian (E_{2b}); this is further confirmed by the presence of *Tylonautilus nodiferus*.

Measures between the Styford and Whitehouse limestones are known in detail only from boreholes; they appear to be mainly argillaceous in character, although a thin bed of sandstone is generally present at or near the top of the sequence. In the Throckley Borehole*, two very thin limestones and one thin coal were recorded at this level. The ammonoid *Homoceras* cf. *henkei*, indicative of the R_{1a} zone, was recorded at a level between the horizons of the Styford and Whitehouse limestones of the Woodland Borehole (Hull, 1968; Mills and Hull, 1976; Ramsbottom et al., 1978). This suggests that throughout north-eastern England, strata of the Arnsbergian (E_2) and Kinderscoutian (R_1) stages are separated in many places by as little as 5 m of poorly dated beds. There is no evidence for a regional unconformity at this level, and strata of the intervening Chokierian (H_1) and Alportian (H_2) stages have been inferred from palynological evidence (Owens and Burgess, 1965; Hull, 1968; Owens, 1972; Ramsbottom et al., 1978).

The **Whitehouse Limestone (Dipton Foot Shell Bed)** (Figure 12), a hard grey argillaceous limestone up to 1 m thick, and the overlying shelly mudstones up to 6 m

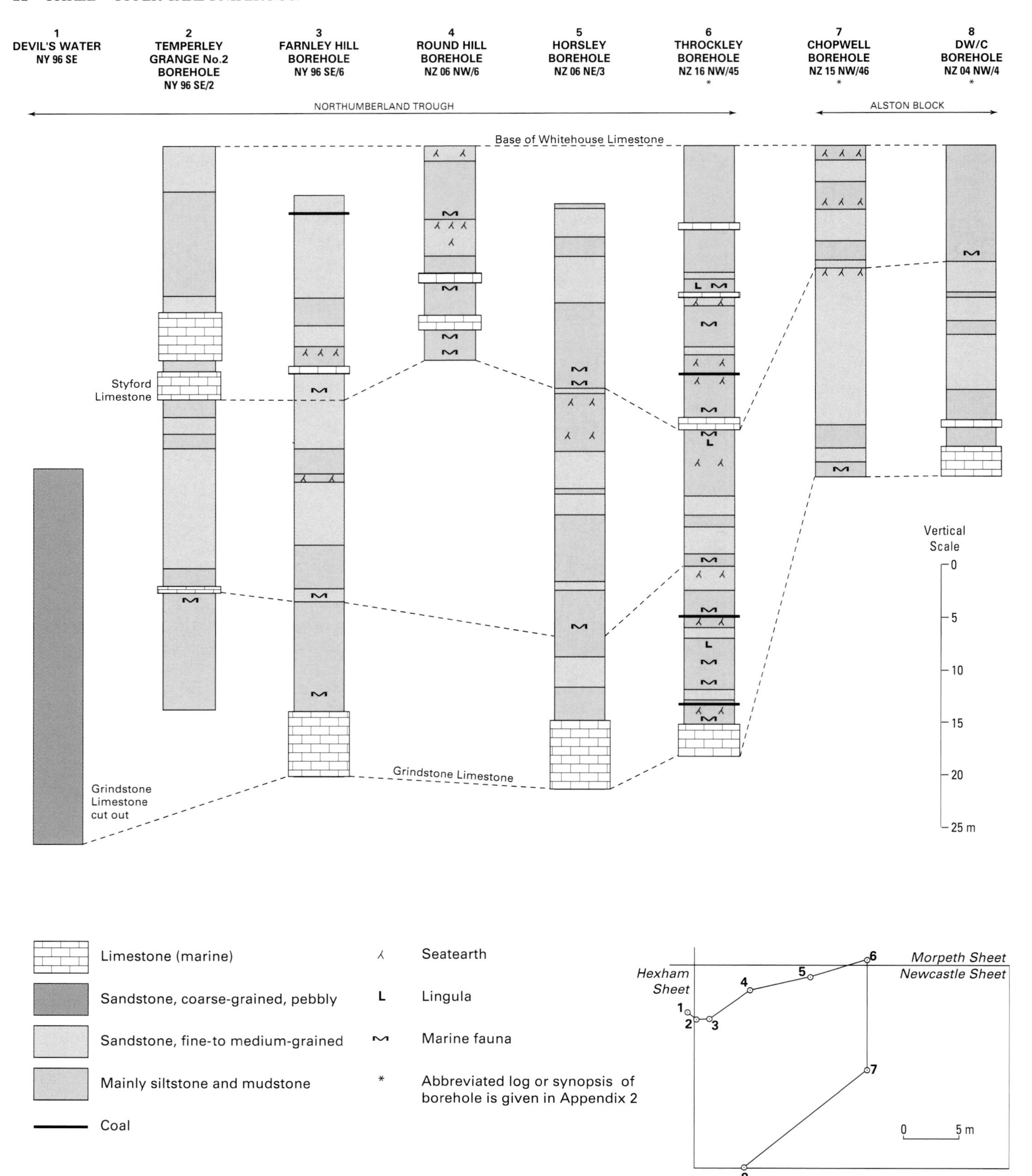

Figure 11 Comparative sections of Stainmore Group strata: Grindstone (Newton) Limestone to Whitehouse Limestone (Dipton Foot Shell Bed).

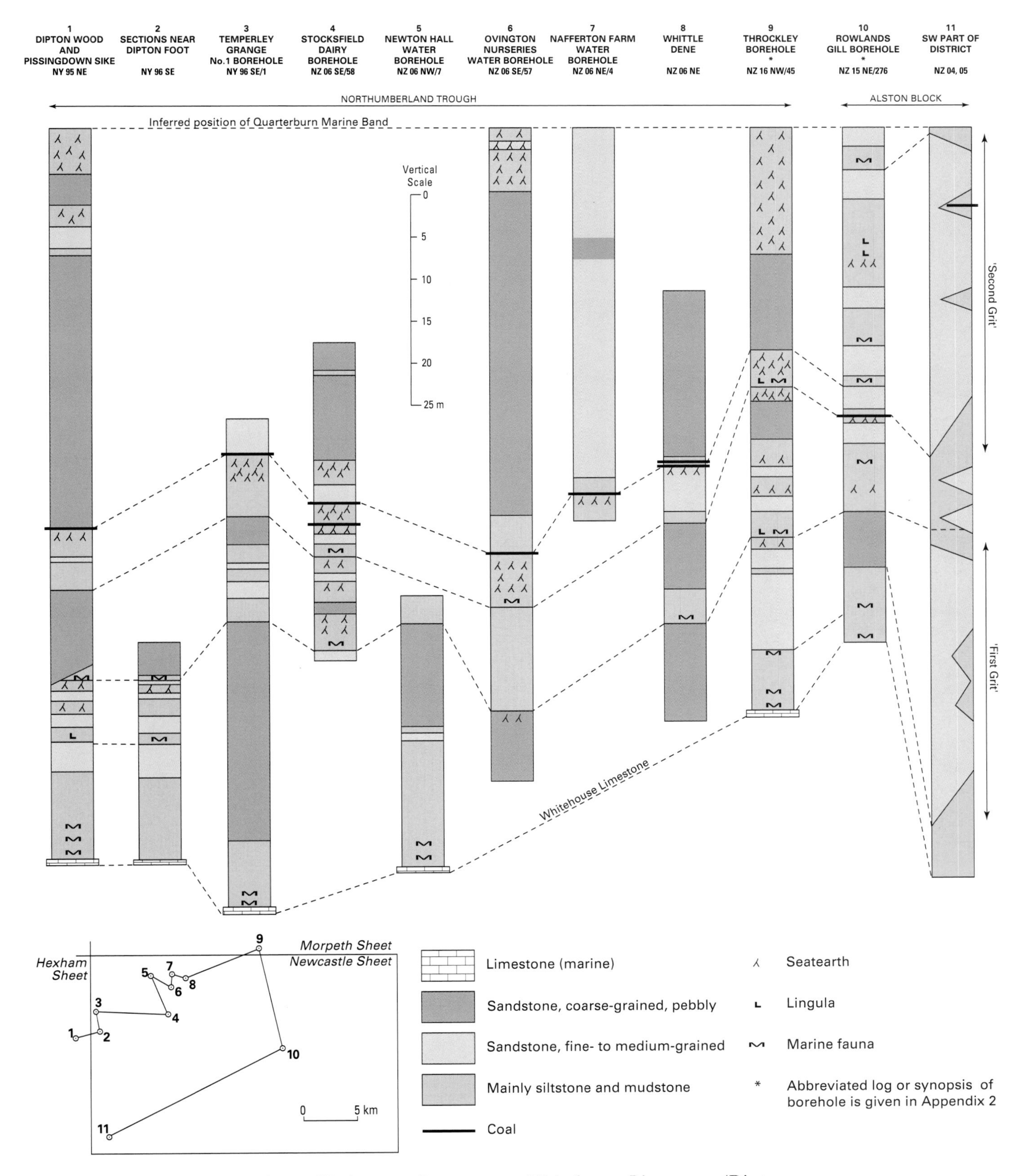

Figure 12 Comparative sections of Stainmore Group strata: Whitehouse Limestone (Dipton Foot Shell Bed) to Quarterburn (Subcrenatum) Marine Band.

thick, form a prominent stratigraphical marker in this and adjacent districts. Strictly, the Alston Block term Whitehouse Limestone (Mills and Hull, 1976) refers only to the thin basal limestone, generally the youngest Upper Carboniferous limestone encountered, whereas the Northumberland Trough name Dipton Foot Shell Bed (or Dipton Foot Marine Band) (Hedley, 1931) includes both the limestone and the overlying fossiliferous mudstones and shales. The ammonoid *Reticuloceras stubblefieldi*, indicative of the R_{1b} zone, has been found at this horizon at several localities in this and other districts (Ramsbottom et al., 1978). The fauna collected from March Burn [NY 9960 6044] is a typical, though richer than average, R_{1b} age fauna for northern England (Pattison, 1980). It is comparable to that found at this horizon on the Alston Block (e.g. Mills and Hull, 1976), and with those of the Peterel Shell Beds of the Hexham district and the Burnfoot Shale of the Brampton district (Trotter and Hollingworth, 1932).

The thickness of strata between the Whitehouse Limestone and the base of the Quarterburn Marine Band (Figure 12) ranges between about 62 and 90 m. A few metres of strata above the Whitehouse Limestone and its associated shelly mudstones generally consist of siltstone and mudstone with rare marine intercalations, but the remainder of the interval is much more arenaceous and corresponds to the lower and middle part of the 'Millstone Grit' of former classifications, including the 'First' and 'Second' Grits of the Primary Survey. Prior to the present resurvey, this part of the sequence was mapped as containing two main beds of 'grit', or medium- to coarse-grained sandstone, commonly pebbly, separated by subordinate argillaceous beds. Evidence has accumulated over a number of years that this is an oversimplified picture, and that the previously held assumption of lateral continuity in the 'grits' can no longer be sustained. While in general terms there is little doubt that many sections through this part of the sequence demonstrate a high arenaceous content, in detail the major sandstones are locally impersistent and discontinuous, and themselves include shale and mudstone beds and partings. Study of the borehole data, and of discontinuous surface exposures, supports the interpretation that the argillaceous rocks, with marine bands at some levels, are much more persistent and extensive than was hitherto supposed. In the detailed description below, the terms 'First Grit' and 'Second Grit' are retained for ease of reference, but only a broad indication of relative stratigraphical position is implied. The sandstones of this part of the Stainmore Group commonly have sharp erosive bases, but there is rarely any evidence for significant associated downcutting.

The highest sandstones of the Stainmore Group form bold, well-marked scarp features in the south-western part of the district (Plate 2). In general terms, two major laterally variable sandstones are present, each with up to two argillaceous intercalations. The sandstones are separated by up to 25 m of siltstones and mudstones, locally with marine intercalations. Towards the north-east, the same overall pattern appears to be maintained,

Plate 2 Derwent Reservoir [015 525] with Edmundbyers Common behind. Gently dipping strata of the Stainmore Group with 'Second Grit' forming prominent escarpment. L2536.

although it is less clear due to the occurrence of widespread drift.

The argillaceous rocks in this part of the Stainmore Group locally contain marine faunas, some restricted to *Lingula*. Faunal and floral evidence is meagre, but is sufficient to indicate the existence of both R_1 and R_2 zones with reasonable certainty (Hull, 1968; Ramsbottom et al., 1978). Some of the marine beds appear to be persistent over a significant area (Figure 12), and may equate with the Woodland Shell Beds and Spurlswood Shell Beds of the Barnard Castle district (Mills and Hull, 1976).

Great Limestone to Crag (Oakwood) Limestone (Figure 8)

These beds, including the Little Limestone, have only been proved within the district in the Chopwell Borehole* drilled in 1897 (Simpson 1904) in the Derwent Valley, 2 km west-south-west of Rowlands Gill. In the meagre record that follows, the term 'post' indicates sandstone.

	Thickness m
CRAG (OAKWOOD) LIMESTONE (top at 255.27 m)	
Limestone, blue, shelly	2.18
Post, grey; shaly towards base (?FIRESTONE SILL)	6.79
Shale, dark grey	1.03
Post, grey; shale partings	1.58
Shale, dark grey, soft	0.48
Post, grey, strong	11.97
Shale, grey	2.69
Post, grey; shale partings (?WHITE SILL)	7.77
Shale, grey	1.38
Post, grey	0.60
Shale, dark grey	3.81
Post, grey; shale partings (?PATTINSON SILL)	2.98
Shale, dark grey; iron pyrites	5.76
LITTLE LIMESTONE	
Limestone, blue	3.71
Post, grey; shale partings (?WHITE HAZLE)	1.65
Shale, grey; post girdles (sandstone layers)	5.39
Post, grey, hard; shale partings (?SNOPE BURN BAND at top)	6.14
Coal	0.10
Post, grey	0.38
Shale, grey; pyrites	6.78
GREAT LIMESTONE (base at 349.96 m)	
Limestone, blue	2.09
Shale, dark grey ('Tumbler Beds')	4.11
Limestone, blue	15.32

In borehole DW/C* at Lambshield Moss in the south-west of the district, the Crag Limestone (at 200 m) consists of 2 m of grey, fine-grained, medium- to thick-bedded, generally argillaceous limestone with calcareous mudstone bands at the base; much comminuted shell debris is present. The limestone rests on 0.54 m of fossiliferous mudstone, coaly at the base, which overlies 1.93 m+ of siliceous sandstone, correlated with the Firestone Sill.

Strata between the top of the Crag (Oakwood) Limestone and the base of the Lower Felltop (Corbridge) Limestone (Figure 9)

The full sequence has been recorded in only two boreholes (Chopwell* and Lambshield Moss*) within the district. The lower part of the interval, including the Crowhall Coals and Plankey Shell Bed of the Hexham district, is only known from these boreholes and from adjoining areas. The higher parts are indifferently exposed in the north-western part of the district, and in immediately adjacent areas.

Around 5 m of fine-grained flaggy sandstone with silty and argillaceous partings, underlying or near the base of the Aydon Shell Bed, are exposed near Aydon Castle [0002 6621 to 0039 6653]. Nearby boreholes indicate the presence of numerous impersistent beds of seatearth, limestone and marine shale in the sandstone. In addition, near the base of the Aydon Shell Bed, there is a persistent coal seam which crops out [0018 6633] in Aydon Dene where it is 0.30 m thick. Paynological studies (Owens, 1978a) indicate a Pendleian (E_1) age for these exposures.

The Aydon Shell Bed is also exposed in Aydon Dene. At the type locality [0018 6633] it is composed of 4.65 m of sparsely shelly sandstone, with thin bands of limestone, mudstone and siltstone. At nearby localities [0018 6621 and NY 9962 6591] 0.5 m of shelly sandstone and 0.3 m of very sandy grey limestone, respectively, form the upper part of this bed. Similar lithological variations have been found in nearby boreholes.

Strata overlying the Aydon Shell Bed, including coarse-grained pebbly sandstones and argillaceous beds, in total about 20 m thick, are sporadically exposed in Aydon Dene [NY 9952 6588 to 0054 6673]. These beds also include thin coals, locally up to 0.45 m thick; the Corbridge Coal of Hedley and Waite (1929) lies about 5 m below the assumed position of the Belsay Dene Limestone.

The Belsay Dene Limestone is a widely recognised, but not well-exposed, marker bed in the district. The best exposure is in Aydon Castle Quarry [0115 6725] where 1.8 m of siliceous or sandy crinoidal limestone is overlain by 0.2 m of weathered calcareous mudstone. Nearby to the east, overgrown exposures in Aydon Dene are no longer visible, except for 0.7 m of shelly sandstone [0020 6620], which is at or just below the limestone base. Hedley and Waite (1929) appear to have mistaken this limestone for the Lower Felltop Limestone, despite the marked difference in lithology and bedding characteristics between the two limestones.

The measures between the Belsay Dene and Lower Felltop limestones around Corbridge consist of a thick-bedded coarse-grained pebbly sandstone overlain by argillaceous strata. In Milkwell Lane [NY 9915 6543] 2 m of this sandstone, coarse-grained, feldspathic, and showing cross-bedding directed to the west-south-west, are visible in the road cutting. This sandstone and overlying argillaceous strata are capped by a fireclay which underlies the Corbridge Limestone. The fireclay was worked until quite recently. At Hook Hill Colliery [NY 998 656], the 'seggar' or seatearth was stated to be 1.02 to 1.83 m thick and separated from the Lower Felltop Limestone by 0.30 m of brown sandstone. In Deadridge Quarry [NY 9960 6564], a short distance to the west, 0.5 m of seatearth-mudstone exposed beneath the limestone was formerly worked.

To the south-east and south, on the Alston Block, the full sequence of strata between Crag and Lower Felltop limestones has been proved in the Chopwell* and DW/C* boreholes, although in neither case is it possible to correlate individual beds with certainty. The borehole sequences are considerably thinner than those of the Northumberland Trough. In the Chopwell Borehole the interval is 61.75 m thick and consists largely of sandstones, which may in part be the equivalent of the Grit Sills and Slate Sills of the Alston Block. A bed of grey sandstone and limestone, 0.41 m thick and 36.09 m above the top of the Crag Limestone, may represent the Aydon Shell Bed, while a very thin coal some 14 m below the Lower Felltop Limestone may indicate the approximate position of the Belsay Dene Limestone. In the DW/C Borehole the interval is 61.99 m thick and, although sandstone also predominates, the following tentative correlations are suggested: Plankey Shell Bed, 0.76 m of calcareous sandstone at 186.73 m (equivalent of the Knucton Shell Beds); Aydon Shell Bed, 0.25 m of calcareous sandstone at 166.36 m. In the Rowlands Gill Borehole* the Belsay Dene Limestone may be represented by 2.77 m of limestone containing crinoid and shell fragments, 10.35 m below the Lower Felltop Limestone.

Lower Felltop (Corbridge) Limestone (Figure 10)

The limestone, formerly well exposed in and around Corbridge is thought to be a correlative of the Lower Felltop Limestone of the Alston Block. The best exposures are now restricted to Deadridge Quarry [NY 9964 6549 to NY 9967 6569] where 6 m of grey, fine-grained, bioclastic limestone are visible (Plate 3). The limestone has numerous apparently persistent argillaceous partings, and thin mudstone interbeds towards the top. Sporadic exposures of the limestone are also visible in the now largely overgrown Aydon Quarry [0100 6625] and in a small digging near Greenleighton [0241 6718]. Smith (1910) noted that the fauna of the Corbridge Limestone in this area commonly includes *Chaetetes, Composita ambigua, Latiproductus latissimus* and *Edmondia sulcata*, and these observations have been confirmed, as have Hedley and Waite's (1929) comments on the general absence of corals, other than chaetetids of uncertain affinities, and the abundance of *Buxtonia* (Pattison, 1980). Palynological studies (Owens, 1978b) here only confirm an early Namurian age.

The Lower Felltop Limestone also has been recorded in the following boreholes: Chopwell*, limestone, grey, 4.77 m thick (at 193.52 m); Rowlands Gill*, limestone, grey, thin- and medium-bedded, argillaceous at top, abundant crinoid and shell fragments, sparse brachiopods, 4.63 m thick (at 212.88 m); Bore DW/C*, limestone, grey, sandy, thin-bedded, fine- to medium-grained, shell debris, 0.42 m thick (at 136.04 m).

Strata between the Lower Felltop and Upper Felltop limestones (Figure 10)

In and around Corbridge, a thick sandstone generally overlies the Lower Felltop Limestone, although a thin mudstone parting is normally present between them, as in Deadridge Quarry [NY 996 656] where it is 0.6 m thick. In the adjacent Aydon road cutting [0031 6534] on the A69, 15 m of fine- to medium-grained, flaggy, cross-bedded sandstone with two mudstone beds were recorded, above which a thin, poor quality coal and 0.4 m of seatearth-sandstone were seen. Palynological studies (Owens, 1978b) indicate an E_2a to E_2b age. Borings around this level have proved a number of coals, one of which, the Chapel House Coal, was sufficiently thick to have been worked north of the district.

Close above these coals is the Pike Hill Limestone, which was formerly exposed at the eponymous locality [076 706] to the north of the district. It typically comprises between 3.5 and 5 m of limestone, argillaceous limestone, and shelly calcareous mudstone containing ironstone nodules. Borings west of Corbridge suggest that the Pike Hill Limestone passes into the Coalcleugh Shell Bed, which consists of shelly mudstone with abundant ironstone nodules. Although noted by the Primary Surveyors, other previous workers (e.g. Hedley and Waite, 1929; Hedley, 1931) did not recognise the Pike Hill Limestone and assigned exposures and borehole provings of the bed to other limestones. Loose dark grey crinoidal limestone, poorly exposed in Brockhole Burn [0280 6612], is now correlated with the Pike Hill Limestone, but was previously equated with the Lower Felltop (Corbridge) Limestone. Just over 4 m of limestone,

Plate 3 Corbridge Limestone (Stainmore Group) at the type locality, Deadridge Quarry [NY 995 656]. The adit is excavated into the underlying fireclay. L1721.

proved in the Aydon South Farm Borehole [0083 6607], is considered to be the Pike Hill Limestone, as is limestone formerly visible at Kip Hill [029 678] in the Morpeth district.

In the north-western part of the district, strata overlying the Pike Hill Limestone are not well exposed. In Brockhole Burn, 0.75 m of mudstone overlain by flaggy sandstone are exposed [0275 6591]. Several metres of coarse-grained gritty sandstone, probably forming a channel infill, are visible at Crags [034 672] and in a quarry [0350 6693] near Shildonhill. A thin coal close to the base of the Upper Felltop (Thornbrough) Limestone was formerly seen near Thornbrough [008 647] and other thin coals at this level have been recorded in a number of boreholes.

Several kilometres to the east-south-east, the full sequence has been proved in boreholes at Chopwell and Rowlands Gill. In the Chopwell Borehole* these strata are 38.87 m thick and include a high proportion of sandstone, especially in the lower part. A grey limestone, 0.82 m thick and roughly midway in the sequence, is equated with the Pike Hill Limestone, or Coalcleugh Shell Bed of the Alston Block. In the Rowlands Gill Borehole* strata are 34.13 m thick, and largely consist of interbedded sandstone and siltstone. However, a medium grey thin-bedded laminated limestone containing a fauna of corals and productoid brachiopods, 2.55 m thick and 21 m below the Upper Felltop Limestone, is considered to be the equivalent of the Pikehill Limestone and Coalcleugh Shell Bed. Also in this borehole, two very thin coals were recorded, one near the top and one near the base of the interval.

In the south-west of the district, strata underlying the Upper Felltop Limestone are visible in the incised meander of the River Derwent between 1 and 2 km south-east of the hamlet of Muggleswick. The valley margins hereabouts are locally very steep, wooded, and in part landslipped, but a composite section seen on the south (County Durham bank) of the river, west and north-west of Combfield House [0589 4915], revealed the following:

	Thickness m
Upper Felltop Limestone	—
Mudstone, dark grey	0.63
Sandstone, ganisteroid; passes laterally into mudstone	0.43
Mudstone, dark grey, shaly at base	2.56
Sandstone, thin-bedded at base, massive at top	3.04
Mudstone, dark	0.12
Sandstone, thin-bedded	1.34
Mudstone, grey, silty; siltstone beds and lenses	1.30+

Broadly similar sections are visible along the banks and slopes of the Derwent Valley in this general vicinity. In Bore DW/C* these strata are 31.18 m thick and are largely argillaceous. They include a thin sandstone, possibly the High Grit Sill, overlying the Lower Felltop Limestone and two thin sandstones, probably the Hipple Sill of the Alston Block, underlying the Upper Felltop Limestone. A shelly mudstone some 5 m above the Lower Felltop Limestone is tentatively equated with the Coalcleugh Shell Bed. In the Derwent–Wear Tunnel (Appendix 3) the Coalcleugh Shell Bed is thought to be represented by about 1.5 m of dark grey, muddy, shelly limestone at chainage 2100–2165 m. The limestone overlies thin argillaceous beds with 0.05m of coal at the base, resting on sandstones referred to the Grit Sills.

Upper Felltop (Thornbrough) Limestone (Figure 10)

The Upper Felltop Limestone was extensively exposed in old workings in the Thornbrough area, east and north-east of Corbridge, but many of these are now overgrown. The limestone is generally between 3 and 6 m thick, and typically composed of medium- to coarse-grained crinoidal limestone, pale grey towards the top, and passing down to grey, fine-grained limestone. Well-marked bedding planes, commonly with mudstone partings, and locally showing palaeokarstic phenomena, are a typical feature of this limestone. It is characterised by abundant *Aulophyllum fungites*, bryozoa and brachiopods, including several dictyoclostid and gigantoproductoid forms. The limestone can be seen in the River Tyne at Thornbrough Wood [0057 6353] and in old diggings around Thornbrough [0092 6428; 0100 6447], but the best exposure in the north-west of the district is seen alongside the A69 Corbridge by-pass road [0098 6466] and in quarries towards Aydon [007 655].

In the Chopwell Borehole* the limestone was 5.99 m thick, with shale partings, and in the Rowlands Gill Borehole* it was 7.34 m thick, medium and dark grey, thin and medium bedded, with argillaceous laminae towards the top; common crinoid debris and shell fragments were noted.

In the south-west of the district the Upper Felltop Limestone is seen on both banks of the River Derwent in the incised meander between 1 and 2 km south-east of Muggleswick. Calcareous mudstones and limestones are underlain by well-bedded limestone, the whole totalling some 4 m. In a borehole [0493 4990] north-east of Calf Hall, near Muggleswick, 4 m of earthy crinoid limestone are referred to this horizon. A borehole [0229 5089] sunk prior to the construction of the Derwent Dam recorded 2.67 m of grey massive limestone, sandy at the top; the lowest part of the bed was dark grey and crinoidal. Exposures of the limestone are lacking south of Edmundbyers, but the approximate position of the bed can be deduced by surface indications along the valley margins of the Eudon and Feldon burns [NY 990 482 and NY 999 482]. In Borehole DW/C* the limestone, 2.04 m thick, is dark grey, muddy, thin to medium bedded, and fine grained with abundant comminuted fossil debris; at the base it is shaly.

Strata between the Upper Felltop (Thornbrough) and Grindstone (Newton) limestones (Figure 10)

The strata in this interval are not well exposed in the north-west of the district. Part of the succession is visible in the cutting on the A69 road near Thornbrough [0110 6445 to 0122 6430], where the following section was recorded:

	Thickness m
Siltstone with sandstone beds and laminae	7.60
Mudstone, silty with siltstone laminae	1.25
Sandstone, fine-grained	1.20
Siltstone with sandstone laminae	0.95
Sandstone, fine- to medium-grained (probably about 5 m above the Upper Felltop Limestone)	4.90

Mudstones and shales at a somewhat higher level are sporadically exposed in Brockhole Burn [0238 6513 to 0217 6484].

In the Chopwell Borehole* these measures, some 17 m thick, include 10.04 m of grey very coarse-grained sandstone, the probable equivalent of the Grindstone Sill, immediately underlying the inferred position of the Grindstone Limestone. In the Rowlands Gill Borehole* the strata, 32.58 m thick, largely consist of sandstone and interbedded siltstone. The assumed Grindstone Sill forms a buff/grey medium- and thick-bedded, mostly coarse-grained, feldspathic sandstone 8.26 m thick; siltstone clasts and quartz pebbles were present in the basal 2 m.

In the south-western part of the district these strata are poorly exposed at the top of steep banks forming the incised meander of the River Derwent, between 1 and 2 km south-east of Muggleswick. The following composite section was recorded in the south (County Durham) bank of the river, west and north-west of Combfield House [0589 4915].

	Thickness m
Sandstone	c.3.04
Mudstone; sandstone beds	4.57
Mudstone, passing into sandstone towards the south-west	2.31–3.65
Mudstone	0.40–0.91
Upper Felltop Limestone	—

Broadly similar sections are visible in the general vicinity but are largely inaccessible. A borehole [0229 5089] sunk prior to the construction of the Derwent Dam recorded:

	Thickness m
Shale, grey to brown, sandy at base (near presumed position of Grindstone Limestone)	6.33
Sandstone, brown-grey, cross-bedded (Grindstone Sill)	7.24
Shale, grey, sandy, micaceous	1.98
Sandstone, greyish brown, argillaceous, massive	1.22
Shale, grey; fine-grained sandstone laminae; some slumping	8.99
Shale, dark grey	5.48
Upper Felltop Limestone	—

On the south side of the Derwent Reservoir, north-north-east of Ruffside, the Grindstone Sill is partially exposed [NY 9930 5204] and consists of nearly 3 m of ferruginous, coarse-grained, rotted calcareous sandstone containing brachiopods. It overlies yellow seatearth-mudstone. A fauna collected from this locality by Mr K Anderson of Struthers Farm, Edmundbyers, includes *Tylonautilus nodiferus* (E_2b), stroboceratid nautiloids, a smooth nautiloid, brachiopods and crinoid debris. Hereabouts, the Grindstone Sill appears to be split into two leaves. Some distance farther to the north-east, and again on the margin of the reservoir, the base of the upper leaf is exposed [NY 9969 5213] and consists of 1 m of flaggy fine-grained sandstone, on argillaceous sandstone and mudstone.

In Feldon Burn, and just south of the district, pyritic black shale [NY 9974 4695] from just below the Grindstone Sill contained ammonoids, gastropods, *Lingula*, small bivalves, byozoa and crinoids. Quite the most interesting specimen obtained from the locality, was, however, the graptolite *Pseudodictyonema heyi* sp. nov. Collected by Mr K Anderson, this is the last-known British graptolite (Chapman et al., 1993, p.314).

In Borehole DW/C* these strata are 21.50 m thick; the Grindstone Sill is a muddy, impure, fine-grained, well-bedded sandstone 7.28 m thick, predominantly grey, but locally tinged green or mottled brown and purple. The remainder of the sequence in this borehole consists almost entirely of mudstone. In the Derwent–Wear Tunnel (Appendix 3) the Grindstone Sill was encountered between Chainage 325 and 440 m, and possibly from 640 to 700 m.

Grindstone (Newton) Limestone (Figure 11)

In the north-west of the district the Grindstone Limestone, generally between 3 and 10 m thick, is a grey, fine-grained bioclastic bed, commonly argillaceous and nodular, with mudstone partings at some localities. In this area the fauna of the limestone is very distinctive, yielding corals and latissimoid productoid brachiopods. The most extensive outcrop occurs to the west and north of Newton, with three principal exposures [0296 6492; 0339 6540; 0357 6558]. The limestone is also exposed in Brockhole Burn [0239 6449 and 0245 6445], was formerly seen in shallow diggings near Farnley Hill [NY 9980 6295], and was proved in shallow bores at Styford Gravel Pit [013 634] and along the line of the Corbridge bypass [0155 6398]. Corals suggestive of this limestone were recorded by Green (1954) and Whittaker (1963) in old workings north of Newton Fell House [0325 6665]. Farther east the outcrop of the limestone is obscured by thick drift.

In the Chopwell Borehole* 1.85 m of grey shale and shelly limestone occur at the inferred level of the limestone but in the Rowlands Gill Borehole* it is apparently absent, its horizon being represented by mudstone containing *Lingula* and bivalves 5.60 m above the Grindstone Sill.

In the south-western part of the district in the vicinity of Edmundbyers and the Derwent Dam, the Grindstone Limestone is not visible at surface, and indications suggest that it is very thin, or represented by calcareous shelly mudstone. In Feldon Burn [NY 9973 4687], just south of the district, the limestone contains a fauna of brachiopods, crinoids and a nautiloid. In Borehole DW/B [0352 5008] west of Muggleswick, two thin limestone beds, 0.30 and 0.27 m thick and separated by 2.40 m of calcareous sandstone and mudstone, may form this horizon, whereas farther south, in Bore DW/C*, the bed consists of 2.34 m of grey to dark grey argillaceous medium-bedded limestone. In the few other boreholes that penetrate this sequence hereabouts the Grindstone Limestone is absent or was not recorded. In the Derwent–Wear Tunnel (Appendix 3) a dark grey limestone recorded between Chainages 470 and 635 m and between 714 and 975 m is thought to be the Grindstone.

Strata between the Grindstone (Newton) Limestone and the base of the Whitehouse Limestone (Dipton Foot Shell Bed) (Figure 11)

In the north-west of the district, shales and sandstones above the Grindstone Limestone are seen in Brockhole Burn [0237 6376], and in the road cutting [017 639] near Brocks Bushes where the following sequence was recorded.

		Thickness m
Sandstone, fine-grained	(seen)	3.20
Siltstones, sandstones and mudstones, alternating		3.10
Sandstone, fine-grained		1.70
Siltstones, sandstones and mudstones, alternating		7.60+

The fine-grained sandstone at the top of this section was formerly extensively worked near High Barnes [021 633].

Overlying this sandstone or its equivalent is the Styford Limestone. Although proved in several boreholes, it is imperfectly exposed in this part of the district. Between 2 and 3 m of shelly, brachiopod-rich, calcareous mudstone and argillaceous limestone are visible at the type locality on the banks of the River Tyne [0285 6215], and about a metre of limestone and mudstone were seen during the excavations for the A69 road diversion. The Styford Limestone was formerly visible in old quarry workings [NY 996 628] south of Corbridge. Even in the poor exposures available, there is good evidence for much lateral lithological variation, with some or all of the limestones being lenticular.

Strata between the Styford and Whitehouse limestones are not exposed in the north-west of the district. In the valley of the Stocksfield Burn, between Wheelbirks Cottages and Hindley Bridge, about 10 m of grey shale and greyish brown fine-grained sandstone have been recorded [0570 5830] underlying the Whitehouse Limestone.

In the Chopwell Borehole* the strata between the Grindstone and Whitehouse limestones are 29.09 m thick and broadly consist of a thick basal sandstone overlain by largely argillaceous measures with two thin sandstones. The Styford Limestone was not recorded, but its horizon may lie near the top of the basal sandstone. The Styford Limestone was also not recorded in the Rowlands Gill Borehole*. In the south-west of the district, in Borehole DW/C*, these strata are at least 18 m thick and largely consist of arenaceous measures at the base; above, a calcareous mudstone with brachiopods, some 18 m above the Grindstone Limestone, may correlate with the Styford Limestone although no direct equivalent of this bed on the Alston Block has been proved.

Whitehouse Limestone (Dipton Foot Shell Bed) (Figure 12)

The Dipton Foot Shell Bed, and the associated Whitehouse Limestone, are exposed [NY 9960 6044] at the type locality in March Burn, south-east of Dipton Foot, where they comprise at least 2.4 m of shelly mudstone overlying 0.5 m of argillaceous shelly limestone. The rich fauna here includes the ammonoid *Reticuloceras*, the bryozoan *Rhombopora*, a wide range of brachiopod species, notably *Crurithyris*, *Productus carbonarius* and *Punctospirifer northi*, and the bivalves *Palaeolima simplex* and *Parallelodon regularis*. Some distance to the west, in a stream [NY 9904 6039] near Hollies House, a 1.1 m bed of shelly siltstone and mudstone between beds of fine-grained sandstone may be at broadly the same position in the sequence. The shelly mudstones are also visible at a number of nearby localities [e.g. NY 9903 6040; NY 9720 5960]. Some 2 km to the north, the Whitehouse Limestone was recorded as 0.77 m of grey, argillaceous, fossiliferous limestone in Temperley Grange No. 1 Borehole [NY 9898 6211]. In the valley of the Stocksfield Burn, between Wheelbirks Cottages and Hindley Bridge, the Whitehouse Limestone is possibly represented by 0.2 m of grey-brown fossiliferous limestone, but this locality [0570 5830] is so isolated from other sections and boreholes that it is difficult to be conclusive.

In Acton Burn [NY 9830 5288], on the north side of the Derwent Reservoir and a few metres west of the district boundary, 0.40 m of ferruginous decalcified shelly sandstone, overlain by 0.30 m of shelly mudstone and siltstone, are exposed. The sandstone, thought to be the equivalent of the Whitehouse Limestone, contains the following: productoid debris indet., bellerophontid indet., *Euphemites ureii*, *Tropidostropha* sp., and *Coleolus* sp. The argillaceous beds have yielded crinoid debris, bryozoa, brachiopods, bivalves, cephalopods including the naultiloid *Catastroboceras* sp. and the ammonoid *Reticuloceras* cf. *stubblefieldi* (juv.), together with ostracods and fish debris, tending to confirm the correlation with the Whitehouse Limestone (Riley, 1982).

In the Chopwell Borehole* the limestone, merely described as 'grey', is 0.58 m thick. In the Rowlands Gill Borehole* the limestone and associated shell beds consist of 1.64 m of dark grey, calcareous, laminated siltstone with an 0.05 m bed of limestone at the base; the fauna includes brachiopods and crinoids.

Strata between the Whitehouse Limestone (Dipton Foot Shell Bed) and the base of the Quarterburn (Subcrenatum) Marine Band (Figure 12)

North-western part of district In this area, which forms part of the southern margin of the Northumberland Trough, the only complete sections are those provided by the Throckley Borehole* (Morpeth district) and a composite sequence in the Dipton Wood area [NY 965 606], built up from scattered sections by R G Carruthers during the revision of the Hexham district to the west.

The strata between the Whitehouse Limestone and the base of the 'First Grit' range between about 8 and 20 m in thickness, and usually comprise argillaceous beds, locally with thin sandstones. A poorly exposed sequence of mudstone, sandstone and seatearth is visible at this level in Riding Mill Burn [NY 9720 5960; 9903 6037; 9979 6093]. The slopes on either side of this stream and for some distance farther downstream [NY 9920 6097 to 9999 6097] are formed by coarse-grained, locally pebbly, sandstone at the base of the 'First Grit'. The erosive base of the sandstone is exposed in places. The sandstone can be up to 40 m thick, but in some areas it appears to be split into two or three leaves, especially near the top. The more important additional localities at which major exposures of the 'First Grit' are visible include the following:

	Thickness m
Prospect Hill [NY9921 6236]	
Sandstone, fine-grained	0.75
Seatearth-mudstone	0.75
Sandstone and mudstone	0.50
Sandstone, fine-grained	0.75
Sandstone and siltstone	1.0–1.5
Sandstone, cross-bedded, coarse-grained (position near top of a lower leaf of the sandstone)	7.0+

Temperley Grange [NY 985 623]: cross-bedded sandstone (Plate 4)

Newton [0345 6449]: sandstone, coarse-grained

Whittle Dene [0728 6587 to 0735 6560]: sporadic exposures of coarse-grained gritty sandstone

Horsley bypass [094 662]: sandstone, locally cross-bedded, coarse-grained, up to 8 m thick

North Dunslaw [0855 6640]: sandstone, coarse-grained, feldspathic

The strata between the 'First' and 'Second' grits are largely argillaceous, and include a coal, seatearths and (locally) a marine band. In Temperley Grange No. 1 Borehole [NY 9898 6211] the interval included dirty coal 0.19 m, on grey sandy fireclay 3.25 m, on sandy shale 4.32 m, on 3.71 m of coarse-grained kaolinitic sandstone, an upper leaf of the 'First Grit'.

The 'Second Grit' also ranges up to 40 m in thickness, and may similarly be split into two leaves. Localities at which it may be seen include:

The River Tyne at Bywell [0498 6177]: sandstone, coarse-grained, cross-bedded, up to 6 m thick, is exposed on both banks of the river; in a nearby borehole [0576 6168], east of Stocksfield Hall, the sandstone was recorded as 13.77 m thick.

Stelling Hall [0511 6576 and 0566 6177]: sandstone, coarse-grained, massive to flaggy, up to 7 m.

Whittle Dene [0753 6521]: sandstone, rusty brown, cross-bedded, fine- to coarse-grained, sharp erosive base, 6 m+ thick.

Planetree Banks [0393 6349]: sandstone, coarse-grained, in overgrown quarry.

The strata overlying this sandstone, up to the inferred position of the Quarterburn Marine Band, were exposed in the A68 road cutting [0204 6059] on Shilford Bank. The section recorded was as follows:

	Thickness m
Quarterburn Marine Band (inferred position)	—
Sandstone, rootlets at top	2.13

Plate 4 'First Grit' (Stainmore Group) at Temperley Grange [NY 985 623]. Coarse-grained, cross-bedded sandstone. L1726.

	Thickness m
Siltstone, with coal 0.23 m at centre	3.28
Sandstone	1.52
Siltstone, with coal 0.08 m at base	1.91
Sandstone	1.83+

Healey area In the extreme west of the district, west and north-west of Hedley Hall [0040 5780], in the middle and higher reaches of March Burn and its tributaries, poorly exposed sections are visible. At Healeyburn Wood [NY 9971 5884] the following sequence was recorded:

	Thickness m
Sandstone, thick-bedded, medium-grained	1.6
Shale	0.07
Shale, ferruginous, fossiliferous	0.23
Seatearth and shale	0.3
Coal	0.02
Seatearth	0.3
Sandstone	2.0
Sandstone, shaly, micaceous	0.6

The fossiliferous shale may also be seen farther north [NY 9953 5920 and NY 9945 5935], and appears to lie just below the 'Second Grit', which is present both to the east and west of March Burn. It may be the Woodland Shell Beds of the Barnard Castle district (Mills and Hull, 1976). The sandstone is predominantly medium to coarse grained and is exposed at various localities farther south in the higher reaches of the burn.

Stocksfield Some 6 km to the east-north-east, strata high in the Stainmore Group are exposed in the upper reaches of the Stocksfield Burn between Wheelbirks Cottages and Hindley Bridge, and also in the lower reaches of an associated tributary, the Lynn Burn. The section recorded is as follows:

	Thickness m
Inferred position of Quarterburn Marine Band	—
Shale, locally ferruginous	1.0
Sandstone, grey, knobbly, fine-grained	0.6
Shale, grey	0.5
Seatearth, brownish grey	0.5
Sandstone, brown, cross-bedded (?'Second Grit')	10.0
(Section obscured)	1.0
Shale, grey, and limestone, sandy; fossiliferous [0531 5924 and 0508 5846]	0.4
Sandstone, grey, fine-grained	0.25
Seatearth, greyish brown	0.5
Sandstone, brown, fine-grained, rooty	1.0
Shale, grey [0570 5830]	0.2

The sandy limestone and fossiliferous shale may equate with the marine beds in the sections described above in March Burn, and may also correlate with a sequence in the higher reaches of Reaston Gill [NY 9883 5721]:

	Thickness m
Sandstone, medium- to coarse-grained	12.0
Shale, shelly, especially in lower part	2.0
Seatearth	1.6
Sandstones and seatearth	1.0
Shales and mudstones, shelly at base	2.5+

Broadly similar sections are present in the other nearby streams running into Stocksfield Burn. However, because the exposures are poor and the strata laterally variable, firm correlation is not possible.

North and north-east of the Derwent Reservoir In the area of Pithouse Fell [NY 996 539] and Manor House Farm [035 539], a sequence of thick sandstones separated by argillaceous measures is much obscured by drift. The following section was recorded in Acton Burn [NY 9830 5288], a few metres to the west of the district.

	Thickness m
Sandstone, flaggy and shaly	0.60+
Mudstone and siltstone	4.26

Coal and seatearth, sandy; coal in two leaves 0.02 and 0.15 m thick	0.27
Mudstone and siltstone	1.05
Sandstone, thick-bedded, medium-grained	0.30
Siltstone	1.53
Siltstone and mudstone, shelly	0.30
Sandstone, decalcified, muddy, ferruginous; fossiliferous (?Whitehouse Limestone)	0.38

The major sandstone overlying this exposure is probably the equivalent of the lower leaf of the 'First Grit.' This sandstone is exposed some distance upstream and forms a thick, cross-bedded, coarse-grained pebbly bed.

Winnowshill Common [NY 986 534] is formed by the lower leaf of the 'Second Grit', here at least 13 m thick composed of thick-bedded medium- to very coarse-grained pebbly sandstone with sporadic mudstone layers. The sandstone is also exposed in Winnowshill Quarry [NY 9815 5388] just to the west.

On Pithouse Fell, flaggy coarse-grained sandstone is exposed [NY 9941 5413] in old diggings. Disused quarries north of Cronkley [0175 5290], and near Birkenside [0307 5235], Carterway Heads [0475 5182] and Fine House [0470 5310], all expose flaggy or cross-bedded medium- to coarse-grained pebbly and feldspathic sandstones referred to the 'Second Grit'. Argillaceous strata within this sandstone-dominated sequence are exposed at Calfclose Hall [0390 5230], where blocky fissile shales 3 m in thickness are overlain by 2 m of blocky grey silty shales capped by 0.5 m of flaggy sandstone. In a disused quarry [0350 5222] the same shales have yielded fish fragments. Disused adits nearby mark the position of a thin impersistent coal.

The high, largely drift-covered ridges around Moor Game [0285 5385] and to the north-east of Manor House [0445 5250] are capped by sandstones close to the top of the Stainmore Group.

Long sections in the 'First' and 'Second' grits were proved in the Letch House-Airy Holm-Derwent Tunnel (Appendix 3). The 'First Grit', a grey and dull yellow sandstone, commonly cross-bedded, medium- to coarse-grained and feldspathic, contained irregular masses of highly decomposed and rotted rock. Sections through various parts of the 'First' and 'Second' grit sequence were also provided by boreholes such as Airy Holm No. 18* and Airy Holm No. 22*.

North side of Derwent Valley from Allensford to near Ebchester The north bank of the River Derwent and its tributaries opposite Consett and Shotley Bridge provide many sections through strata high in the Stainmore Group. The following sequence is traversed in an eastward dipping succession in Mere Burn.

	Thickness m
Quarterburn Marine Band [0887 5485]	—
Seatearth-mudstone, pale grey, silty; grades to	0.15
Seatearth-sandstone, pale brown, fine-grained, micaceous; coarser towards base	1.00
Sandstone, pale brown, flaggy to cross-bedded, medium- to coarse-grained, micaceous ('Second Grit'); exposed to confluence with river [at 0937 5498]	25.00

Seatearth-mudstones overlying flaggy sandstones in Snow's Green Burn, downstream from Snow's Greenburn Bridge [0921 5341], are thought to equate with the seatearths below the Quarterburn Marine Band in Mere Burn. In Burnhouse Gill, farther to the south-west, coarse-grained, cross-bedded, feldspathic sandstones are exposed in the stream bed [0808 5276] immediately upstream of a fault that brings the Stainmore Group into contact with the Lower Coal Measures. For about 0.5 km south of Shotley Bridge [0907 5275], the River Derwent flows through a steep-sided ravine (Plate 5) incised into a thick sandstone ('Second Grit') which underlies the Quarterburn Marine Band. The sandstone, probably over 15 m thick, is flaggy to cross-bedded, medium- to coarse-grained, micaceous, and locally pebbly and feldspathic. Impersistent beds of silty shale also occur.

Wallishwalls Burn enters the River Derwent north of Allensford Bridge [0769 5024] where the following composite section is recorded:

	Thickness m
Sandstone, brown to white, cross-bedded, medium- to coarse-grained, conglomeratic towards base [0810 5046]	4.0
Siltstone, green-grey, argillaceous, micaceous; eroded top	0.95
Sandstone, pale brown to white, flaggy and cross-bedded, medium- to coarse-grained; conglomeratic towards base (base of 'Second Grit')	12.00
Mudstone, dark grey, silty; thin beds of fine-grained sandstone, eroded top	3.25
Sandstone, green-yellow, coarse-grained	1.25
Mudstone, dark grey, silty	1.85
Sandstone, pale brown, cross-bedded, medium-grained [0769 5024]	0.75

In an isolated section in Fairley May Gill, some 8 km to the north-north-west, a small area of high Stainmore Group strata occurs on the south side of the Ninety Fathom Fault. The following sequence was recorded underlying the inferred Quarterburn Marine Band.

	Thickness m
Shales, seatearths and thin sandstones	?8.0
Sandstone, brown, cross-bedded, medium- to coarse-grained	10.0
Shale, grey, fossiliferous [0447 5746]	0.6
Seatearth, brownish grey	0.3

Derwent Valley near Rowlands Gill In the Chopwell Borehole* the strata between the Whitehouse Limestone and the Quarterburn Marine Band form a relatively condensed sequence 35.63 m thick, of which the upper part is largely sandstone. The lower part consists mainly of interbedded sandstone and shale. In the Rowlands Gill Borehole* these measures are 61.93 m thick, much of the sequence consisting of interbedded sandstone and siltstone. Except for 6.53 m of coarser-grained, locally pebbly, sandstone near the base, the coarse arenaceous facies so characteristic of this part of the sequence is largely absent. Marine fossils were noted at six separate levels.

Edmundbyers, Muggleswick and south side of River Derwent to Allensford In this area, and extending to the southern margin of the district, the sandstones tend to form bold escarpments and ridge features, with the argillaceous beds forming intervening hollows, or 'slacks'. Exposures are poor, however, and detailed information is largely wanting. The total thickness of strata probably ranges up to 90 m and consists of two main sandstones, themselves locally split, separated by up to 20 m of

Plate 5 'Second Grit' (Stainmore Group) in River Derwent at Shotley Bridge [088 522]. Coarse-grained, cross-bedded sandstone. L1770.

argillaceous strata (Figure 12, column 11). On Edmundbyers Common [NY 9953 4908] over 8 m of cross-bedded coarse-grained sandstone, locally rotted, ('First Grit') are exposed. Similar exposures occur along the length of the ridge feature formed by the sandstone.

The most important exposures east of Burnhope and Feldon Burns, and including the Muggleswick Park area, are listed below.

Cathouse Quarry (disused) [0180 4896] ('Second Grit'): sandstone, flaggy, cross-bedded, medium- to coarse-grained, 1.52 m, on sandstone, well-bedded, medium- to coarse-grained, 2.13 m; basal 0.43 m soft, brown, rotted.

Exposure formerly recorded [0132 4866]: shale, fireclay and coal in old diggings.

Black Cleugh Crags [0288 4882]: sandstone, thick-bedded, coarse-grained, occasionally pebbly ('Second Grit'), 2.43 m+ thick.

Old digging south of Calf Hall [0455 4943]: sandstone, pinkish and dull yellow weathering, coarse-grained, pebbly ('Second Grit)' about 3 m thick.

Exposure near Calfclose Hall [0452 4853]:

	Thickness m
Sandstone, ('First Grit'), coarse-grained, becoming more silty towards base	2.43
Shale	0.60
Not exposed	3.04
Sandstone, silty, micaceous	1.82
Sandstone, medium- to thin-bedded	3.04+

Coalgate Burn [0480 4848]:

	Thickness m
Sandstone, massive becoming flaggy towards base, coarse-grained	3.65+
Shale	0.60
Not exposed	3.04
Sandstone, silty, micaceous	4.87+

Hisehope Burn at Hisehope Bridge [0479 4830]: shelly sandstone, near level of the Whitehouse Limestone.

Horsleyhope Burn, east bank [0621 4866 to 0624 4839]: interbedded sandstone, mudstone and shale containing three shelly sandstone bands, c. 18.28 m thick.

Healeyfield Quarry [0658 4835] formerly recorded:

	Thickness m
Sandstone, coarse-grained, pebbly ('Second Grit')	2.13
Sandstone, flaggy, micaceous	1.21
Shale, dark, carbonaceous	0.91
Sandstone, coarse-grained, friable	2.43
Ganister (not seen)	—

A prominent bed of sandstone, referred to the 'First Grit', is visible along much of the south bank of the River Derwent between the Healeyfield Fault [0656 4975] in the west and to between Wharnley Burn [0748 5002] and Allensford [0770 5024] in the east. The sandstone is well seen at Wharnley Burn itself where the lower part of the bed forms a prominent waterfall, while slightly farther west at Ravens Crag [0717 4988] about 30.48 m of massive to cross-bedded gritty sandstone, coarse-grained and often pebbly, are exposed. Micaceous carbonaceous mudstones and siltstones with thin, sometimes calcareous, sandstone ribs are indifferently exposed underlying the base of the sandstones in this vicinity, and these are assumed to occupy a position between the base of the 'First Grit' and a few metres above the Whitehouse Limestone. Some distance farther to the east, a section [0779 4992] at West Fines Wood formerly recorded the following:

	Thickness m
Sandstone, cross-bedded, mainly coarse-grained ('Second Grit')	9.14
Mudstone, silty, shelly at base	1.98
Mudstone, calcareous, shelly	0.30
Not exposed	0.15
Limestone	0.30
Sandstone and seatearth	0.91
Sandstone	3.04

The limestone here may be at the horizon of the Woodland Shell Beds (Mills and Hull, 1976), but the occurrence of limestone at a stratigraphical level above the Whitehouse Limestone is unusual.

In the eastern part of the district, west of Gateshead, a borehole [1988 6167] sunk from the base of Swalwell Engine Pit proved 51 m of strata underlying the inferred position of the Quarterburn Marine Band. However, the record is old (1839) and lacks detail. An 18 m-thick bed of sandstone logged towards the bottom of the borehole may be equivalent to the 'Second Grit'. In Delta Ironworks Borehole* at Gateshead, 33.5 m of strata underlying the assumed position of the Quarterburn Marine Band were recorded. This included 3.81 m of sandstone thought to form a leaf of the 'First Grit', underlain by shales containing *Lingula, Productus* and *Orbiculoidea.*

COAL MEASURES

The generalised succession of the Coal Measures in the district, based on data from the whole Northumberland and Durham Coalfield, is shown in Figure 13, together with the current classification and nomenclature. The total preserved thickness of the group within the district is 480 m. Biostratigraphical zonation is based on the non marine bivalves ('mussels') but the stage boundaries are drawn at marine bands, which, though relatively thin, are commonly extensive and in some instances permit international correlation.

Generalised descriptions of the Coal Measures of the district and the wider region appear in accounts by Winch (1817), Buddle (1831a, 1831b), Hurst (1860), Lebour (1878), Brown (1888), Murton (1892), Kirsopp (1907), Woolacott (1913), Carruthers et al. (1931), Hickling and Robertson (1949), Johnson (1970), Taylor et al. (1971) and Jones and Magraw (1980). Comprehensive accounts of the Coal Measures of adjacent areas include those by Smith and Francis (1967), Land (1974), Lawrence and Jackson (1986, 1990) and Smith (1994). Investigations into the lithology and sedimentary structures of rock types, more especially with reference to sandstones, have been carried out by Haszeldine (1983a, 1983b, 1984a), Haszeldine and Anderson (1980), Fielding (1982, 1984a, 1984b, 1984c, 1986) and Fielding and Johnson (1986). Richardson and Francis (1971) have studied the occurrence of fragmental clayrocks in this and other districts. Considerable attention has been paid to Coal Measures palaeontology, both flora and fauna, in this district and a wider area. The floras were studied by Howse (1890a, 1890b), Kidston (1922) and Bolton (1926), and faunal studies include work by Hopkins (1927–1934) and Calver (1968a, 1968b) on nonmarine bivalves, and Pollard (1966, 1969) on ostracods. A large amount of unpublished palaeontological data on the Coal Measures is held in BGS archives, and has been used in the compilation of the memoir. However, many borehole records submitted to BGS include fossil names which are not backed up by collected material, and which are therefore unverified.

The former Geological Survey classification (Table 2) placed the boundary between the Millstone Grit and Coal Measures at the Ganister Clay Coal, which lies near the top of the 'Third Grit' of the Primary Survey. However, following the formal division of the Coal Measures of England and Wales by Stubblefield and Trotter (1957), the base of the Coal Measures in north-east England was lowered to the base of the Quarterburn Marine Band, the inferred position of the Subcrenatum Marine Band which defines the base of the Westphalian Series (Mills and Hull, 1968, 1976; Land, 1974; Ramsbottom et al., 1978). The base of the Coal Measures was thereby lowered by about 50 m. Following Stubblefield and Trotter (1957) and current practice in nearby districts (Smith and Francis, 1967; Land, 1974; Mills and Hull, 1976; Smith, 1994), the Coal Measures of the district are now divided into Lower and Middle divisions at the base of the Harvey Marine Band, the local correlative of the Vanderbeckei Marine Band (Ramsbottom et al., 1978), rather than at the Brockwell Coal as in earlier classifications (Figure 13; Table 2). The Coal Measures of the district are of Westphalian A to C age (Ramsbottom et al., 1978). New names for these divisions, introduced since the 1:50 000 map was compiled, are used in this memoir (Table 2; Figure 13). No strata of the Upper Coal Measures, as currently defined, occur within the district.

The conditions of deposition and lithology of the Coal Measures have been outlined above. The sequence consists of interbedded shales, mudstones, siltstones, sandstones, coals and seatearths, with a few thin marine bands. Although there is a wide local variation in the proportions of the different lithologies present, a cyclical pattern of sedimentation is evident throughout, and the overall thickness is remarkably constant. The lowest part of the sequence, below the Ganister Clay Coal, contains only thin and sporadic coals and is characterised by a high proportion of sandy strata. These were formerly referred to collectively as the 'Third Grit', thereby emphasising the similarity between this part of the Coal Measures and the highest beds of the underlying Stainmore Group. The remainder of the sequence, above the Ganister Clay Coal, includes widely persistent coals, although seams thicker than 0.9 m are largely confined to the beds between the Brockwell and High Main coals. Twenty coals (if split seams are counted as single seams) have been worked,

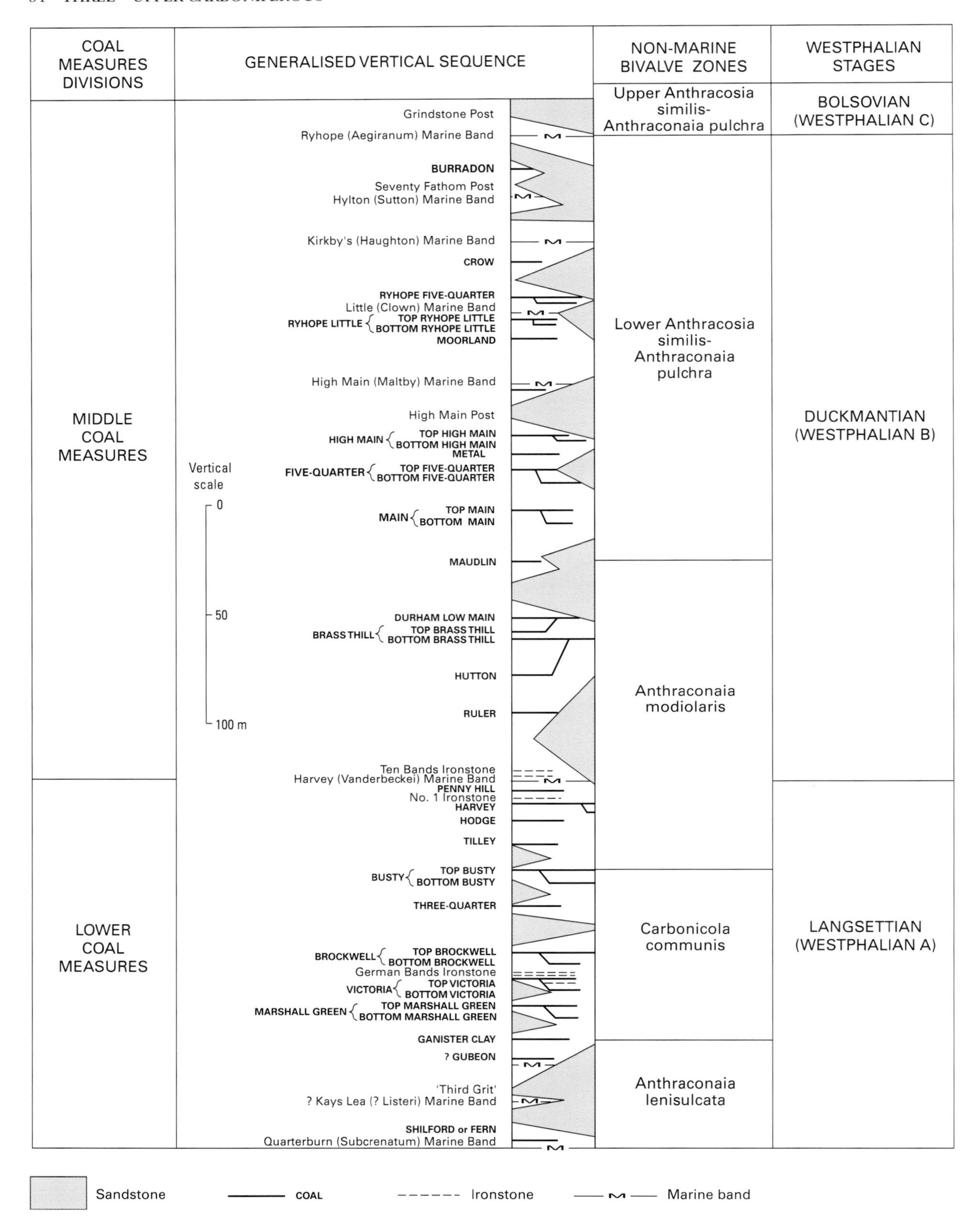

Figure 13 Coal Measures: generalised succession, nomenclature and classification.

three to only a limited extent. Above the Brockwell Coal the thicker and more widespread sandstones tend to occur between the Brockwell and Tilley coals, above the Durham Low Main Coal, above the High Main Coal, and above Kirkby's Marine Band.

The Coal Measures occupy the whole of the central and eastern part of the district, and also extend up to the western margin in a narrow tongue on the interfluve between the Tyne and Derwent valleys. The structure is comparatively simple, so that the lower measures generally crop out around the western fringes of the main coalfield, whereas the higher strata are present around Stanley and Annfield Plain, east of Jesmond Dene in the extreme north-east, and in the higher areas of Gateshead along the eastern margin of the district.

Exposure of the Coal Measures is generally poor due to the drift cover, being largely restricted to sandstones on the higher ground and steeper slopes, and to isolated and discontinuous stream sections. The lack of surface exposure is offset by the numerous borehole logs, shaft records and mine plans, which provide a wealth of data on these strata.

Seam correlation

Traditionally, each colliery applied its own set of seam names. This led to much confusion and to a proliferation of local names (Table 3). Thus, not only were individual seams given numerous local names, but where the same name was used in several collieries it was commonly applied to different coals. Later, standard sets of coal seam names were established separately in County Durham and Northumberland, but only in a few instances (notably in the lower measures) were similar names used for the same coals. This led to further confusion, a situation made worse by the frequent miscorrelation of seams within and between the two counties. The National Coal Board then assigned standard names and index letters, and it is now accepted that these largely hold good throughout the coalfield. The standard names are used in this account, but the local names are shown as well as the standard names and index letters in Table 3.

Stratigraphy

Lower Coal Measures

The base of the Lower Coal Measures in England and Wales is drawn at the Subcrenatum Marine Band, which also defines the base of the Langsettian Stage (Westphalian A) (Figure 13; Table 2). The position of the **Quarterburn Marine Band**, the inferred correlative of the Subcrenatum Marine Band, is generally determined from meagre borehole evidence, and from the proven or mapped outcrops of key stratigraphical markers such as the Ganister Clay or Marshall Green coals and the underlying 'Second Grit'. In the few localities where the marine band is exposed, it comprises up to 5 m of mudstone and silty shale, generally associated with silty seatearth, and an irony rib (or nodules) together with a knobbly and rooty, commonly calcareous, sandstone containing the trace fossil *Zoophycos*. Scattered *Lingula* fragments are commonly present in the mudstone. At some localities a more prolific fauna has been obtained, including brachiopods, molluscs, echinoderms, sponge spicules, crinoid columnals, fish and plant debris. At one of these localities, a tiny nodose fragment of the ammonoid *Gastrioceras?* was collected, but no definite occurrence of *Gastrioceras subcrenatum* has been recorded.

The **strata between the Quarterburn Marine Band and the Ganister Clay Coal** (Figure 14) range up to 60 m in thickness, and are largely dominated by sandstones in which thin impersistent ganisters are locally recorded, more especially in the south-west. The sandstones that dominate this part of the sequence were formerly referred to collectively as the 'Third Grit', the top part of the 'Millstone Grit' of the Primary Survey. The name 'Third Grit' is useful for general descriptive purposes, although up to four main beds of sandstone of widely varying thickness may be present, and only rarely is the bulk of the sequence made up of a single bed. The sandstones range from fine to coarse grained (locally pebbly), and are cross-bedded in places.

Several argillaceous partings, which generally comprise mudstone, shale, coal and seatearth, occur within the sequence. These are commonly cut out by, or pass laterally into, sandstone. Up to four thin impersistent coals have been recorded. The most notable are the local **Shilford** or **Fern Coal** not far above the Quarterburn Marine Band, a coal 15 to 20 m above the marine band, and another coal between 15 and 20 m below the Ganister Clay. Due to lack of evidence, doubt must remain over the correlation of these coals. It is possible (Figure 14) that the thin seam up to 20 m above the marine band may be equivalent to the **Saltwick Coal**, while that between 15 and 20 m below the Ganister Clay may be the **Gubeon Coal**. The Saltwick Coal, exceptionally up to 0.6 m thick, has been proved in the Morpeth and Tynemouth districts (Lawrence and Jackson, 1986, 1990; Land, 1974). A marine band recorded in the roof of this coal in the Rowlands Gill* and Throckley* boreholes is tentatively equated with the **Kays Lea Marine Band** of the Barnard Castle district (Mills and Hull, 1976). This is the inferred equivalent of the Listeri Marine Band, a marker widely recognised in British coalfields and elsewhere in north-western Europe (Ramsbottom et al., 1978). The supposed Gubeon Coal has been recorded in a number of bores but the overlying **Gubeon Marine Band**, present in the Morpeth, Tynemouth and possibly Sunderland districts (Lawrence and Jackson, 1986, 1990; Land, 1974; Smith, 1994), has not been proved.

Throughout the district, the **Ganister Clay Coal** (Figure 15) forms a thin impoverished seam seldom more than 0.2 m thick; only rarely does it appear to be absent. The boundary between the *Anthraconaia lenisulcata* and *Carbonicola communis* nonmarine bivalve zones is taken at this coal (Figure 13).

The **strata between the Ganister Clay and Marshall Green coals** (Figure 15) generally range between 8 and 15 m in thickness, and are commonly characterised by a high proportion of sandstone. The horizon of the Well

Table 3 Coal seam nomenclature.

This memoir	Index letter	Co. Durham (includes Gateshead area)	Northumberland (includes Newcastle upon Tyne)	Local Names of coal seams frequently in previous use (where different from this account)				
				N. side of Tyne Valley (including Newcastle)	S. side of Tyne Valley (including Gateshead)	Chopwell, Medomsley and Consett	Burnopfield, Annfield Plain & Stanley	Low Fell, Birtley and Chester-le-Street
Ryhope Five-Quarter		Ryhope Five-Quarter	Ryhope Five-Quarter				Crow; Charlaw	Three-Quarter
Ryhope Little		Ryhope Little	Ryhope Little	70 Fathom			Charlaw; 70 Fathom; Crow No. 2	
Moorland	DE1	(Moorland)	Moorland					
High Main	E	High Main	High Main		Bensham		Shield Row	
Metal	FI	Metal	Metal					
Five-Quarter	F2	Five-Quarter	Five-Quarter	Stone			Metal & Five-Quarter	Stone; Yard; Main
Main	G	Main	Yard[1]	Five-Quarter; Yard; Bentinck[1]			Brass Thill; Yard	Yard; Bensham
Maudlin	H	Maudlin	Bensham					
Durham Low Main	J	Durham Low Main	Durham Low Main	Benwell Main; Low Main	Grand Lease; Main; Six-Quarter; Bensham	Hutton; Pontop Hutton Hutton	Hutton; Pontop Hutton; Maudlin	Six-Quarter; Maudlin
Brass Thill	K	Brass Thill	Northumberland Low Main		Five-Quarter; Crow	Little	Low Main; Little; Little Hutton	Five-Quarter; Low Main
Hutton	L	Hutton	Hutton	Grove; Low Main	Five-Quarter; Old Five Quarter; Candle	Main; Old Five-Quarter	Main; Pontop Main	Low Main
Ruler	M	Ruler	?Cheeveley, Plessey	Plessey; Little; Tyne Level; Kitty's Drift	Little; Plessey	Sun	Plessey	
Harvey	N	Harvey	Beaumont	Engine	Towneley; Barlow Fell; Redheugh; Whickham	Towneley No 1; Barlow Fell or Barlowfield	Beaumont; Towneley	Beaumont
Hodge	O	Hodge	Hodge			Foot		
Tilley	P	Tilley	Tilley				Hand	
Busty	Q	Busty	Busty	Main	Five-Quarter; Six-Quarter; Main; Stone	Five-Quarter; Stone; Yard; Six-Quarter		
Three-Quarter	R	Three-Quarter	Three-Quarter		Yard	Pasture Drift		
Brockwell	S	Brockwell	Brockwell	Denton Low Main	Tyne	Hownes Gill		
Victoria	T	Victoria	Victoria					
Marshall Green	U	Marshall Green	Marshall Green					

1. The Bentinck Seam between the Yard and Five-Quarter seams in Northumberland is sometimes called the Top Main.
Sources: Land, 1974; Lawrence & Jackson, 1986, 1990; British Coal (formerly National Coal Board).

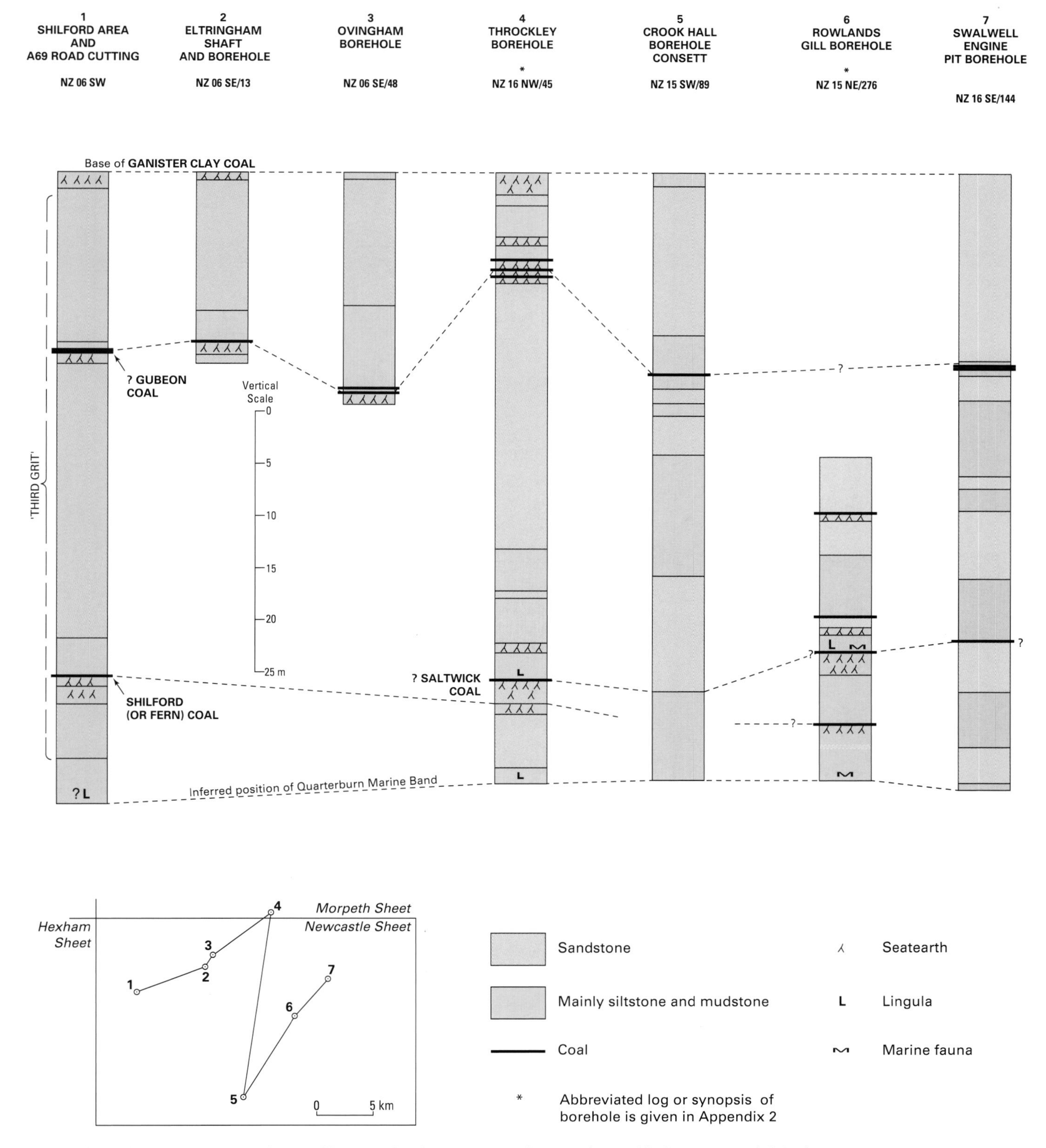

Figure 14 Comparative sections of Lower Coal Measures: Quarterburn (Subcrenatum) Marine Band to Ganister Clay Coal.

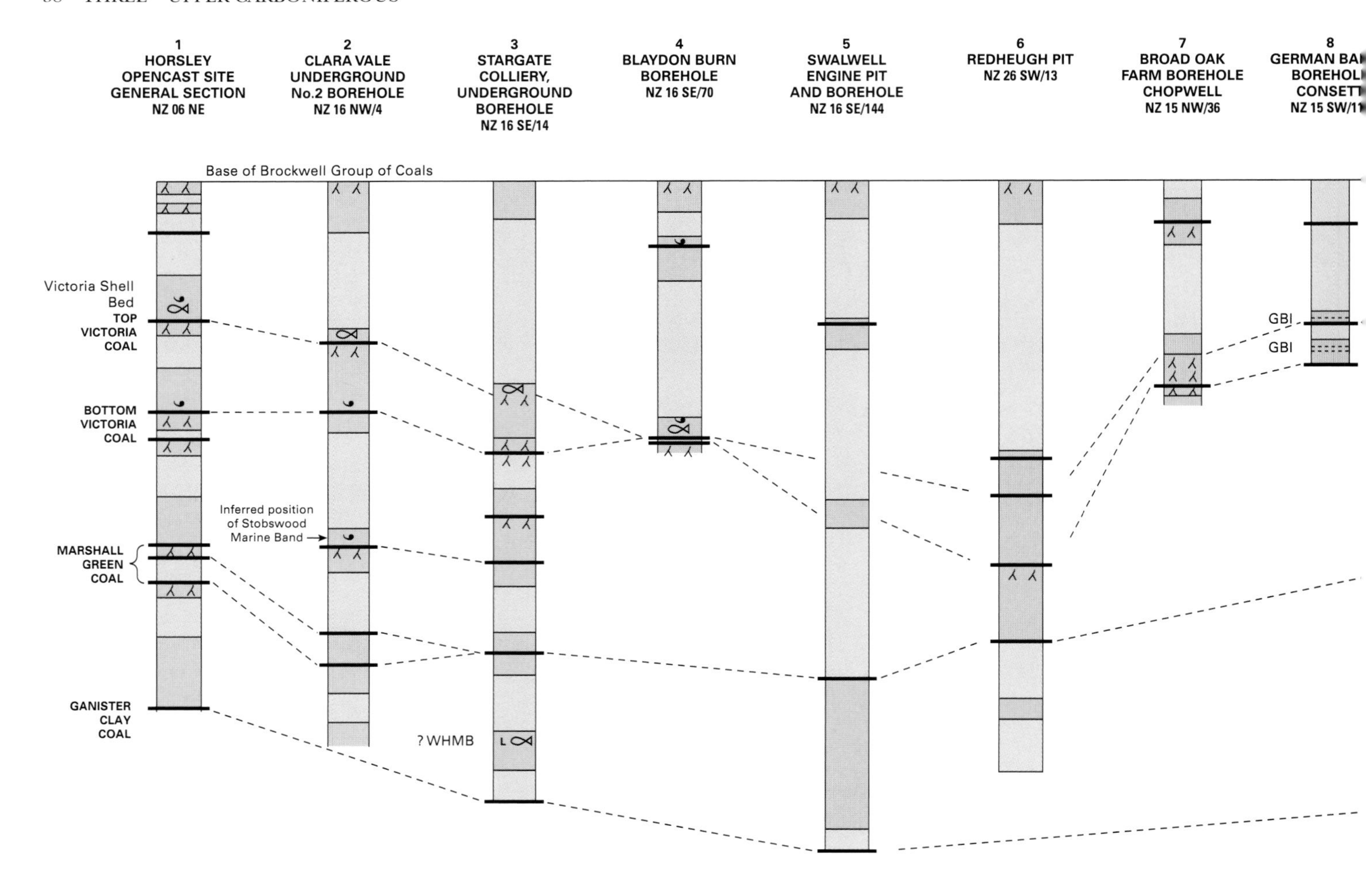

Figure 15 Comparative sections of Lower Coal Measures: Ganister Clay Coal to Brockwell Coal.

Hill Marine Band of the Morpeth district may be represented by a band containing fish fragments present (in places) in the middle of the interval.

The **Marshall Green Coal** (Figure 15), rarely over 0.5 m thick, is commonly split into top and bottom leaves separated by up to 3.0 m of argillaceous strata. In the southern and south-eastern part of the district, the few records suggest that it is generally thinner than in the north and again split into two leaves.

The **strata between Marshall Green and Victoria group of coals** (Figure 15) range between 6 and an exceptional 37 m in thickness. A large proportion of this interval consists of sandstone or sandy strata, more especially in some of the thicker sections. Up to two thin coals have been recorded, the lower of which is inferred to underlie the horizon of the **Stobswood Marine Band**. The upper thin coal is regarded as a split off the Bottom Victoria Coal.

The **Victoria Coal** (Figure 15), ranging up to 0.6 m thick, is usually split into two seams, **Top** and **Bottom Victoria,** separated by up to 11 m of largely argillaceous strata with a thin sandstone locally near the top. The Bottom Victoria Coal, itself locally split into top and bottom leaves, in places contains a fauna of small mussels and fish remains in its roof. In the south-western part of the district, notably in the area west and south-west of Consett, impersistent thin beds of ironstone and ironstone nodules, the **German Bands Ironstone**, are present in the measures between the Bottom and Top Victoria coals, and also in the beds overlying the Top Victoria Coal. The Top Victoria Coal, again locally split, is less persistent than the Bottom Victoria; where absent, its position is generally marked by a seatearth.

The **strata between the Victoria and Brockwell groups of coals** (Figure 15) generally range between 10 and 20 m thick, a high proportion of the sequence consisting of sandstone or beds of a sandy character. However, argillaceous strata are generally present overlying the Top Victoria Coal or its assumed horizon, and are commonly very shelly. This bed, the **Victoria Shell Bed**, contains a distinctive fauna dominated by the mussel *Carbonicola pseudorobusta* and also including *Naiadites* sp., *Planolites ophthalmoides* and fish debris. In the south-western part of the district the German Bands Ironstone also occurs in the immediate roof of the Top Victoria Coal (see above). Argillaceous measures, often associated with a thin coal, are also generally present in the upper part of the

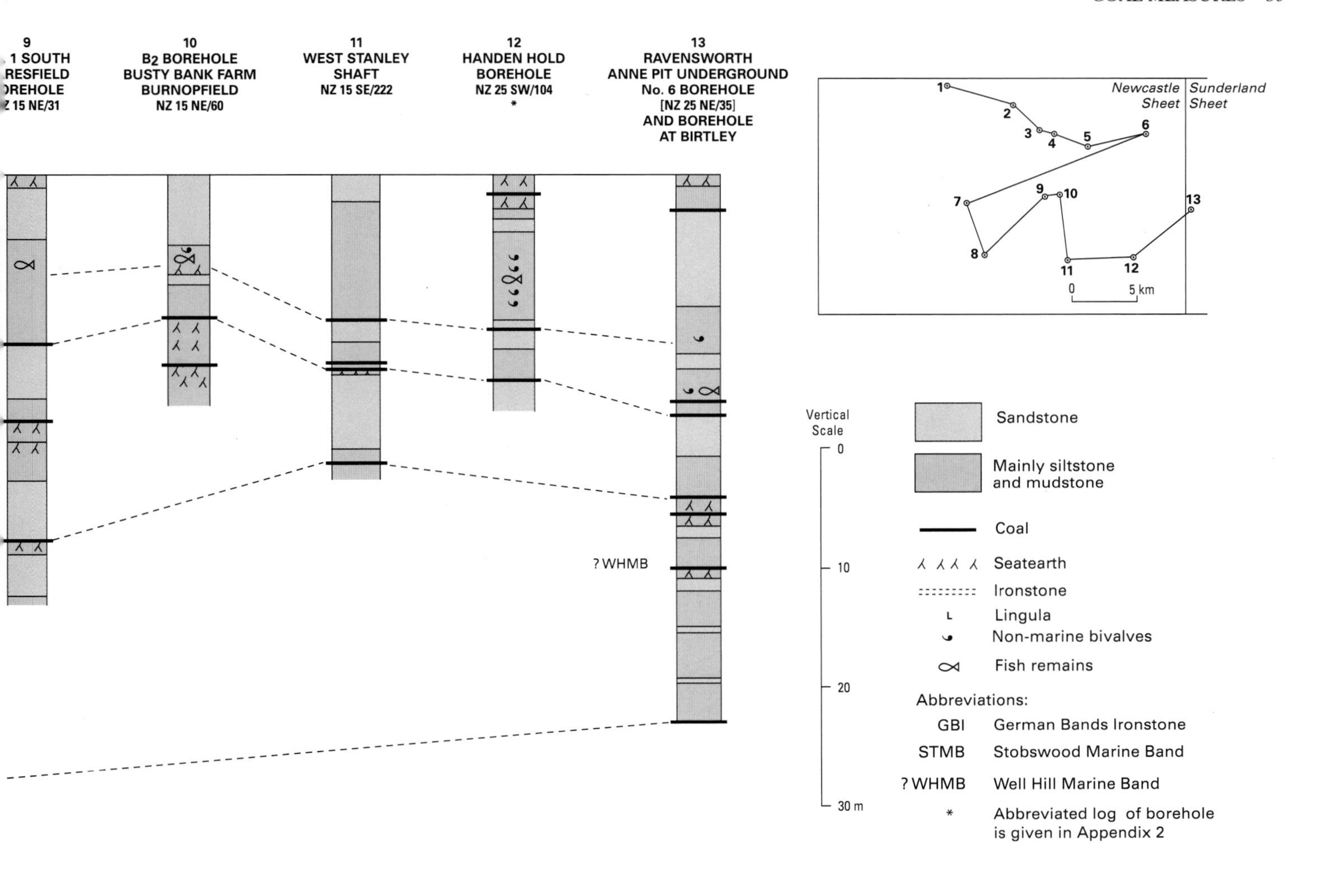

sequence, comprising some 7 to 10 m of strata below the Brockwell Coal. A fauna of small mussels has been recorded at the inferred horizon of this coal. It is probable that this is the equivalent of a thin seam recorded at a high level in the Victoria–Brockwell interval of the Morpeth and Tynemouth districts (Land, 1974; Lawrence and Jackson, 1986).

The **Brockwell Coal** or **Brockwell group of coals** (Figures 16 and 17) forms the lowest widely worked seam in the district. In many areas, more particularly in the west, centre, south and south-east, the coal is split into upper and lower leaves, the **Top** and **Bottom Brockwell coals**, whose total aggregate thickness is generally about 1.15 m. The thickness of beds separating the leaves varies widely, but rarely exceeds 5 m. In the south-east, both the Top and Bottom coals may themselves be split. Only in parts of the north and the north-east does the coal generally form a single seam.

The **strata between the Brockwell and Three-Quarter coals** (Figures 16 and 17) range between 3.6 and 20 m in thickness, the minimum separation occurring in the south-east and in adjacent parts of the Sunderland district (Smith, 1994). A high proportion of the sequence commonly comprises sandstone, particularly overlying the Brockwell Coal itself. Three thin impersistent coals and seatearths have been recorded. In some boreholes, mussels and fish debris occur in the roof measures of the Brockwell Coal, and at the former Horsley Opencast Site the **Brockwell Ostracod Band** was recorded in fragmental clayrock and mudstones immediately overlying the seam (Richardson and Francis, 1971). Mussels, ostracods and fish fragments have also been recorded from a bed between 3 and 7 m below the Three-Quarter Coal. In both instances, there are close similarities with the equivalent strata in the Tynemouth district (Land, 1974). It appears likely that the higher fossiliferous bed is just above the horizon of a coal seam equated with the New Seam of Stobswood Colliery in the Morpeth district (Land, 1974, p.29).

The **Three-Quarter Coal** (Figures 16 and 17) is widely recognised throughout the district and generally ranges between 0.4 and 0.7 m thick. In the south-east, the coal is split into Top and Bottom leaves, the latter approaching to within 4 m of the Brockwell Coal.

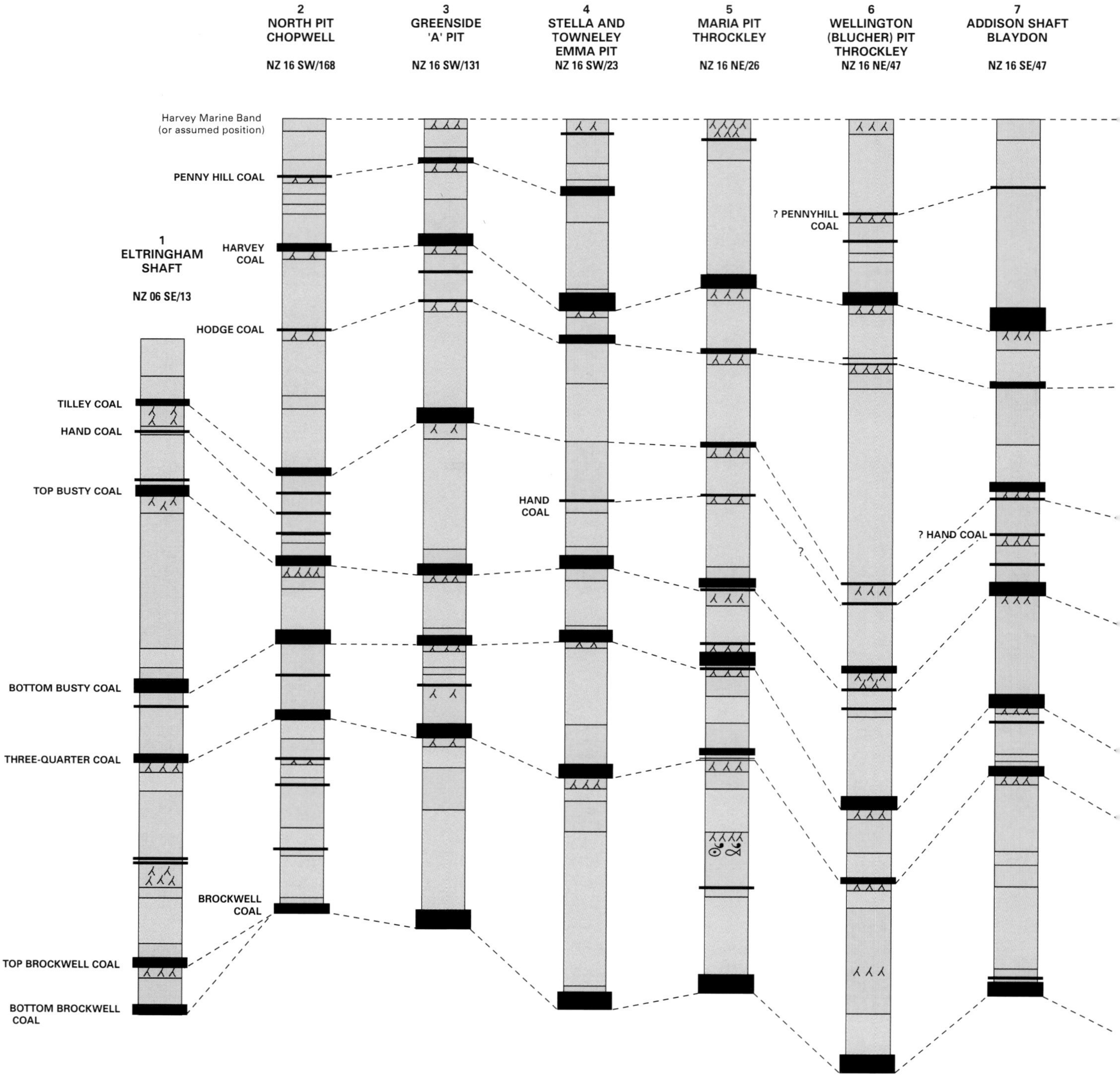

Figure 16 Comparative sections of Lower Coal Measures: Brockwell Coal to Harvey (Vanderbeckei) Marine Band (northern part of district).

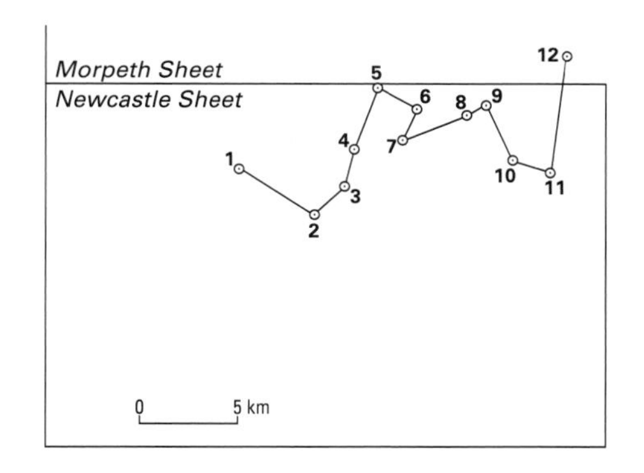

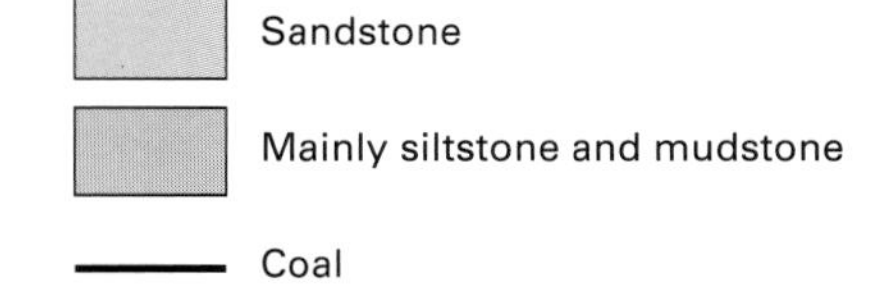

Sandstone

Mainly siltstone and mudstone

Coal

Seatearth

Fish remains

Ostracods

Non-marine bivalves

* Abbreviated log of borehole is given in Appendix 2

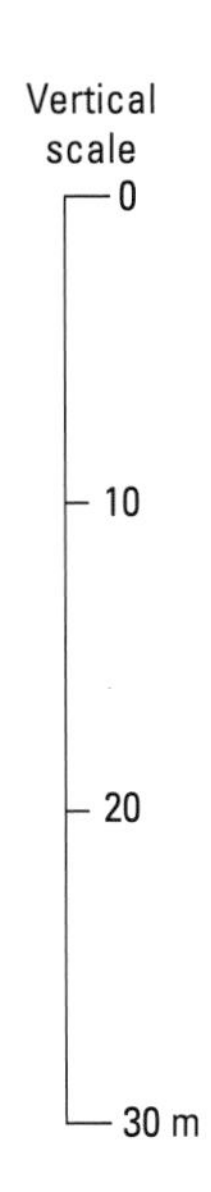

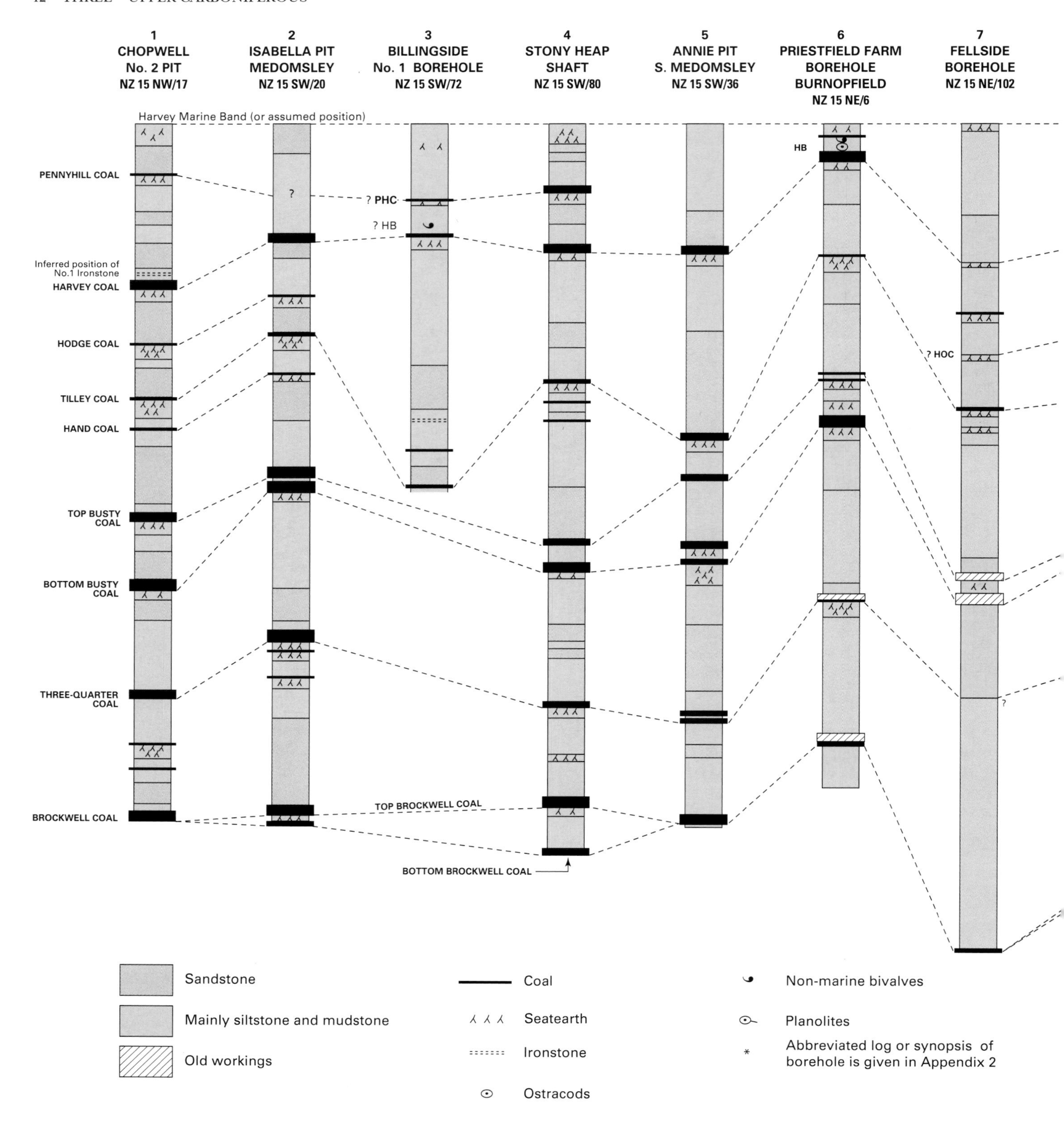

Figure 17 Comparative sections of Lower Coal Measures: Brockwell Coal to Harvey (Vanderbeckei) Marine Band (southern part of district).

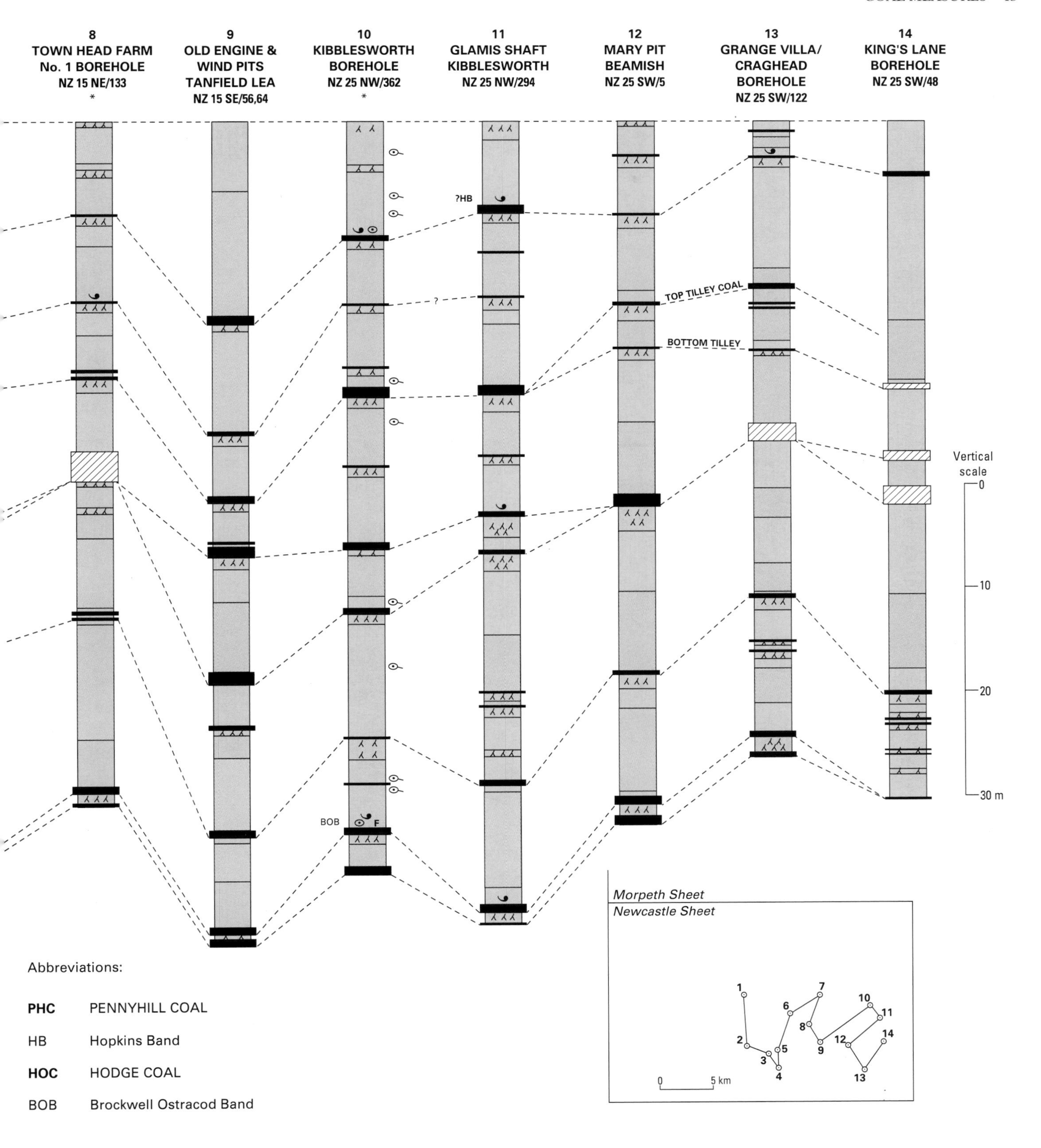

Abbreviations:

PHC	PENNYHILL COAL
HB	Hopkins Band
HOC	HODGE COAL
BOB	Brockwell Ostracod Band

The **strata between the Three-Quarter and Bottom Busty coals** (Figures 16 and 17) range between 4.5 and 22 m in thickness. Considerable local variations in lithology characterise this interval and, although a high proportion of sandstone commonly is present, it is not as widespread as in the Tynemouth and Sunderland districts (Land, 1974; Smith, 1994). Up to three thin impersistent coals and seatearths are present. At one locality, a fragmental clayrock has been recorded in the roof of the Three-Quarter Coal, and a bed in the overlying mudstones locally contains mussels, ostracods and fish remains.

The **Busty Coal** (Figures 16 and 17) is usually split into Top and Bottom leaves, a single composite Busty Coal being restricted to very limited areas in the district, most notably in the south-east. The occurrence of a split seam closely equates with that recorded in the Morpeth, Tynemouth and Sunderland districts (Land, 1974; Lawrence and Jackson, 1986, 1990; Smith, 1994). The **Bottom Busty Coal**, ranging generally between 0.76 and 1.2 m in thickness, is separated from the **Top Busty Coal**, 0.6 to 1.0 m thick, by up to 16 m of strata in which sandstone or sandy beds tend to predominate (Figures 16 and 17). A thin coal is locally present in the interval. The Busty Coal, more particularly the Bottom Busty, has been widely worked. The boundary between the *Carbonicola communis* and *Anthraconaia modiolaris* bivalve zones is taken at the top of the Top Busty Coal (Figure 13).

The **strata between the Top Busty and Tilley coals** (Figures 16 and 17) range between about 3 and 22 m in thickness, the thicker sequences being developed more especially in the south-east. Commonly, a high proportion of the sequence is sandstone, although the pattern is rather irregular. Thin, largely impersistent coals are recorded in the interval, one of them, the **Hand Coal**, being traceable with some certainty, especially in the western part of the district. Mussels, including *Carbonicola* sp., have been recorded from the roof of the Top Busty Coal.

The **Tilley Coal** (Figures 16 and 17) shows much variability and ranges from a single seam, generally between 0.5 and 0.7 m in thickness and largely confined to the western part of the district, to a group of four thin coals separated by up to 6 m of strata, the bulk of these being sandy. Despite such variation and complexity, there is no evidence to assume that either the Hand Coal, below, or the Hodge Coal, above, are splits from the Tilley coals in this district. The Tilley Coal or coals have been worked locally in many parts of the district, but nowhere to the same extent as the Busty or Harvey coals. The patchy nature of the workings reflects the lateral variability of the seam.

The **strata between the Tilley and Harvey coals** (Figures 16 and 17) usually range between 9 and 12 m in thickness and show much variability. Locally they contain a high proportion of sandstone, but elsewhere argillaceous strata predominate. Up to three, largely thin and impersistent, coals and seatearths are present. One of these is equated with the **Hodge Coal**, a seam that can be traced with some confidence in the north and west of the district, but less certainly elsewhere. Mussels and ostracods have been recorded in the roof of the coal or at its inferred horizon.

The **Harvey Coal** (Figures 16 and 17), widely worked in many parts of the district, forms a good quality coal usually ranging between 0.55 and 1.14 m thick. Locally, a thin argillaceous band is present in the lower part. The seam is somewhat thinner in the south-eastern part of the district.

The **strata between the Harvey Coal and the base of the Harvey Marine Band** (Figures 16 and 17) generally range between 11 and 19 m in thickness, the extremes being 3 and 24 m. Although wide variations occur, the sequence is typically argillaceous, or of mixed character with up to four thin sandstones and two thin impersistent coals or seatearths. One of these, the **Penny Hill Coal**, is recognisable over a wide area in the north-west. The Harvey Coal itself is commonly overlain by a thin kaolin-rich fragmental clayrock, which in turn is succeeded by the **Hopkins Band,** a widespread and reliable marker. This last consists of mudstone with a characteristic mussel–ostracod fauna including *Geisina arcuata* (Hopkins, 1927, 1928; Carruthers, 1930; Armstrong and Price, 1954; Pollard, 1966, 1969; Richardson and Francis, 1971). However, the boreholes which penetrate this part of the sequence record little palaeontological detail, although the characteristic lithology is widespread. Thin ironstones or beds of ironstone nodules occur in the roof measures of the Harvey Coal and, in the Consett area, were worked as the **No. 1 Ironstone** in the last century. A thick seatearth, locally overlain by a thin impersistent coal, is present at the top of the interval, underlying the Harvey Marine Band. On the published 1:50 000 scale map of the district, the No. 1 Ironstone is shown incorrectly as the Ten Bands Ironstone; the latter in fact occurs in the measures overlying the Harvey Marine Band.

Quarterburn (Subcrenatum) Marine Band (Figures 12 and 14)

In the north-western part of the district, several boreholes are inferred to have penetrated the Quarterburn Marine Band, but its precise position has nowhere been proved with certainty. A mudstone containing doubtful *Lingula* at the supposed position of the marine band was formerly exposed in the A68 road cutting [0204 6059] at Shilford.

In the Healey area [downstream from 0007 5858], low cliffs alongside the burn expose up to 6 m of sandstone thought to be at the top of the Stainmore Group. Strata overlying this sandstone, and sporadically exposed in the twin headwaters of Healey Burn, comprise about 5 m of shale, silty shale and mudstone with seatearths, the latter containing ferruginous nodules; a thin coal seam (?Shilford or Fern) is present at the top of the sequence. These beds are presumed from evidence to the west of the area to contain the Quarterburn Marine Band. Shales, locally nodular, crop out [0086 5668 to 0088 5656] in Healey East Burn, south of the Ninety Fathom Fault; their position relative to the inferred crop of the Ganister Clay Coal to the south suggests that they may contain the Quarterburn Marine Band. To the east-south-east in Fairley May Gill [0441 5723], the presumed horizon of the marine band is included in 6.10 m of grey, rubbly, fossiliferous mudstone overlying shales, seatearths and thin sandstones, a sequence very similar to that encountered in Healey Burn.

In the valley of the Stocksfield Burn, between Wheelbirks Cottages and Hindley Bridge, and also in the lower reaches of Lynn Burn, a distinctive rubbly sandstone, 1.0 m thick, contains the marine band. The sandstone is greyish brown, fine-grained, knobbly, rooty, sometimes calcareous, and contains abundant *Zoophycos* [at 0534 5958]. It lies within a sequence which includes shales below and seatearth above. The marine band is thought to be in a sequence in Shotleyfield Burn, upstream from Hammermill Bridge [0649 5299], where the following section was recorded:

	Thickness m
Lower Coal Measures	
Mudstone, blue-grey, fissile; fossiliferous	0.36
Ironstone nodule bed; fossiliferous	0.03
Mudstone, blue-grey; fossiliferous including *Lingula*	0.31
Ironstone nodule bed	0.03
Mudstone, blue-grey, silty, fissile; fossiliferous including *Lingula*	0.33
Sandstone, green and grey, fine- to medium-grained, argillaceous, with micaceous laminae; fossiliferous including abundant *Zoophycos* and rare *Lingula*	0.10
?Stainmore Group	
Seatearth-mudstone, pale grey, silty	0.60
Siltstone, pale grey, flaggy, argillaceous	0.70
Mudstone, grey, silty	1.25
Sandstone, dark blue, ganisteroid, very hard; *Zoophycos* on upper surface	0.15
Seatearth-mudstone, clayey at top, silty at base	0.60
Mudstone, grey, silty; impersistent sandstone and siltstone beds, on sandstone ('Second Grit')	0.52

The fauna collected from the upper part of this section [at 0625 5309], and identified by Dr N J Riley, included *Lingula mytilloides, Orbiculoidea bulla, O.* cf. *cincta,* O. cf. *elongata,* orthotetoid sp. (indet), *Aviculopecten dorlodoti, Myalina pernoides, Palaeoneilo* cf. *taffiana, Sanguinolites interruptus, Schizodus antiquus, Euphemites* sp., *Retispira striata,* fish debris indet, and *Zoophycos* sp.

Broadly similar, though less complete, sections are visible in Mere Burn [0937 5498], Snow's Green Burn downstream from Snow's Greenburn Bridge [0921 5341] and Mereburn Bridge [0859 5488]. In Mere Burn the sequence is distinctly more sandy than that in Shotleyfield Burn, and is extremely rich in molluscs. The fauna collected [0886 5485] in Mere Burn, and identified by Dr Riley, included sponge spicules, echinoderm plates and spines, crinoid columnals, *Lingula mytilloides, Antiquatonia* cf. *hindi, Aviculopecten* sp., *Cypricardiella* cf. *parallela, Edmondia* spp., *?Leiopteria* sp., *Nuculopsis scotica, Palaeolima* sp., *?Phestia attenuata, Sanguinolites interruptus,* S. cf. *spinulosum, Schizodus antiquus, Coleolus reticulatus, Bellerophon* sp., *Donaldina* sp., *Euphemites* cf. *anthracinus, Glabrocingulum* sp., *Hemizyga* sp. nov., *Hesperiella* cf. *loudini, Tanthinopsis* sp., *Platyconcha hindi, Retispira exilis, Soleniscus* sp., *Yuania* cf. *parva, Solenchilus* sp., *Gastrioceras* sp., *Serpuloides stubblefieldi, Bairdiacypris* sp., *Elonichthys* sp., and *Zoophycos* sp. Beds of hard, dark blue-green sandstone with the trace fossil *Zoophycos* are seen in sporadic exposures in Burnhouse Gill downstream from Shotley Bridge Road [0768 5290]. This sandstone resembles the *Zoophycos*-bearing post of Mere Burn and Shotleyfield Burn and is thought to be close to the Quarterburn Marine Band.

In the south-west of the district, outliers of the marine band are inferred at Edmundbyers Common [NY 999 506] and Muggleswick Park [030 492]. The middle reaches of Wharnley Burn [070 488], just east of the Healeyfield Fault, expose strata near the base of the Coal Measures. Five boreholes penetrate this horizon elsewhere in the district, notably Crook Hall Borehole [c.1186 5068] near Consett, Chopwell Borehole*, Swalwell Engine or Axwell Pit Borehole [1988 6167], near Gateshead, Delta Ironworks Borehole* and Rowlands Gill Borehole* (Figure 14). In the last-named, the presumed Quarterburn Marine Band is represented by 1.77 m of grey laminated siltstone containing fragments of a bivalve and the trace fossil *Planolites,* and in the Delta Ironworks Borehole * by 1.62 m of shale containing *Lingula.* In the other three boreholes the position of the marine band is not precisely located.

Strata between the Quarterburn Marine Band and the Ganister Clay Coal (Figure 14)

In the north-west of the district these strata are thought to range between 45 and 60 m in thickness. They include a high proportion of sandstone, generally split into three or four leaves and referred to the 'Third Grit'. The sequence also contains several argillaceous beds with up to four thin coals. In the A68 road cutting [0204 6059] at Shilford the following strata were recorded overlying the inferred position of the Quarterburn Marine Band.

	Thickness m
Sandstone ('Third Grit')	6.10+
Mudstone	3.35
Coal	0.38
Mudstone	2.36
Sandstone	4.88+

The coal, locally known as the Shilford or Fern Coal, ranges up to 0.53 m thick and was worked to a limited extent east of the A68 road, notably at Shilford Wood [028 609] and Juniper Hill Wood [033 609]. Several exposures are referred to the 'Third Grit' in this part of the district. Quarries at North Acomb [0493 6427] revealed 4 to 5 m of coarse-grained sandstone, and similar rocks were recorded in Cushatbank Wood [0448 6479]. In old quarries [0554 6548] at Old Nafferton, medium- to coarse-grained sandstone is present forming an upper leaf to the sandstone. A quarry [0535 6405] near Bearl exposed up to 4 m of coarse-grained sandstone. South of High Shilford [021 605], and contained within an arcuate faulted area, four thin sandstones make up the 'Third Grit', the intervening argillaceous measures containing at least three thin coals. One of these is tentatively equated with the Saltwick Coal of the Morpeth district.

A similar sequence also occurs farther to the south and south-west, and on the north side of the Ninety Fathom Fault between the Healey area and Wheelbirks [044 584], but beds low in the Coal Measures are generally very poorly exposed. The ground rises south to High Fotherley [020 572] and the sequence, largely drift-covered, consists of alternating sandstones, which form strong topographical features, and argillaceous strata, which give rise to slacks or relatively lesser slopes. The sandstones are generally of a flaggy, fine- to medium-grained character but more massive coarse-grained lenses and beds are present, as for example on Eastwoodhouse Fell [003 598], west of Broomleyfell Plantation. Two thin coals are present. The lower, tentatively equated with the Saltwick Coal and estimated to be some 20 m above the Quarterburn Marine Band, thickens from 0.03 m near Healey Hall [004 579] to 0.60 m in Broomleyfell Plantation [017 596], where a small area was worked opencast, and to between 0.3 and 0.6 m near Fotherley Gill. The higher seam, possibly the Gubeon Coal, is marked by a few trials but is more difficult to trace. It was formerly visible in a pipeline trench [0244 5787; 0239 5830] near Low Fotherley as thin streaks of coal in up to 1 m of coaly shale. In Stocksfield Burn [062 599] a composite section of strata, 13 m thick and some 14.5 m below the assumed position of the Ganister Clay Coal, records fine-grained sandstone, largely flaggy, with interbedded thin shaly beds and a coal 0.15 m thick, perhaps the Gubeon.

Coal Measures below the Ganister Clay Coal, north-west of Minsteracres [025 556], include three sandstones. The lower part of this sequence is exposed in Healey East Burn [010 564]:

	Thickness m
Sandstone, brown, flaggy, medium-grained	3.0+
Shale, grey	1.0
Sandstone, grey, rooty; sandy seatearth at base	0.8
Shale, grey-brown, sandy	1.6
Shale, grey; thin ripple-marked sandstone beds	5.0
Sandstone, grey-brown, fine-grained, hard	0.1
Seatearth, grey-brown, sandy; some rooty sandstone	1.6
Sandstone, brown, flaggy, medium-grained	2.5+

A similar sequence is visible in Fairley May Gill [0393 5657], some 4 km to the east-north-east, where a coal 0.08 m thick in the middle part of the exposed sequence is thought to be at a level between the Ganister Clay and ?Gubeon coals.

The largely drift-covered country in the neighbourhood of Shotleyfield [060 534] is very similar to that south of Shilford; sandstones form positive features and intervening slacks are formed by argillaceous strata with thin impersistent coals. The following section is exposed in Small Burn [0609 5307] south of Shotleyfield.

	Thickness m
Sandstone, brown, coarse-grained, cross-bedded, micaceous; conglomeratic and feldspathic towards base (base of 'Third Grit')	4.00+
Seatearth-sandstone, brown, medium-grained; shaly siltstone intercalations; eroded top	1.20
Siltstones, shaly; thin beds of fine-grained, rippled and cross-laminated sandstone	2.00
Shale, silty; fine-grained micaceous sandstone ribs	1.50
Quarterburn Marine Band	—

A broadly similar sequence is visible in Mere Burn downstream from Mereburn Bridge [0859 5488]. The upper reaches of Mere Burn [0778 5445; 0720 5433] and Yecklish Burn [0697 5492] near Newlands Grange expose mainly medium- to coarse-grained flaggy to cross-bedded sandstone below the inferred position of the Ganister Clay Coal.

East of the River Derwent, the following strata are exposed in Snow's Green Burn upstream from Snow's Greenburn Bridge [0921 5341].

	Thickness m
Sandstone, brown, cross-bedded, coarse-grained to conglomeratic, feldspathic; ganisteroid beds	15.0
Seatearth-sandstone and mudstone	0.40
Sandstone, green-brown, coarse-grained	0.40
Sandstone, dark grey to brown, shaly, fine-grained, locally friable; poorly fossiliferous including *?Myalina* sp.	1.30
Sandstone, orange-brown, fine-grained	0.10
Coal, bright (? Saltwick)	0.15
Seatearth-sandstone, capped by ganister 0.10 m	0.78
Sandstone, grey-brown, ripple-marked, cross-laminated, medium-grained	0.30
Sandstone and siltstone with shaly beds and laminae; more shaly downwards	3.30
Mudstone, silty, dark grey	1.20
Quarterburn Marine Band	—

The same coal can be seen at the base of a low cliff [0951 5397] on the east bank of the River Derwent. The bed with *Myalina* may be the marine band near the roof of the Saltwick Coal, or a lower marine band not recognised elsewhere.

Sandstone up to 13 m thick, and referred to the 'Third Grit', was formerly quarried [0770 4942] north-west of Castleside. The sequence consists mainly of dull yellowish grey to brown, flaggy, locally cross-bedded, fine- to medium-grained sandstone, often highly decomposed, with layers up to 0.75 m thick of harder, medium- and coarse-grained, locally pebbly sandstone. Discontinuous layers of ganister up to 1.2 m thick were formerly visible near the base of the section, and passed laterally into brown earthy sandstone. Shale laminae, and a shaly coal 0.10 m thick, are present near the top of the sandstone, which is overlain by up to 3.65 m of micaceous carbonaceous mudstone and siltstone. In the south-west of the district, sandstone low in the Coal Measures caps Edmundbyers Common [NY 999 506] and Muggleswick Common [030 492]. There are few exposures, but most of the surface debris consists of blocks of flaggy medium- to coarse-grained sandstone.

Within the main coalfield area to the east, a number of boreholes have been sunk to just below the Ganister Clay Coal, but as noted above only five penetrate the full sequence to below the Quarterburn Marine Band; these are at Crook Hall [c. 1186 5068], Chopwell*, Swalwell Engine or Axwell Pit [1988 6167], Delta Ironworks*, and Rowlands Gill*. In the last named, nearly 29 m of strata overlying the Quarterburn Marine Band were largely argillaceous, with two sandstones near the top. Four thin coals were proved. One of these (?Saltwick), 10 m above the basal marine band, is overlain by a mudstone-siltstone phase containing a sparse fauna of *Lingula*, turritellid gastropods and shell fragments. The other boreholes are difficult to interpret, and the suggested correlations (Figure 14) are tentative.

Ganister Clay Coal (Figure 15)

The Ganister Clay Coal forms a thin impoverished seam throughout the district, and is rarely absent. In the north-west, boreholes have recorded the coal as up to 0.40 m thick, but more generally it ranges between 0.05 and 0.20 m. Farther south, and in the watershed area between the rivers Tyne and Derwent, boreholes indicate that the coal is between 0.13 and 0.20 m thick. In this general area, the Ganister Clay Coal has only rarely been identified at outcrop, its position usually being inferred by reference to higher coals. Exceptions to this are in the tributary [0970 5308] of the Snow's Green Burn near Snow's Greenburn Bridge, where coal 0.15 m thick has been observed; in a gas pipeline trench [0224 5624] in the northern part of Minsteracres Park, where coal 0.20 m thick was seen; and in Newfield Burn [037 563], where coal 0.15 m thick was recorded. The few boreholes that have penetrated this level farther east, in the main coalfield area, indicate the existence of a thin seam rarely more than 0.20 m thick. Locally, in the extreme southern part of the district, in the neighbourhood of Burnhope and Edmondsley, thicknesses of up to 0.45 m have been recorded.

Strata between the Ganister Clay and Marshall Green coals (Figure 15)

In the north-western part of the district these strata range between 8 and 12 m in thickness, and in most cases contain a high proportion of sandstone or sandy 'mixed beds'. Some local exceptions to this, where the strata are wholly argillaceous, were proved in a borehole from the base of the Eltringham Shaft [0845 6298] and in bores in the Spital area [080 667]. Boreholes drilled to prospect the Horsley Opencast Site [090 670] proved an interval about 10 m thick, the upper part mostly consisting of sandstone; no marine bands were recorded.

Farther east, along the Tyne Valley, in Stargate Colliery No. 1 Borehole* near Ryton, fish debris and *Lingula?* were proved in a bed 1.27 m thick approximately midway between the Marshall Green and Ganister Clay coals, possibly at the horizon of the Well Hill Marine Band of the Morpeth district. Short sequences below the position of the Marshall Green Coal, in sandy mudstone and mudstones with thin ironstone ribs, are exposed [1737 5877 and 1718 5885] on the banks of the River Derwent north-east of Rowlands Gill. At the Redheugh Pit [2443 6263], 10 m of strata largely consisting of two sandstone beds were proved immediately underlying the Marshall Green Coal.

In the south-western part of the district, these strata occur at or near surface between the Derwent Reservoir and Minsteracres, and more particularly along the line of the watershed between the rivers Tyne and Derwent. The outcrops, largely concealed by drift, are much affected by small-scale faulting. In a gas pipeline trench just outside the western boundary of Minsteracres Park, a rubbly sequence consisting of shales and mudstones with thin interbedded sandstones, and overlain predominantly by sandstone up to the assumed position [0219 5578] of the Marshall Green Coal, were recorded. In Kellas Plantation, some 2.5 km farther west, sandstones above the Ganister Clay Coal are visible in both Healey East and West burns; they are mainly medium grained and well bedded. Farther east, at Greymare Hill and Shotley Low Quarter, boreholes in these strata have proved sandstone, or alternations of sandstone with subordinate argillaceous strata. Sandstones immediately overlying the inferred crop of the Ganister Clay Coal cap ridges and crop out extensively to the south and south-east of Shotley Low Quarter. Sandstones occur at both Pikehill Quarry [0735 5392] and at Blue House [0771 5345], where up to 5 m of pale brown medium-grained cross-bedded sandstone have been recorded.

In the Letch House–Airy Holm–Derwent Tunnel (Appendix 3) nearly 600 m were driven in grey, fine- to medium-grained cross-bedded sandstone underlying the seatearths of the Marshall Green Coal.

In the Newlands area, north of Consett, the following section is exposed along Mill Dene.

	Thickness m
Shale, grey; with thin sandstones	2.5
Sandstone, brown, medium- to coarse-grained [0914 5536]	3.0
Shale, grey-brown, sandy; interbedded thin shaly sandstones	3.3
Sandstone, brown, flaggy	1.0
Shale, grey	1.3
Sandstone, ochreous, medium- to coarse-grained, passes laterally into thinner beds	1.6
Sandstone, brown, flaggy, cross-bedded	3.0
Fault — downthrows to the north-east	—
Inferred position of Ganister Clay Coal	—

Upstream of the bridge [0873 5563] there are a few exposures of medium-grained flaggy sandstone which are thought to overlie the Ganister Clay Coal. South of Consett, in Hown's Gill [097 489] and farther to the south-east, medium- to coarse-grained sandstone underlies the Marshall Green Coal and forms the bulk of the lower part of the steep-sided glacial overflow channel in this vicinity.

In the southern part of the district, in the main coalfield area, a few boreholes penetrate to just below the Marshall Green Coal and generally prove sandy measures.

Marshall Green Coal (Figure 15)

In the north-western part of the district this is a single seam ranging between 0.38 and 0.51 m thick, or is split into two leaves, Top and Bottom Marshall Green, separated by between 2.4 and 3.0 m of argillaceous strata. At the Horsley Opencast Site [090 670] the generalised section was Top Marshall Green 0.71–0.84 m (locally itself split), shale and sandstone 1.0–1.65 m, Bottom Marshall Green 0.33–0.35 m. Few boreholes penetrate this level farther east in the northern part of the district, but indications suggest that it is generally present as a single seam up to a maximum of 0.35 m thick, and with dirt partings.

In the south-western part of the district, several disused adits along Woodhouse Burn record former workings in the Marshall Green Coal, which was proved in a borehole [0023 5455] to be 0.40 m thick. Nearby, however, and some distance farther north in the vicinity of Healey West and East burns, no trace of the coal was found. In the area of Greymare Hill and Shotley Low Quarter, the Marshall Green Coal, usually a single seam up to 0.50 m thick, has been proved in a number of boreholes. The coal crops out beneath drift at Wood House [0590 5220], south of Shotleyfield. In the Consett and Shotley Bridge areas the outcrop is largely concealed by made ground and drift, but the Marshall Green Coal, thought to range up to 0.30 m thick, was formerly worked from old surface adits in the vicinity of Hown's Gill [097 491]. In the Newlands area the Marshall Green Coal, 0.60 m thick, is visible at the water's edge [0873 5563] near Newlands Bridge.

Little is known about this coal in the southern part of the district, in the main coalfield area east of Consett, but the few records suggest that it ranges from a single seam up to 0.60 m thick, to a split seam consisting of a Bottom Marshall Green up to 0.20 m thick and a Top Marshall Green up to 0.35 m thick, separated by up to 1.22 m of argillaceous strata.

Strata between the Marshall Green and Victoria coals (Figure 15)

In the northern part of the district, these strata range between 9.75 and 17 m in thickness, a large proportion generally being made up of sandstone, shaly sandstone or sandy 'mixed beds'. Sandstone tends to dominate the roof measures of the Top Marshall Green Coal, but this is not always the case. At the former Horsley Opencast Site [090 670] the separation between the two seams was generally just over 9 m, and largely comprised argillaceous measures together with a thin median shaly sandstone. In Clara Vale Colliery No. 2 Borehole [1407 6501], 0.12 m of coal, some 7 m above the Top Marshall Green Coal, is overlain by a grey sandy mudstone containing sporadic poorly preserved mussels including *Carbonicola* sp. and *Curvirimula* sp. This mussel band is the assumed equivalent of the Stobswood Marine Band, the approximate position of which is probably marked in some other boreholes by a thin coal or seatearth. A higher thin coal, sometimes regarded as a split off the Victoria or Bottom Victoria Coal, has been recorded in some boreholes, but its occurrence is sporadic. The coal was proved in some of the boreholes for the Horsley Opencast Site [090 670].

In the south-western part of the district, in the vicinity of Minsteracres [025 556], and also north and west of Newlands

Grange [066 548], sandstones form well-marked topographical features between the crops, or inferred positions, of the Marshall Green and Victoria coals, although bedrock is rarely visible at surface.

The northern 686 m of the Letch House–Airy Holm–Derwent Tunnel (Appendix 3) was driven in a cross-bedded fine- to medium-grained siliceous sandstone, which rested directly on the Marshall Green Coal or was separated from it by up to 2 m of argillaceous strata.

At Hown's Gill [097 489], south of Consett, sandstone appears to form the bulk of the sequence between the Marshall Green and Victoria coals. On the eastern flank of Hown's Gill, sandstone was worked for building stone and roof tiles in a galleried quarry supported by massive stone columns (Plate 6). The section [0975 4893] is as follows:

	Thickness m
Sandstone, thin-bedded and flaggy; mudstone and siltstone beds	3.66+
Mudstone and siltstone; thin shaly sandstone beds	1.83–3.05
Sandstone, yellow and grey-brown, massive to flaggy, often cross-bedded; variably textured, but mainly medium-to coarse-grained; micaceous and carbonaceous; central portion of face consists of up to 6.10 m of good flagstone (main part of quarried stone)	up to 6.10
Sandstone, thick-bedded, with occasional large elongate ferruginous concretions (lower part of section only partly visible)	24.4+

The lowest bed immediately overlies the Marshall Green Coal. The sandstone at this level occurs in the valley sides for some kilometres down Hown's Gill, Knitsley Burn, Back Gill Burn and the upper reaches of Smallhope Burn. The sandstone is also exposed [1110 4862] at Knitsley, and at Knitsley Mill Quarry [1200 4818]; at both localities over 9 m of light grey medium- and coarse-grained beds are visible. Medium- to coarse-grained sandstone overlying the Marshall Green Coal is exposed [1526 5708] along the south bank of the River Derwent 1 km south-west of Rowlands Gill.

There are few records of this part of the sequence east of Consett, in the main coalfield area, but evidence suggests that the interval rarely exceeds 10 m in thickness, and includes a high proportion of sandstone (Figure 15). The Stobswood Marine Band horizon has not been recorded.

Victoria coals (Figure 15)

The Victoria coals are variable in the northern part of the district, and are rarely present as a single seam. At the former Horsley Opencast Site [090 670] the Bottom Victoria Coal, between 0.50 and 0.63 m thick, was separated by a maximum of 11 m of argillaceous strata from the Top Victoria Coal, referred to locally as the Bays Leap Coal, ranging between 0.05 and 0.20 m thick. A fauna of small mussels and fish debris was obtained locally from the roof of the Bottom Victoria Coal. Farther east in the Newburn and Ryton area, the Top Victoria Coal is absent, very thin, or represented by a black cannelloid shale. In Blaydon Burn No. 2 Borehole [1724 6275] the Victoria Coal is a virtually united seam 0.61 m thick. At Lady Pit [2138 6631], Montagu Main Colliery, the Top and Bottom Victoria coals were 0.12 m and 0.15 m thick respectively, separated by some 5 m of 'mixed beds'. In the Swalwell Engine or Axwell Pit [1988 6167] the Victoria Coals were not present in a predominantly arenaceous sequence proved in a borehole from

Plate 6 Underground workings in the sandstone between the Marshall Green and Victoria coals (Lower Coal Measures), Hown's Quarry, Consett [097 489]. L1750.

the base of the shaft. By contrast, in the Redheugh Pit Shaft [2443 6263], the coals were 0.23 and 0.30 m thick respectively.

In the western part of the district, and about 2 km north of the Derwent Reservoir, the Victoria (or Bottom Victoria) Coal was proved in several boreholes, in one of which [0084 5461] it was recorded as 0.36 m thick. In the Greymare Hill [046 555] area, extensive opencast extraction has taken place. Boreholes hereabouts showed the following generalised sequence:

	Thickness m
Top Victoria Coal	
Coal	0.34
Shale	0.28
Coal, shaly and carbonaceous	0.24
Mudstone, siltstone and seatearth; median 2.28 m sandstone	5.36
Bottom Victoria Coal	
Coal, shaly and inferior at base	0.27
Seatearth-mudstone and siltstone	5.36
Coal	0.09

The upper coal, which showed minor variation in thickness, was subsequently worked opencast. Old ironstone workings, recorded during previous surveys [0531 5521] west of Newlands Grange, indicated local extraction of the German Bands Ironstone, which occurs impersistently in this general area above both the Victoria coals. In the Consett area, both Victoria coals are present and are separated by up to 2.74 m of largely argillaceous strata, locally with a thin sandstone bed. The German Bands Ironstone, comprising thin ironstone beds, or beds of large clay ironstone nodules, was extracted from the strata above both coals. It was mined from adits and bell pits, more particularly west of Consett, and on the eastern side of Hown's Gill. All trace of former extraction has now disappeared, but the original discovery of the 'bands' is said to have been made south-west of Delves in the vicinity of Knitsley [110 488], near Back Gill Burn, on the extension of Hown's Gill.

Around Chopwell, Medomsley, Burnopfield, Tanfield Lea, Stanley, and Pelton, the Bottom Victoria Coal is up to 0.43 m thick and locally includes a thin mudstone band. It is overlain by up to 5.0 m of predominantly argillaceous measures succeeded by a very thin Top Victoria Coal or alternatively by a seatearth-mudstone underlying its inferred position. In the Handen Hold Borehole*, West Pelton, the Bottom Victoria Coal, 0.23 m thick, is separated from the Top Victoria Coal, 0.03 m thick, by 4.7 m of strata which are sandy at the top and argillaceous at the base. A broadly similar sequence was proved in the Kings Lane Borehole [2469 5360] where the Bottom Victoria Coal, itself split, was overlain by mudstone containing mussels and fish debris and succeeded by a thin sandstone. The Top Victoria Coal was absent in this borehole.

Strata between the Victoria and Brockwell coals (Figure 15)

At the former Horsley Opencast Site [090 670] these strata ranged between 10 and 13 m in thickness, with an intermediate seam 0.28 to 0.40 m thick in the upper part of the sequence. The beds overlying the Top Victoria Coal consist of argillaceous measures, at their base containing the Victoria Shell Bed, a shale up to 2.74 m thick with large mussels including *Carbonicola pseudorobusta*, *Planolites ophthalmoides* and fish debris. The sequence is otherwise largely dominated by sandstone or beds of a sandy character, although a thin seatearth is locally present between the intermediate seam and the Brockwell Coal.

The Victoria Shell Bed was also recognised farther east, at Clara Vale and Stargate collieries near Ryton, and the overlying interval, up to 20 m thick, is composed of sandstone, commonly coarse-grained, or beds of a sandy nature. In Blaydon Burn No. 2 Borehole [1724 6275], a united Victoria Coal is overlain by 1.67 m of mudstone containing mussels and fish debris. In the few records of this sequence immediately west and south-west of Newcastle, the interval, up to an exceptional 22 m in thickness, is mostly made up of sandstone (Figure 15). In the Redheugh Pit [2443 6263], a coal 0.30 m thick is present some 2.74 m above the inferred Top Victoria Coal.

In the south-west, these strata occupy some of the higher ground north of the Derwent Reservoir between Barley Hill [022 549] and the western margin of the district, but surface outcrops are rare. In the Greymare Hill [046 555] area, these measures are in the general range of 14 to 16 m thick, the lower part generally being argillaceous with a thin coal, 0.2 m thick, near the top. The upper part, to within a metre or two of the base of the Brockwell Coal, is largely arenaceous. This thin seam was worked opencast at Greymare Hill with the removal of other coals. In the Consett area, and farther south-west towards Hown's Gill, ironstones in the roof of the Top Victoria Coal (part of the German Bands Ironstone) were mined from adits and shallow pits.

In the Chopwell, Medomsley, Burnopfield, Stanley and Pelton areas, these strata range in thickness between 7 and 16 m and contain a high proportion of sandstone or sandy 'mixed' beds (Figure 15). At most localities the Victoria Shell Bed, containing mussels and/or fish debris, is present in the roof of the Top Victoria Coal. Locally, a thin intermediate coal, 0.20 to 0.38 m thick, is present in the upper part of the sequence. Strata recorded in the Handen Hold Borehole*, near West Pelton, are particularly fossiliferous in the 6 m of argillaceous measures overlying the Top Victoria Coal, containing in upward succession *Naiadites* sp., *Carbonicola* sp., fish spines, *Carbonicola* sp., *Curvirimula* sp. A record of *Lingula* in the original log of the borehole is almost certainly based on a misidentification of *Curvirimula*. Four thin coals, totalling 0.39 m in thickness, are present nearly 1 m below the Bottom Brookwell Coal.

Brockwell Coal (Figures 16 and 17)

In the north-western part of the district, the Brockwell Coal is typically split into top and bottom coals. At the former Horsley Opencast Site [090 670], the general section recorded was Top Brockwell Coal 0.76 to 0.81 m, largely argillaceous strata 3.04 to 4.57 m, Bottom Brockwell Coal 0.17 to 0.35 m. A broadly similar sequence has been recorded in the Prudhoe area as, for example, in the Eltringham Shaft [0845 6298] (Figure 16). East of Prudhoe and Wylam, the two parts of the seam become progressively closer and unite, though thin argillaceous partings are common. In addition, the seam tends to thin, with the aggregate total thickness generally being in the range 1.01 to 1.14 m. In the Newcastle area, the seam, only occasionally split, is generally about 0.76 to 1.06 m thick. In the extreme north-east, however, it becomes split into several thin coals separated by mudstone.

Outliers of Brockwell Coal extend virtually to the western margin of the district in the area north of the Derwent Reservoir. The coal has been worked from adits and, at one locality at Barley Hill, was subsequently opencast; aggregate thicknesses of coal of up to 1.9 m were recorded, separated by up to 1.0 m of sandstone and shale. At the former Greymare Hill Opencast Site [046 555], the general section recorded was Top Brockwell Coal 0.68 m, shale 0.28 m, Bottom Brockwell Coal 0.22 m. East and south-east of Consett, the Brockwell Coal rarely forms a single seam and is usually split into Top and Bottom coals, up to

0.48 and 0.63 m thick respectively, separated by up to 5 m of argillaceous strata.

East of Stanley, the Brockwell Coal is generally split into two or even more seams, with an aggregate thickness ranging between 0.5 and 2.0 m, but individual leaves rarely exceed 0.9 m in thickness (Figure 17). Both main leaves have been worked locally, but the majority of workings are in the Bottom Brockwell Coal. In the south-east of the district the seams are commonly further split, a typical section being Top Brockwell Coal (coal 0.35 m, fireclay and mudstone 1.50 m, coal 0.15 m) on up to 5.0 m of mudstone with thin beds of seatearth, on Bottom Brockwell Coal 0.38 to 0.90 m thick, commonly with dirt bands and in two leaves. By contrast, around Burnhope and Edmondsley on the southern margin of the district, the Brockwell Coal is generally a single seam up to 0.83 m thick, but is locally impoverished.

Strata between the Brockwell and Three-Quarter coals (Figures 16 and 17)

In the northern part of the district, these strata range between about 8.5 and 20 m in thickness, and generally contain a high proportion of sandstone. A sandstone, exceptionally up to 14 m thick, is generally present in the roof of the Brockwell Coal or is separated from the seam by only a few metres of argillaceous strata. This sandstone is locally thick bedded and medium to coarse grained. At the Horsley Opencast Site [090 670], Richardson and Francis (1971) recorded a thin fragmental clayrock containing mussel fragments in the roof of the Brockwell Coal, overlain by a mudstone and ferruginous coaly layer with mussels and fish remains. The fauna is comparable to that described in the Brockwell Ostracod Band by Pollard (1969, p.244). A higher sandstone, sometimes split into two, is generally present in the middle and upper parts of the sequence. Up to three thin coals, none exceeding 0.15 m thick, have been recorded locally in this interval, more especially in the west, but commonly they are absent or their horizon indicated by a seatearth. In the Maria Pit [1569 6726] at Throckley, mussels including *Curvirimula* sp., together with ostracods and fish fragments, were recorded in grey and blue shales up to 8 m below the Three-Quarter Coal (Figure 16). In Blaydon No. 2. Borehole [1724 6275], at approximately the same horizon, poorly preserved mussels were also recorded in dark grey mudstone 4.57 m below the Three-Quarter Coal.

In the central and southern parts of the district there is greater lithological variation in this interval. The strata range from as little as 3.6 m in the south-east to over 18.3 m elsewhere. Although sandstone tends to predominate, more especially in the lower and middle parts of the interval, it is locally thin or absent. Sandstone is particularly abundant in the area around Burnopfield and Tanfield Lea. For example, in the Fellside Borehole [1899 5830], north-east of Burnopfield, some 25 m of white, fine- to medium-grained sandstone, possibly a channel-infill, were recorded overlying the Brockwell Coal, and the Three-Quarter Coal is absent, perhaps cut out. In the extreme south-east of the district, the thickness of strata between the Brockwell and Three-Quarter coals is much reduced, and consists largely of mudstones and siltstones with seatearths. Thin impersistent coals, nowhere exceeding 0.15 m in thickness, occur at several levels in the sequence. Mussels, including large *Carbonicola* sp., have been recovered locally from mudstones in the roof of the Brockwell Coal, and in the Kibblesworth Borehole*, north-west of Kibblesworth, grey and dark grey mudstones containing mussels, abundant ostracods and ?fish debris (Brockwell Ostracod Band) were also recorded at this horizon (Figure 17). In a borehole [0973 5767] much farther west, at the former Whittonstall New Drift, a fauna of small mussels, together with ostracods, was obtained from the immediate roof measures of a thin coal 3.66 m below the Three-Quarter.

Three-Quarter Coal (Figures 16 and 17)

This coal is widely present, and is generally between 0.40 and 0.71 m thick. Thin dirt partings and bands are locally recorded. In the south-east of the district, the coal tends to thin and split into Top and Bottom coals, separated by up to 2.13 m of argillaceous strata. Both Top and Bottom coals are themselves locally split into top and bottom leaves. On account of the thinness and variable quality of the seam, there are many areas in which it has been worked to a limited extent only.

Strata between the Three-Quarter and Busty coals (Figures 16 and 17)

The strata in this interval range between 4.57 and 22 m in thickness. Although a large proportion of the interval commonly comprises sandstone or sandy beds, considerable variation is present and there is no clear relationship between thickness and lithology. In some places a thick, commonly cross-bedded, medium- to coarse-grained sandstone occupies virtually the whole interval between the two coals, with only thin argillaceous measures at top and base. For example, north of Burnopfield, in Leapmill Burn, sandstone is exposed [1752 5779] on the hill slope due east of Busty Bank Farm, and locally rests directly on the roof of the Three-Quarter Coal. Elsewhere, thinner sandstones are common at both the base and top of the interval. More rarely, the interval is dominantly argillaceous, with ironstone ribs, beds or nodules. Up to three thin impersistent coals or seatearths have been recorded, and where present generally occur in the lower or middle part of the interval. Richardson and Francis (1971) recorded a thin, unfossiliferous fragmental clayrock in the roof of the Three-Quarter Coal at the former Horsley Opencast Site [090 670], overlain by mudstone with the following fauna: *Spirorbis* sp., *Carbonicola* spp., *Curvirimula* sp., *Carbonita claripunctata, C. humilis, C. pungens, Geisina arcuata.* Fish remains and abundant ostracods were present in the basal layers of the mudstone.

Busty coals (Figures 16 and 17)

The Busty Coal characteristically comprises two separate leaves, and only locally forms a single composite seam, though in several areas the Top and Bottom Busty approach to within 2 or 3 m of each other. For example, in the vicinity of the Beaumont Pit [2266 6332], in the Elswick area of Newcastle, 4.62 m of strata include 1.87 m of coal in six leaves, with interbedded seatearth. North-east of Consett, in the neighbourhood of Medomsley, near Burnopfield and towards Sunniside, the Busty Coal locally forms a united composite seam. In the south-eastern part of the district, east and south of Beamish, a single seam is commonly present, generally ranging between 1.65 and 2.08 m in thickness. The Busty coals have been widely worked, more especially in the west and central parts of the district. Mining in the north-east, and down the eastern fringes of the district, has been much more selective.

In the northern part of the district, the Bottom Busty Coal is generally in the range 0.76 to 1.21 m thick (Figure 16), the thicker seam sections often being marked by the presence of argillaceous partings. Locally the seam is split into a thick median coal underlain and overlain by thin coals and accompanying seatearth, as at Maria Pit [1569 6726], Throckley. In the central and southern parts of the district, the Bottom Busty Coal generally ranges between 0.76 and 1.0 m in thickness (Figure 17), and only locally contains dirt bands. The coal is subject to local washout and impoverishment, more especially in the southern part of the district.

The thickness of strata between the coals ranges up to a maximum of 16 m, but is more generally between 5 and 8 m (Figures 16 and 17). The thickness varies sharply from area to area. Sandstone tends to form much of the interval, particularly where the thickness is relatively large; it generally lies in an intermediate position but locally forms the roof of the Bottom Busty Coal. A flaggy, medium- to coarse-grained sandstone is widely present in the roof of the Bottom Busty Coal around Consett, and was formally exposed in the vicinity of the Old Fell Coke Works [102 601]. Where more 'mixed' strata predominate, an impersistent median coal up to 0.17 m thick has been recorded locally. In the area east of Consett the strata between the coals is characterised in places by ironstone beds ranging up to 0.17 m thick.

In the northern part of the district the Top Busty Coal is generally between 0.76 and 1 m thick, although it tends to deteriorate in both thickness and quality towards the north-east, where thicknesses in the range 0.50 to 0.70 m are more common (Figure 16). Locally, as in the south-west part of Newcastle, it deteriorates to a thin seam between 0.20 and 0.50 m thick. A greater range of thicknesses is encountered in the central and southern parts of the district; 0.60 to 0.91 m is a common thickness range in the more westerly areas around and to the east of Consett, but farther east thicknesses are rarely over 0.60 m. Locally, as in Priestfield Farm [1575 5663] and Tanfield Lea [1700 5257] boreholes, the seam deteriorates into two or three thin leaves separated by mudstone and seatearth. In the southern part of the district the Top Busty Coal is washed out, or impoverished, over an extensive area around Greencroft.

Strata between the Busty (Top Busty) and Tilley coals (Figures 16 and 17)

In the northern part of the district (Figure 16) these strata range between about 8 and 16 m in thickness, the higher values tending to occur towards the east. The strata are rather variable in character, with a high proportion of sandstone, commonly medium- to coarse-grained, or sandy 'mixed beds'. Up to three impersistent coals, none over 0.22 m thick, and seatearths, are present. One of these coals, the Hand Coal, which tends to occupy a median position in the sequence, appears to be reasonably consistent in the west but its wider correlation eastwards is less certain.

A similar pattern is present in central and southern parts of the district (Figure 17), but the thickness of the interval varies more widely and irregularly, from as little as 3 m near Medomsley to nearly 22 m locally in the east and south-east. A large proportion of the sequence comprises sandstone, commonly coarse-grained, or is of a sandy character, with sandstone locally resting directly on the roof of the Busty/Top Busty Coal. The Hand Coal can be traced as a thin, impersistent bed largely in the west of the area. In the Tanfield Lea Borehole [1700 5257], dark grey cannelloid mudstone, 0.12 m thick and containing abundant *Carbonicola* sp., was recorded from the roof measures of a split Top Busty Coal. Similarly in the Glamis Shaft [2432 5619], Kibblesworth, mussels were present in a shale bed, 0.71 m thick, just over 0.30 m above the Top Busty Coal.

Tilley Coal (Figures 16 and 17)

There is little doubt about the lateral continuity of the Tilley Coal, or Tilley group of coals, in this district, but considerable local variation is present. In general, a single seam, up to 0.99 m thick but more generally 0.50 to 0.71 m, is found in the western part of the district. When traced towards the east the coal splits locally into two or more leaves separated by thin mudstones or seatearths. Exceptionally, the Tilley seam comprises three or four thin coals separated by up to 6 m of strata, mainly seatearths and sandstone. More rarely the coal, or group of coals, is thin or absent, perhaps due to washout. Workings in the Tilley Coal, and in particular the lower leaf, occur widely throughout the coalfield, but are localised on account of the thinness of the seam.

Strata between the Tilley and Harvey coals (Figures 16 and 17)

These beds range between 7.3 and 24 m in thickness, but more generally the interval is 9 to 12 m thick. The sequence locally includes one or more sandstones, which tend to occur more commonly in the upper and lower parts. More rarely, sandstone occupies virtually the whole sequence. Elsewhere, argillaceous rocks with only thin sandstones or sandy beds, and locally with thin ribs and nodules of ironstone, form the interval. A section in these strata is exposed [1397 5439] on the south side of Pontop Burn, 2 km east of Medomsley, where some 21 m of interbedded sandstone, siltstone and mudstone have been recorded; argillaceous rocks predominate towards the base, where there is a pronounced bed containing large subcircular and elongate ironstone nodules. Other sections are also visible at Oaky Bank and Oaky Burn [1400 5450], a little distance to the north-east.

Up to three coals or seatearths, largely thin and impersistent, are present. The median coal is equated with the Hodge Coal of west Durham (Figures 16 and 17). This coal can be traced with some confidence in the northern and western parts of the district, where it attains a local maximum thickness of nearly 0.76 m, though more generally it is in the range 0.20 to 0.40 m. Eastwards the position of the coal becomes less certain, and it appears locally to be absent. In the Town Head Farm Borehole*, south of Burnopfield, poorly preserved mussels were recorded in grey mudstone 0.71 m thick overlying the Hodge Coal, and in the Tanfield Lea Borehole [1700 5257], some 3 km to the south, black cannelloid shale with mussels and ostracods was recorded from a similar horizon. Plant remains are particularly common in the argillaceous measures in this part of the sequence.

Harvey Coal (Figures 16 and 17)

In the northern part of the district the Harvey Coal generally ranges between 0.76 and 1.15 m thick (Figure 16). In some of the thicker sections, an argillaceous parting may be present in the lower part of the seam. In the central and southern parts of the district the seam is thinner, generally between 0.55 and 0.88 m (Figure 17). At a few localities the seam is impoverished or washed out. Towards the south-east, the general thickness tends to fall to between 0.45 and 0.56 m. The Harvey Coal has been widely worked in the western part of the coalfield, but workings farther east, though extensive, are more restricted, as under the central and northern parts of Newcastle.

Strata between the Harvey Coal and the base of the Harvey Marine Band (Figures 16 and 17)

These strata range in thickness from as little as 3 m to over 24 m, the thicker sequences being accounted for largely by the local dominance of sandstones, of both sheet and channel facies. More generally, the thickness ranges between 10.66 and 19.24 m. As a whole these beds are characterised by considerable lateral and vertical lithological change. Although sandstone locally predominates, more especially near the top and base, in a more typical occurrence between one and four relatively thin sandstones are present in mainly argillaceous strata, including thin impersistent coals and seatearths.

The strata immediately overlying the Harvey Coal were recorded in an opencast site near Woody Close [1269 5089], east-north-east of Crookhall, the section being as follows.

	Thickness m
Mudstone, laminated, micaceous and carbonaceous; planty	2.00+
Coal	0.09
Mudstone and seatearth	1.96
Coal	0.09
Seatearth-mudstone	0.34
Coal	0.21
Shale, black	0.10
Seatearth-mudstone	0.10
Seatearth-sandstone	0.37
Mudstone; siltstone beds	1.90
Mudstone	0.10
Harvey Coal	—

The Harvey Coal itself is widely overlain by a thin kaolin-rich fragmental clayrock, particularly where the roof to the coal is mudstone or shale. At Ravensworth Anne Pit, the fragmental clayrock was recorded as locally up to 30 mm thick, but passes rapidly into carbonaceous mudstone with coal laminae and pyrites (Richardson and Francis, 1971). The clayrock, 50 mm thick, has also been recorded in the nearby Kibblesworth Borehole*. It is commonly at the base of, and closely associated with, the Hopkins Band, which contains a characteristic mussel–ostracod fauna. The Hopkins Band is believed to be present throughout the district, but has been fully recorded in only a few boreholes. Dark fissile mudstones and carbonaceous shales are commonly recorded in the roof of the Harvey Coal, and boreholes in which the Hopkins Band has been recorded include those at Priestfield Farm [1575 5663], Marley Hill Downover [2078 5760], Kibblesworth*, Glamis Shaft [2432 5619] at Kibblesworth, Ravensworth Anne Pit No. 6 [2792 5640] (just within the Sunderland district) and Grange Villa/Craghead Road [2262 5117].

The argillaceous measures overlying the Harvey Coal locally have a high iron content. In the Consett area, and immediately to the east, the roof measures of the Harvey Coal consist of up to 3 m of dark grey mudstone containing several thin beds of ironstone, or layers of ironstone nodules, which were worked extensively, along with the Harvey Coal, as the No. 1 Ironstone by the Consett Iron Company during the last century. Ironstone ribs and nodules also occur locally at this horizon elsewhere, as at Redheugh Pit [2443 6263], Gateshead. The No. 1 Ironstone is named incorrectly on the published 1:50 000 map of the district as the Ten Bands Ironstone, a bed that lies higher in the succession.

In the overlying beds, the Penny Hill Coal thickens to a maximum of about 0.76 m between Mickley, Prudhoe and the area between Chopwell and Blaydon. To the east and south-east it is thin and impersistent but it has been locally identified (Figures 16 and 17). Isolated boreholes have recorded a few mussels above the Pennyhill Coal or its inferred position. A higher coal or seatearth is very thin and impersistent. The Harvey Marine Band itself is commonly underlain by a thick seatearth-mudstone or seatearth-sandstone underlying a thin coal.

Middle Coal Measures

The **strata from the Harvey Marine Band to the base of the Hutton Coal** (Figures 18 and 19) range between 25 and 55 m in thickness. The **Harvey Marine Band** is the local correlative of the Vanderbeckei Marine Band, the base of which defines the base of the Duckmantian Stage (Westphalian B), and the base of the Middle Coal Measures (Table 2; Ramsbottom et al., 1978). This marine band is an important and widespread stratigraphical marker, but detailed records and collections in the district are poor. The marine band is overlain by the **Harvey Shell Bed**, a dark grey mudstone containing an abundant fauna of mussels, and which is also widely present in the district. More detailed collections have been obtained from both the marine band and shell bed in nearby districts (Smith and Francis, 1967; Land, 1974; Smith, 1994). In the Consett area, this part of the sequence is marked by the presence of thin beds and nodular layers of ironstone, the **Ten Bands Ironstone**. On the published 1:50 000 map of the district this name is incorrectly applied to an ironstone overlying the Harvey Coal (see above).

The remaining, part of the Harvey Marine Band–Hutton Coal sequence consists of sandstones and argillaceous beds, the former comprising the greater proportion of, and locally dominating, the interval; several channel sandstones are present. **The Ruler Coal**, a thin seam generally up to 0.3 m thick, but often split into two leaves, is present in the lower part of the sequence, especially in the west. In the Newcastle area, and in the adjacent Morpeth and Tynemouth districts, the seam is more generally known as the **Plessey Coal**, and is also generally split into two leaves. Two beds well known in the Tynemouth district are not recorded in the present district. They are the Cheeveley Coal, which occurs between the Harvey Marine Band and the Ruler Coal, and the Plessey Shell Bed above the Plessey Coal.

The **Hutton Coal** (Figures 18 and 19) has been widely worked in the district and generally ranges between 1.1 and 1.8 m in thickness, but locally exceeds 2 m. In the north-east, and very locally elsewhere in the east, it rapidly deteriorates and is absent or very thin.

The **strata between the Hutton and Brass Thill coals** (Figures 18 and 19) range between 3 and 18 m in thickness, and vary considerably in lithology . No consistent pattern of sediment type is obvious, although argillaceous strata tend to predominate. In contrast to the Tynemouth and Sunderland districts, there are no records of mussels in the roof mudstones of the Hutton Coal. In one instance fish debris was recorded.

Over most of this district the **Brass Thill Coal** is typically split into a Bottom Brass Thill and a Top Brass Thill, the latter being virtually united or very closely associated with the overlying Durham Low Main to form a composite seam widely termed the Top Brass Thill/Durham Low Main (Figures 18 and 19). Prior to regional standardisation, seam nomenclature and correlation in this part of the sequence was especially confusing (Table 3). **The Bottom Brass Thill** is widely present as a seam ranging between about 0.7 and 1.3 m in thickness, separated from the overlying **Durham Low Main Coal** (Top Brass Thill/Durham Low Main) by largely argillaceous measures up to 20 m thick. This latter coal generally ranges between 1.1 and 2.0 m in thickness, but in parts of the south of the district it forms a composite seam over 3 m thick, including dirt partings.

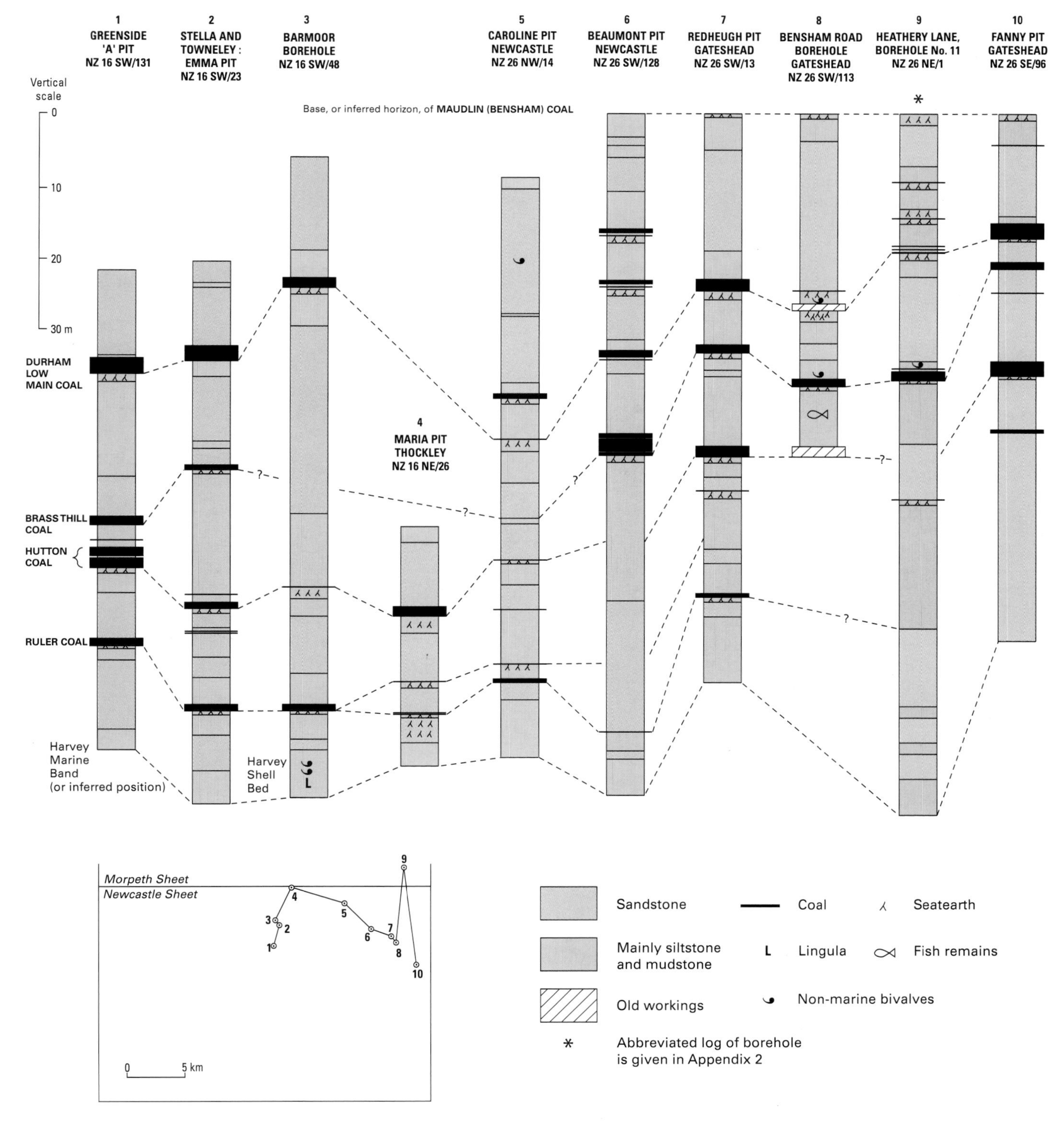

Figure 18 Comparative sections of Middle Coal Measures: Harvey (Vanderbeckei) Marine Band to Maudlin (Bensham) Coal (northern part of district).

The **strata between the Durham Low Main and Maudlin coals** (Figures 18 and 19) generally range between 8 and 35 m in thickness, and commonly include a sandstone (Plate 7), the **Durham Low Main Post**, often in two or three main beds. Locally, the channelised base of this sandstone extends down to, or cuts out, the Durham Low Main Coal. Thin, impersistent coals have been recorded in this sequence, more particularly in the north-east, where a

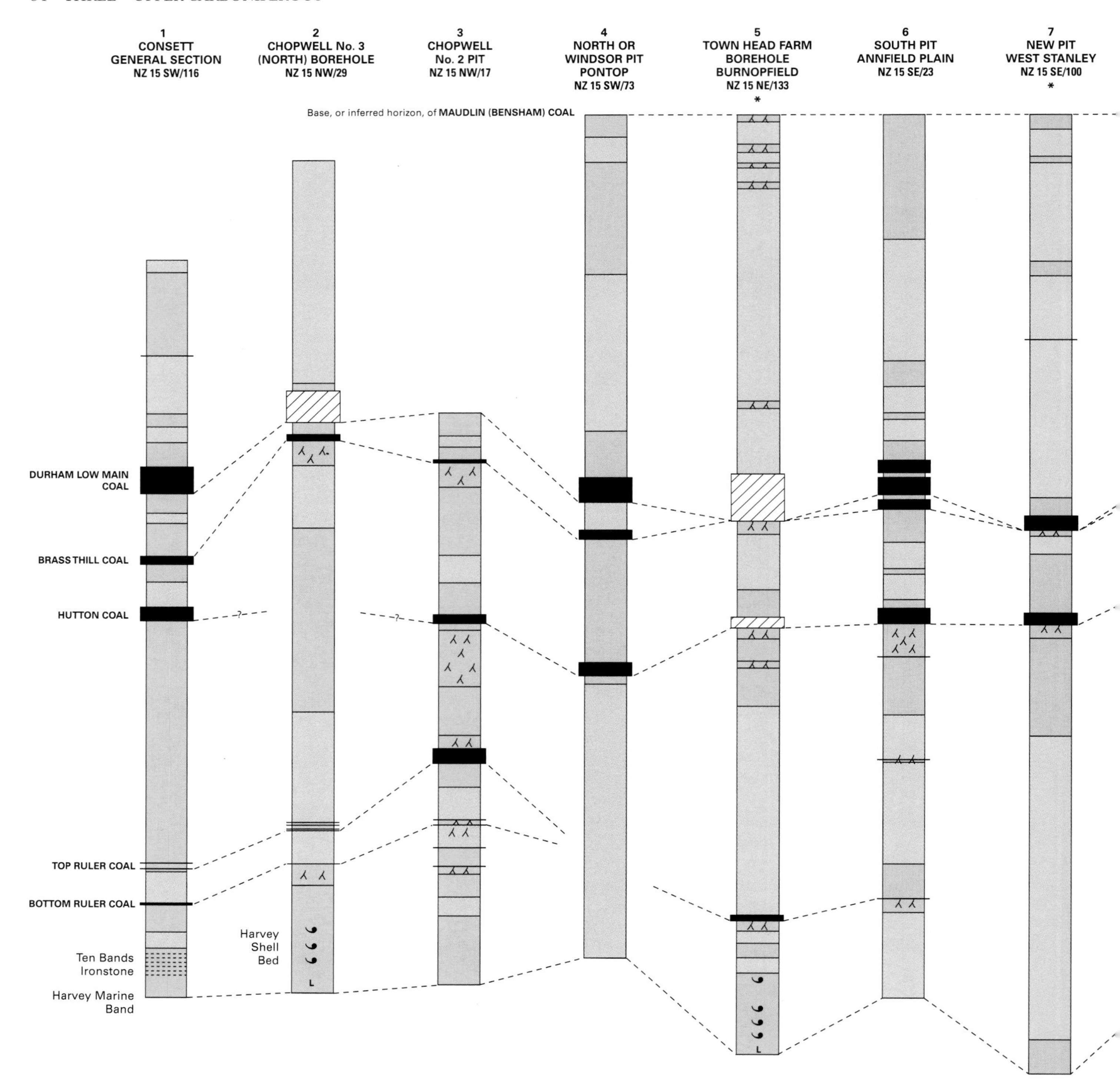

Figure 19 Comparative sections of Middle Coal Measures: Harvey (Vanderbeckei) Marine Band to Maudlin (Bensham) Coal (southern part of district).

higher proportion of argillaceous measures is present. Mussels have been recorded at one locality from the roof measures of the Durham Low Main Coal. In general, these strata contain a higher proportion of sandstone than the equivalent measures in the Tynemouth and Sunderland districts, although in the former district the Table Rocks Sandstone (Land, 1974, p.66), up to about 15 m thick, has many features in common with the Durham Low Main

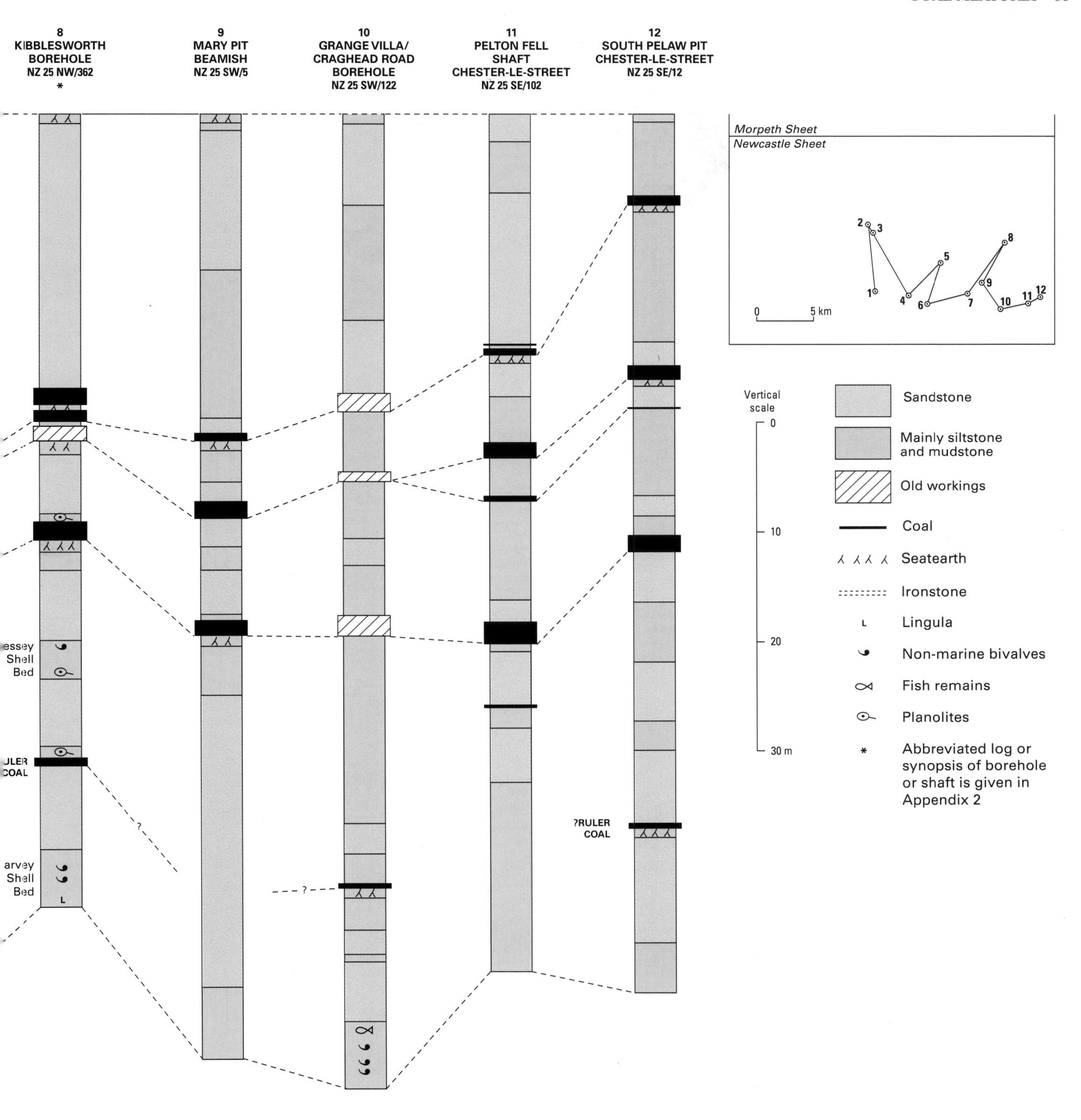

Post. Again, in parts of the Morpeth district (Lawrence and Jackson, 1986), a persistent sandstone between 9 and 23 m thick is present between the two coals.

The **Maudlin (Bensham) Coal** (Figure 20) is commonly thin, impoverished or washed out, and it is only along part of the eastern margin of the district, and in the south-east, that it is thick enough (0.6 to 1.4 m) to have been worked. This contrasts markedly with the Tynemouth and Sunderland districts, where the coal, or its splits, are consistently thicker and have been widely worked. To the north, however, in the Ponteland–Morpeth area (Lawrence and Jackson, 1986), the character of the coal is very similar to that described here. The boundary between the *Anthraconaia modiolaris* and

Plate 7 Flaggy-weathering sandstone (Middle Coal Measures) at Mountsett Quarry, near Tantobie [165 551]. The widened joints are due to collapse into old workings of the Durham Low Main Coal, just beneath. L1759.

the Lower *Anthracosia similis–Anthraconaia pulchra* nonmarine bivalve zones is taken at the top of the Maudlin Coal.

The **strata between the Maudlin and Main coals** (Figure 20) generally range between 13 and 18 m in thickness. The strata tend to be predominantly argillaceous but include up to four thin sandstones. Thin impersistent coals are locally present, and up to three mussel bands have been recorded at various levels. The lowest of these, in the roof measures of the Maudlin Coal, is widely recorded in adjacent districts to east and north-east.

The **Main Coal** (Figure 20) has been extensively worked in the southern part of the district, where it forms a good quality seam ranging between 1.3 and 1.8 m thick. Locally, and more particularly in the extreme east and south-east, it becomes split into Top and Bottom Main coals, the higher coal rapidly deteriorating. In the north-eastern part of the district the Main Coal is again relatively thin. Farther north-east it splits, the upper leaf, a possible correlative of the Bentinck Coal of Northumberland (Land, 1974), being separated by up to 12 m from the lower leaf.

The **strata between the Main and Five-Quarter coals** (Figure 20) generally range between 14 and 18 m in thickness but, exceptionally, decrease to as little as 0.35 m in the west of the outcrop area where the coals are almost united. The strata are largely argillaceous in character although some sandstone is present locally in the south, and a slightly higher proportion of sandstone is also recorded in the north-east. In two boreholes, mussels have been recorded near the top of the interval.

The **Five-Quarter Coal** (Figure 20) forms a single seam generally ranging in thickness from 0.6 to 2.7 m over much of the southern part of the district, but in the north-east and south-east the coal is split into Top and Bottom Five-Quarter coals, separated by up to 12 m of strata, most of which are argillaceous and locally include nonmarine bivalves. The coal was widely worked where it is a single seam, but elsewhere extraction was much more localised due to deterioration both in thickness and quality.

The **strata between the Five-Quarter and High Main coals** (Figure 20) range between 15 and 36 m in thickness, and are predominantly argillaceous. Some thin sandstones are also present, and in the east the proportion of sandstone locally becomes significantly higher. Up to three coals, for the most part thin and impersistent, are locally present. One of these, however, regarded as the equivalent of the **Metal Coal** of the Tynemouth and Sunderland districts, thickens sufficiently in the east and north-east to have been worked very locally in the Gateshead area and in parts of Newcastle.

The **High Main Coal** (Figure 20) is one of the thickest and most consistent seams in the district. The early coal mining industry of the area was largely based on this coal, which is rarely less than 1.5 m thick and is generally over 2 m thick. Locally it is subject to washout. In the south-east of the district it locally becomes split into two leaves, the **Top** and **Bottom High Main** coals, separated by thin argillaceous measures. Small areas of coal are believed to remain unworked in the south-east and beneath parts of Gateshead, but elsewhere the High Main Coal has effectively been worked out.

The **strata between the High Main Coal and the base of the High Main Marine Band** (Figure 20) range between 7 and 24 m thick and occupy a restricted area in this district. Because of the relative thinness and poor quality of the seams in this part of the sequence, borehole and mining records provide only limited information. Equivalent beds to the east are better known (Land, 1974; Smith, 1994). A thick sandstone, the **High Main Post**, commonly makes up most of the interval, locally occupying the whole of it, and in some areas it cuts down to impoverish or wash out the High Main Coal. Above the **High Main Post** a few metres of argillaceous beds, including a seat-earth and a thin impersistent coal, the **High Main Marine Band Coal**, generally underlie the High Main Marine Band. North of the Ninety Fathom Fault, in the Newcastle area, the whole interval is relatively thin and sandstone is subordinate, if present at all. '*Estheria*' and fish scales have been recorded from one bed in this sequence. Equivalent strata in the Tynemouth and Sunderland districts also contain thick sandstones, but show greater diversity of lithology and fauna.

The **High Main Marine Band**, or Maltby Marine Band of the British standard nomenclature (Ramsbottom et al., 1978), has only been recorded in one borehole (Figure 21) in the north-eastern part of the district, although the characteristic lithology of dark grey and black shale and mudstone associated with this horizon has been proved elsewhere. Locally the marine band may be cut out by the overlying sandstone. However, as Land (1974, p.93) has pointed out, the marine band is not present everywhere in the coalfield.

The **strata between the High Main Marine Band and the Ryhope Little Coal** (Figure 21) range between 12 and 33 m in thickness, the separation increasing to the east and north-east. The strata are predominantly argillaceous in character, but include a very much higher proportion of sandstone near the eastern margin of the district, in the south part of the Gateshead conurbation. One or two thin coals are present, the upper one, occurring in the east and north-east, being equated with the **Moorland Coal** of the Tynemouth district (Land, 1974). Faunal assemblages have been proved at three levels in a borehole in the north of the district; the lowest, in the roof of the High Main Marine Band, is the **High Main Shell Bed**. This is regarded as one of the most prolific mussel bands in the coalfield and is probably widespread, but has generally not been recorded.

The **Ryhope Little Coal** (Figure 21) often split into **Top** and **Bottom Ryhope Little** coals, is characterised by widespread and rapid changes in thickness, and by a tendency to split and reunite over comparatively short distances. Also, one or both coals can rapidly die out, while very locally the Bottom Ryhope Little Coal is itself split. Where it forms a single seam, the Ryhope Little Coal is in the general range of 0.6 to 0.7 m thick, but both leaves, either singly or where present together, can attain thicknesses of this order. Locally, as in the eastern part of Newcastle, the single seam attains a thickness of more than 1 m and has been mined. Areas of both Bottom and Top Ryhope Little coals have been worked opencast in the south and south-east.

The **strata between the Ryhope Little and Five-Quarter coals** (Figure 21) range between 8 and 20 m in thickness. In the more western part of the outcrop argillaceous and sandy strata are present in roughly equal proportions but farther east sandstone is more dominant, especially south of Gateshead and in the Jesmond Dene area of Newcastle. Little sandstone is present north of the Ninety Fathom Fault, where the separation between the two coals is least. The Little Marine Band is the local equivalent of the Clown Marine Band of British standard nomenclature (Ramsbottom et al., 1978). It is impersistent in adjacent areas and has not been proved in the district.

The **Ryhope Five-Quarter Coal** (Figure 21) forms either a single seam up to 0.7 m thick, or is split into two leaves, the lower of which appears to be thicker and more consistent. Where it is split, the maximum separation is 3 m, made up of argillaceous strata. The seam has been worked locally in the north-east, and has been worked opencast at a small number of sites in the south.

The **strata between the Ryhope Five-Quarter Coal and Kirkby's Marine Band** (Figures 21 and 22) range up to 35 m thick, and contain a high proportion of sandstone, especially south of Gateshead. Up to three thin impersistent coals are present. The uppermost, locally termed the Crow Coal, may equate with a seam that lies close below Kirkby's Marine Band. Sandstone usually overlies the Ryhope Five-Quarter Coal, and locally occupies the whole interval.

Kirkby's Marine Band is only exposed at one locality in this district, but has been proved in a number of boreholes (Figure 22), and is also exposed on the bank of the River Tyne just to the east of the district. It is one of the most widespread marine bands in the coalfield, and is characterised by an alternation of marine and nonmarine phases. The sequence typically consists of mudstone containing *Lingula* and foraminifera, overlain by grey mudstone with mussels, in turn succeeded by strata containing '*Estheria*', *Planolites,* foraminifera and *Lingula.* The band, discovered by Kirkby (1860, p.412) and named by Armstrong and Price (1954, p.978), has been collected extensively in the Tynemouth and Sunderland districts (Land, 1974, p.99; Smith, 1994, p.40). It is the correlative of the Haughton Marine Band of the British standard nomenclature (Ramsbottom et al., 1978).

Strata overlying Kirkby's Marine Band are largely restricted to the extreme north-east part of the district, their presence being inferred from meagre borehole data and scanty surface information, combined with data from adjacent districts (Land, 1974; Smith, 1994). Figure 22 shows an inferred generalised sequence for the extreme north-east of the district, and the graphic log of a borehole in the Morpeth district. This is the nearest section, some 2 km to the north of the district, which affords a complete record through these higher strata.

The **Hylton Marine Band** is the correlative of the Sutton Marine Band of the standard British marine band nomenclature (Ramsbottom et al., 1978). The **Seventy Fathom Post**, a major sandstone above Kirkby's Marine Band is present in the extreme north-east, mainly under drift. It is thought to cut out or replace the Hylton

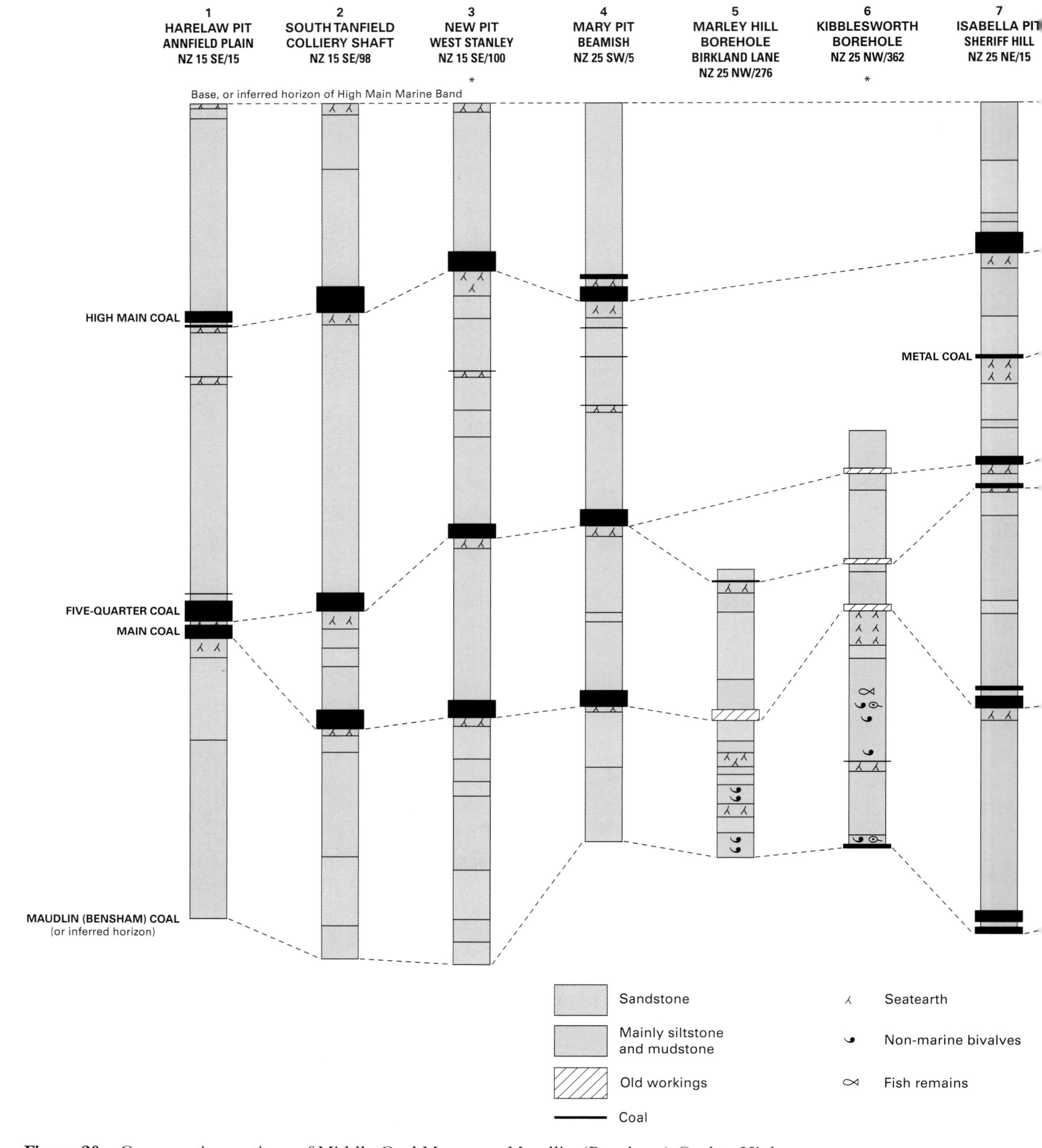

Figure 20 Comparative sections of Middle Coal Measures: Maudlin (Bensham) Coal to High Main (Maltby) Marine Band.

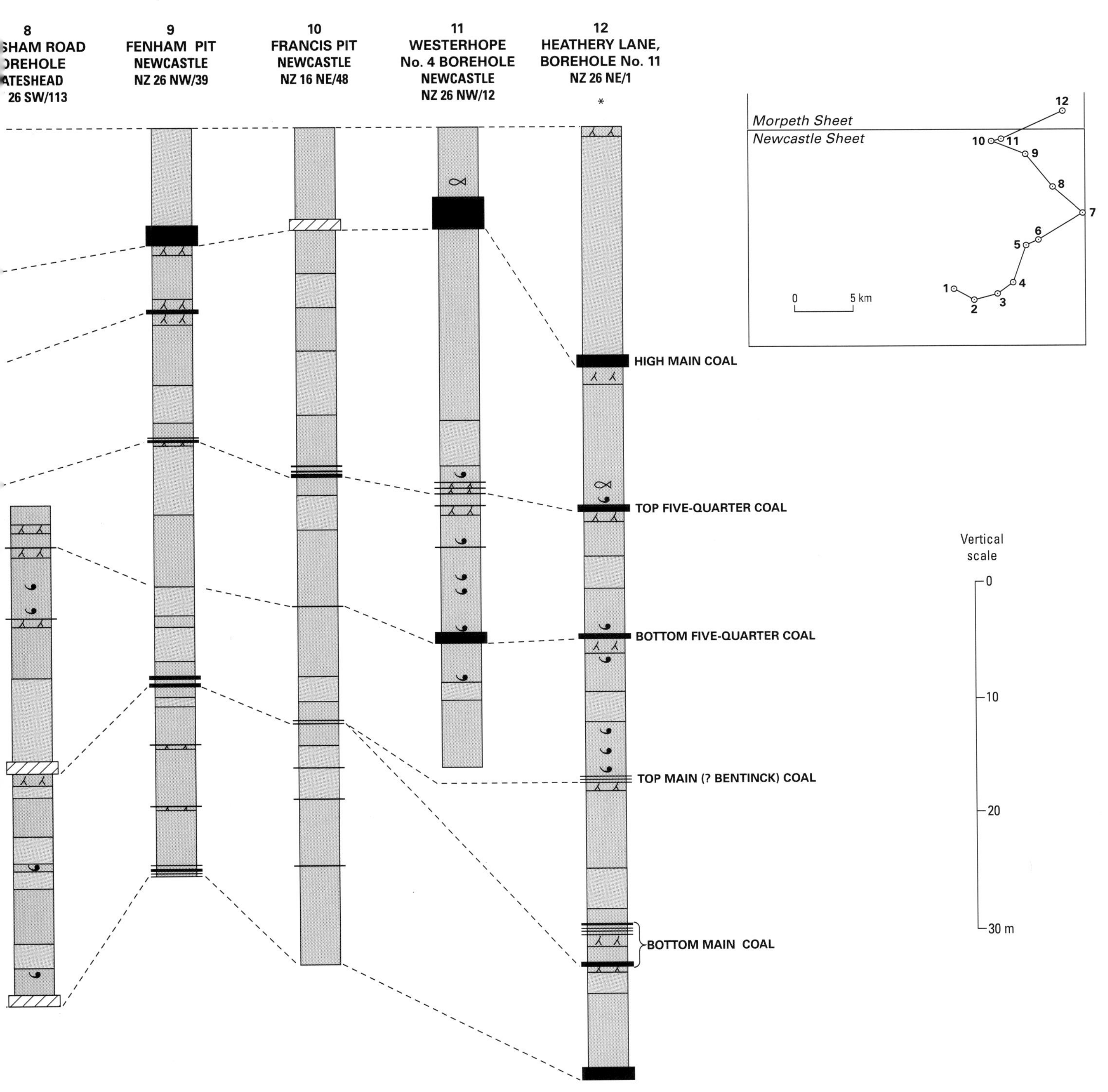

Planolites

* Abbreviated log or synopsis of borehole is given in Appendix 2

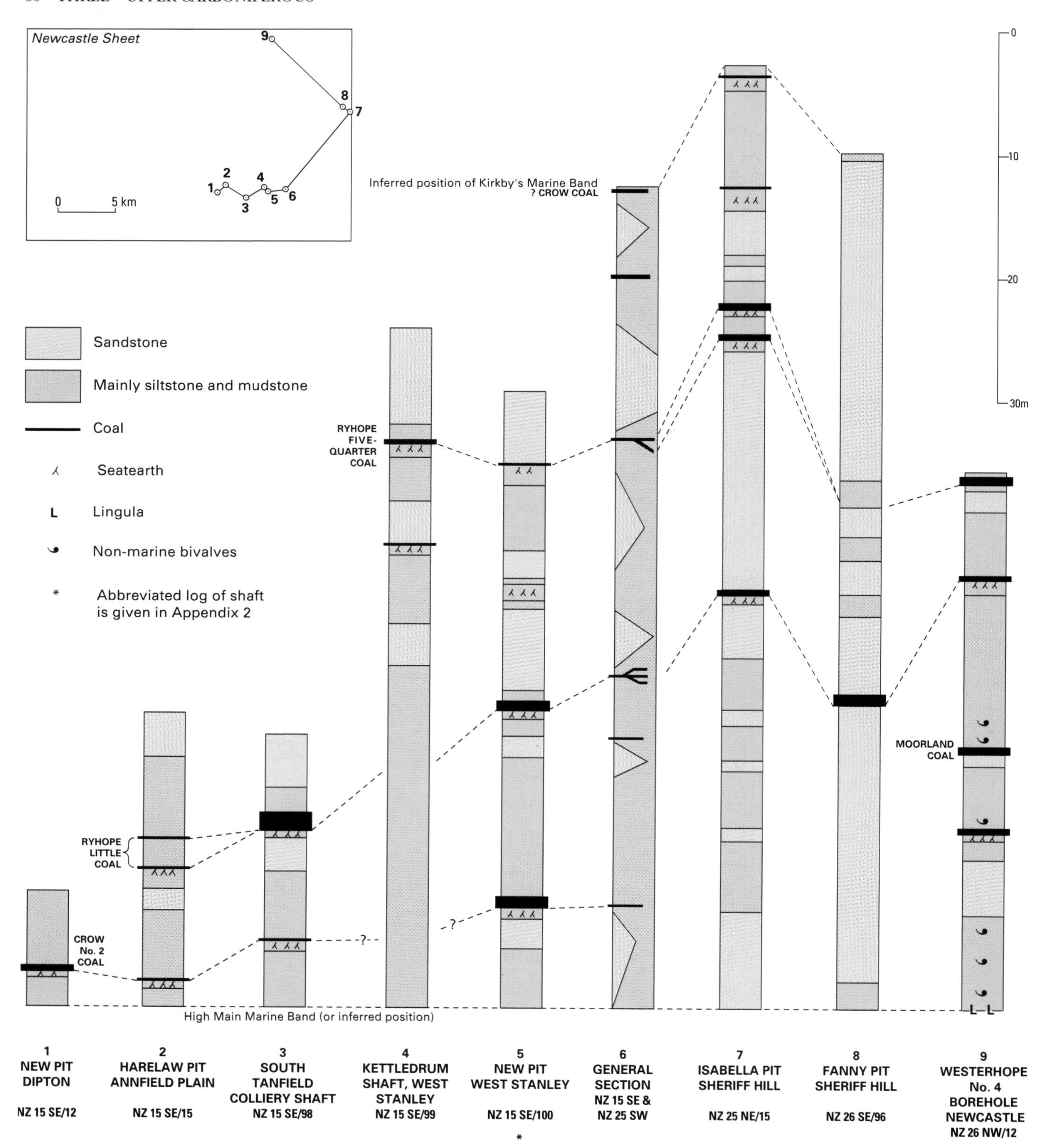

Figure 21 Comparative sections of Middle Coal Measures: High Main (Maltby) Marine Band to Kirkby's (Haughton) Marine Band.

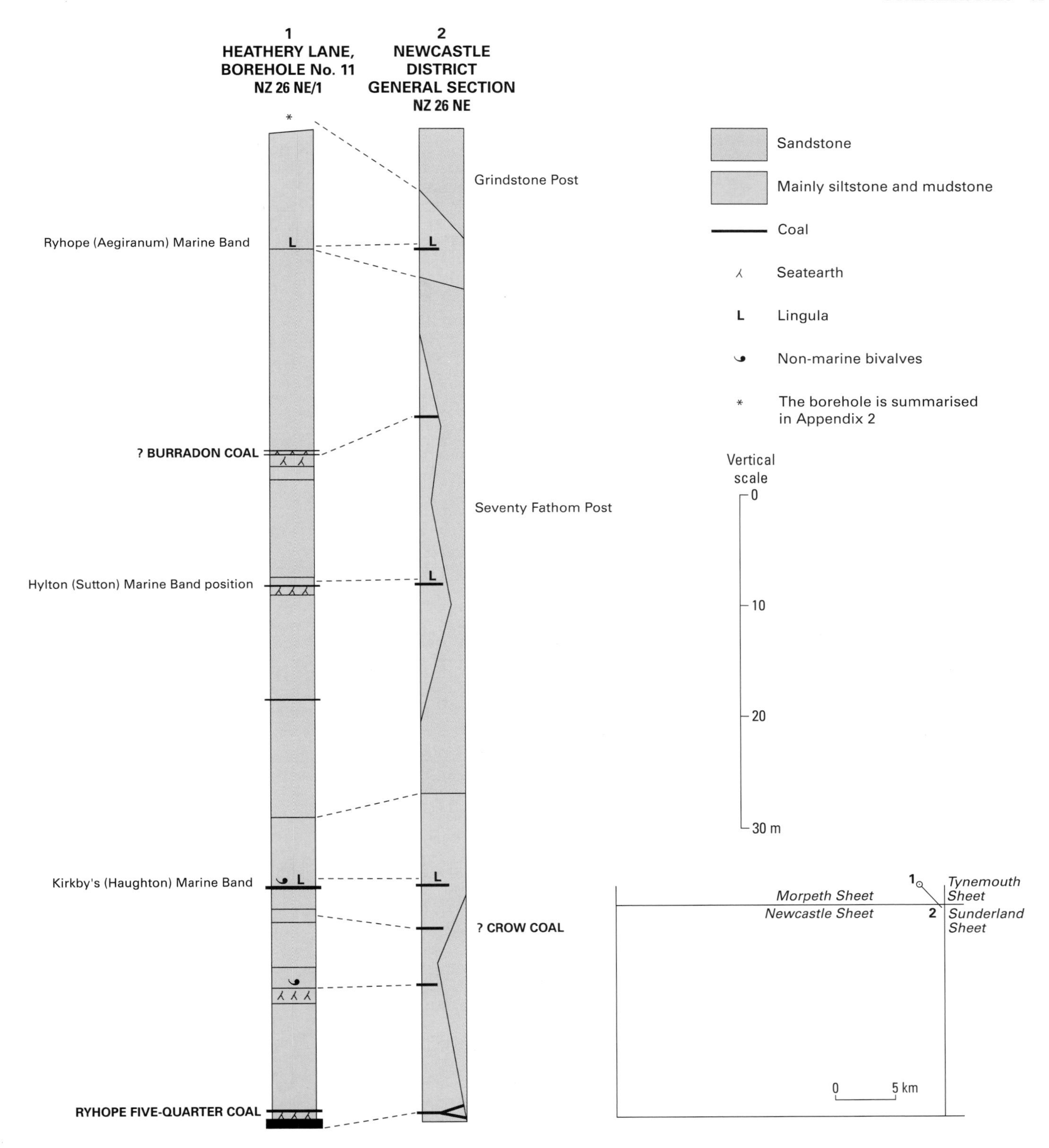

Figure 22 High Middle Coal Measures: north-east part of the district (generalised) and a borehole nearby.

Marine Band and **Burradon Coal,** neither of which has been recorded in this district. The **Ryhope Marine Band**, some overlying argillaceous beds, and the succeeding few metres of the **Grindstone Post**, are the highest strata occurring in the district (Figure 22). They are only a few metres thick and their presence is inferred. The base of the Ryhope Marine Band, which is correlated with the Aegiranum Marine Band, marks the base of the Bolsovian Stage (Westphalian C) (Table 2) and is also the boundary between the Lower and Upper *Anthracosia similis–Anthraconaia pulchra* nonmarine bivalve zones (Ramsbottom et al., 1978).

Harvey (Vanderbeckei) Marine Band and Harvey Shell Bed (Figures 18 and 19)

The Harvey Marine Band (Armstrong and Price, 1954) has been proved in a number of boreholes, especially in the south. The bed is widespread throughout north-east England, and although marine faunas have not been noted everywhere, the characteristic lithology is widely recorded. The Harvey Marine Band generally comprises up to 0.3 m of very dark grey micaceous shale and silty mudstone containing *Lingula,* with some fish debris and sporadic *Planolites.* At most localities it contains only a few specimens of *Lingula,* and lacks the more diverse fauna of the Durham and West Hartlepool district, where marine bivalves, brachiopods and sponge spicules have been recorded.

The marine band is overlain by the Harvey Shell Bed, or is locally separated from it by a thin barren mudstone. The shell bed consists of grey mudstone with ironstone nodules and a fauna of mussels and fish debris. It is widely present in the district, and has been proved in a number of boreholes, although detailed palaeontological collections are largely lacking. *Naiadites* sp., *Carbonicola* sp. and *Anthracosia* sp., together with fish debris and *Spirorbis,* have been recorded from Chopwell No. 3 Borehole [1102 5913]. Mussels including *Naiadites* sp. have been recorded at several levels in 7.19 m of mudstone overlying the Harvey Marine Band in the Town Head Farm Borehole* near Burnopfield. East of Consett, several thin beds of ironstone and ironstone nodules characterise this part of the sequence, and were worked as The Ten Bands Ironstone by the Consett Iron Company in the last century. Locally both the Harvey Marine Band and Harvey Shell Bed are cut out by channel sandstones, as for example in North Pit [1447 5248] at Pontop.

Strata between the Harvey Shell Bed and the base of the Hutton Coal (Figures 18 and 19)

These beds vary widely in lithology and thickness, and include sheet sandstones and more local channel sandstones with associated washouts. They range between 21 and nearly 50 m in thickness, the thicker sections usually including the thicker sandstones. For example, in the New Pit* at West Stanley, and at Mary Pit [2102 5356] at Beamish, sandstones over 26 m thick form most of the interval.

North of a line from Medomsley to the southern outskirts of Gateshead, the Ruler Coal is present generally in the lower to lower-middle part of the interval. Exceptionally recorded as a single seam up to 1.06 m thick, it is more often split into Top and Bottom Ruler coals separated by up to 6 m of strata, and these commonly include sandstone. In many areas, both leaves are full of mudstone partings. In the north-east of the district the Ruler Coal, more generally known here and to the north as the Plessey Coal, is again generally split into two leaves, both of which are thin and impersistent. However, on the western flank of the Team Valley, in the Lobley Hill area, the coal, here thought to be a united seam, is up to a metre thick and has been worked over a limited area. Similarly, the coal and its underlying seatearth may have been worked together beneath Scotswood. Elsewhere the Ruler Coal has only been worked in very limited areas. In the western part of the coalfield a thin, impersistent coal is locally recorded between the Hutton and Ruler seams.

In the Kibblesworth Borehole* a mudstone bed 2.90 m thick, with abundant mussels, was recorded some 9 m below the Hutton Coal. It may represent the only record of the Plessey Shell Bed in this district. Bold sandstone features characterise this part of the sequence in several areas, notably east and north-east [108 543] of Consett, around the higher flanks of the Derwent Valley to the south of Hamsterley, in the Horse Gate [115 597] area to the west of High Spen, around Labourn's Fell [096 587] west of Chopwell, on the high ground south [079 619] and east of Mickley Square, and west of Winlaton. Sandstones overlying a thin Ruler Coal were formerly exposed [1380 4957] at Woodside Common Quarry, 2 km east of Delves, where some 7 m of medium- and coarse-grained sandstone were visible in the immediate roof of the seam.

Hutton Coal (Figures 18 and 19)

This is a widely worked seam, usually of high quality and between 1.06 and 1.82 m thick. Locally it thickens to more than 2.5 m, but then includes thin beds and partings of inferior quality coal. North of the River Tyne, in the Newcastle area, the seam deteriorates very rapidly towards the north-east, rarely exceeding 0.30 m in thickness, and locally is very thin or absent. The coal is also slightly thinner in the south-east of the district, generally in the range 0.63 to 1.29 m.

Strata between the Hutton and Brass Thill coals (Figures 18 and 19)

These strata range from as little as 3m to over 18 m thick, but usually the thickness is about 9 m. The sequence varies considerably and no overall pattern of lithological distribution is apparent. In some localities it is composed almost entirely of argillaceous strata, as at Billingside No. 1 Borehole [1376 5284] at Leadgate, and North Pit [1447 5248], Pontop, whereas in others it is made up almost entirely of sandstone, as in the Kibblesworth Borehole*, in Mary Pit [2102 5356] at Beamish and in 'D' Pit [2549 5414] at Urpeth.

Few fossils have been recorded in this part of the sequence. However, in the Bensham Road Borehole [2482 6223], Gateshead, a few fish scales were recovered from mudstone 6.7 m above the Hutton Coal, and in the Kibblesworth Borehole*, *Planolites* was recorded from the roof of the coal.

Bottom Brass Thill Coal (Figures 18 and 19)

This coal is present as a single seam in much of the district, and generally ranges between 0.68 and 1.27 m in thickness. Along the western fringes of the coalfield, broadly west of a line extending from Throckley to Pontop, the coal tends to be thinner, rarely exceeding 0.60 m. In the north-east it is impoverished, thin, and split by dirt bands or washed out, as at Caroline Pit [2035 6589], Newcastle. In the south-east the coal is usually split into two leaves, up to 0.81 and 0.55 m thick, separated by up to 4 m of argillaceous strata. Although widely worked, the coal is commonly dirty, and there are many areas where workings were patchy. In the north-east of the district it was left unworked over a large area, due to seam impoverishment.

Strata between the Bottom Brass Thill and Durham Low Main coals (Figures 18 and 19)

These strata range up to a maximum of 20 m thick, the greater thicknesses being due to the occurrence of unusually thick sandstone, as in the Barmoor Borehole [1407 6441], near Ryton, where the Bottom Brass Thill Coal is probably washed out. More generally, the strata are thinner, normally less than 9 m, and largely argillaceous. At localities where the Top Brass Thill Coal is close to the Durham Low Main Coal, a thick seatearth-mudstone is locally present near the top of the interval. In the Bensham Road Borehole [2482 6223], a few small mussels in grey mudstone were recovered from the roof of the Brass Thill Coal.

Durham Low Main Coal (Figures 18 and 19)

This, one of the most extensively worked coals in the district, generally ranges between 1.11 and 1.98 m in thickness, locally reaching 2.43 m, although it often contains dirt partings. Along the southern margin of the district the seam section tends to be somewhat thinner, at between 0.85 and 1.20 m. In many areas, more particularly in the south-west, the Top Brass Thill and Durham Low Main coals virtually form a single seam, ranging up to nearly 3.04 m thick, with little intervening seatearth and mudstone. At South Pit [1626 5177], Pontop, the section recorded was: coal 0.83 m, dirt 0.28 m, coal 0.45 m, dirt (brass) 0.02 m, coal 1.07 m, blue metal 0.73 m, on coal 0.60 m (with 0.05 m dirt parting). Broadly similar sections have been proved elsewhere in the south and south-west, notably around Leadgate, east of Consett, around Beamish and near Kibblesworth.

The Durham Low Main is subject to local washout by the overlying sandstone, for example in a broad east-north-east-trending belt to the north of Burnhope, and near Edmondsley in the south of the district. Except for such areas, and in the extreme north-east where it deteriorates and rarely exceeds 0.60 m in thickness, the seam has been widely worked in the district.

Strata between the Durham Low Main and Maudlin (Bensham) coals (Figures 18 and 19)

These strata generally range between 8 and 35 m in thickness, with a recorded maximum of 41 m. The roof measures of the Durham Low Main generally comprise up to 6 m of siltstones and mudstones, the latter commonly containing ironstone nodules. A thin impersistent coal is very locally present at the top of this sequence. In the Bensham Road Borehole [2482 6223], at Gateshead, a few mussel fragments were recorded from the roof of the Durham Low Main Coal, and similar occurrences are recorded [260 530] east of Pelton.

The interval is characterised in most of the district by a thick sandstone, the Durham Low Main Post, which is usually split into three or four beds separated by thin argillaceous strata. The sandstone, which commonly rests on, and locally cuts out, the Durham Low Main Coal, ranges from cross-bedded and medium to coarse grained to more flaggy, fine grained, micaceous and carbonaceous, with argillaceous partings dividing it into a number of posts. The maximum composite thickness of the sandstone occurs in the Pontop, Burnopfield, Stanley and Kibblesworth areas, where values of up to 34 m have been recorded. The Durham Low Main Post caps much of the high ground in the central part of the district, as around Medomsley, east of Leadgate, and most notably on the south side of the Ninety Fathom Fault between Labourn's Fell and Chopwell; here many old diggings and small quarry workings formerly exposed white, grey, yellow and buff sandstone, cross-bedded, medium- to coarse-grained, and with sporadic coarse pebbly beds. In the old Mountsett Quarries [165 551] (Plate 7) near Tantobie, medium- to coarse-grained sandstone up to 21 m thick was formerly exposed overlying the Durham Low Main Coal (Plate 7). Isolated outcrops of this sandstone are still exposed in the grounds of the adjacent crematorium. In Causey Gill, north-east of Tanfield Lea, the stream has cut a gorge-like feature along most of its length, exposing the Durham Low Main Post. On the north side of Broomhill Dene [140 516], east of Leadgate, up to 9 m of sandstone forming the roof measures to the Durham Low Main Coal are exposed. The sandstone is flaggy, with sporadic ironstone concretions. In Town Head Farm Borehole* the main bed of sandstone, predominantly fine to medium grained at the base and coarse grained at the top, is over 31 m thick and is overlain by 0.30 m of rooty mudstone and black shale beneath a thin Maudlin Coal.

Towards the eastern margin of the district, the thickness of the interval between the seams decreases rapidly, if irregularly, and less sandstone is generally present. Two thin impersistent coals occur locally (Figure 19). In the Beaumont Pit [2266 6332], Newcastle, two coals, 1.09 and 0.73 m thick and containing many dirt bands, were present in the lower part of the interval. In the Caroline Pit [2035 6589], Newcastle, 0.50 m of 'shelly' post, mixed with 'coal pipes', was recorded some 16 m above the Durham Low Main Coal, in a total sandstone sequence of at least 27 m.

Maudlin (Bensham) Coal (Figure 20)

This coal is thin or impoverished in the central part of the district, ranging between 0.10 and 0.25 m thick. Locally, the coal is cut out by a thick sandstone which in some places extends down towards, and almost unites with, the Durham Low Main Post. Along the eastern margin of the district, and in the south-east, the seam thickens rapidly and progressively; locally it forms a sequence of up to four thin dirty coals, none of them over 0.45 m in thickness, but elsewhere it is between 0.60 and 1.44 m thick, commonly with a dirt band in its lower part. South of the River Tyne, workings are concentrated near the eastern margin of the district. In the Whickham area, where the seam ranges from 1.2 up to an exceptional 2 m thick, the seam has been almost worked out. North of the River Tyne, workings are very restricted since the seam is largely thin, impoverished or washed out. However, towards the extreme north-east, the coal thickens and approaches 0.83 m.

Strata between the Maudlin and Main coals (Figure 20)

These strata range between 9 and 24 m in thickness, the more general range being about 13 to 18 m. The strata tend to be predominantly argillaceous, with one to four thin sandstones. Where a thicker sandstone is present, it generally lies towards the top of the interval. Up to three thin impersistent coals are known to occur north of the River Tyne, but elsewhere only one has been recorded. In the south of the district, strata overlying the Maudlin Coal are exposed on the south bank [2332 4945] of the Cong Burn, just to the north of Edmondsley. The following sequence has been recorded:

	Thickness m
Sandstone	0.60
Shale, sandy	0.60
Sandstone	0.60
Shale, sandy	0.60
Sandstone	1.52
Maudlin Coal	0.91

Faunas have been obtained from three different levels, namely the roof measures of the Maudlin Coal (or its inferred position), at a level in the lower part of the interval, and also approximately midway between the Maudlin and Main coals (Figure 20). Mussels have been recorded in the roof measures of the Maudlin Coal from a number of boreholes; for example, in the Town Head Farm Borehole* abundant mussels, with *Spirorbis* sp., were recorded in 3.08 m of grey mudstone above the coal, and in the Kibblesworth Borehole* small mussels, with *Planolites* sp., were recorded from 0.73 m of grey mudstone with ironstone ribs at a similar level. This borehole also recorded mussels from the two higher levels. The lower contained two mussel bands in mudstone overlying a coal 0.13 m thick, some 8 m above the Maudlin Coal, and the higher, at a level 13 m above the coal, contained poorly preserved mussels and ?fish fragments. The upper mussel band may correlate with the Blackhall Estheria Band of areas to the east (Magraw et al., 1963; Smith, 1994).

Main Coal (Figure 20)

This is a consistently thick, high-quality seam throughout most of the southern half of the district, where it has been extensively worked; it generally ranges between 1.34 and 1.82 m thick, and is normally comparatively free of argillaceous partings. The thicker sections tend to be in the west. The coal has only been partially worked in the east-central and south-eastern parts of the district; here it generally ranges between 0.75 and 1.82 m and locally splits into Top and Bottom Main coals. The upper coal deteriorates rapidly, and may be separated by more than 3 m of argillaceous strata from the lower coal.

North of the River Tyne, in the Newcastle area, the stratigraphy is more complicated but modern data are scanty. In the north-western part of the city two thin coals at this level, thought to be Top and Bottom Main, are separated by thin beds of mudstone and seatearth. South-eastwards they combine to form a workable seam up to one metre thick in the Heworth area, east of the district boundary. However, to the north-east, and in the nearest part of the Morpeth district, the coal is split into Top and Bottom Main coals separated by nearly 12 m of strata, including a 3 m thick sandstone near the base. The Top Main Coal is locally in two leaves, 0.25 and 0.12 m thick, separated by 0.38 m of mudstone. The Bottom Main consists of leaves 0.71 and 0.22 m thick separated by over 2 m of argillaceous measures. The seam and its splits have not been worked north of the Tyne in the Newcastle area. Here, the maximum recorded thickness of 0.71 m (Top Main Coal only) occurs north of the Ninety Fathom Fault.

Strata between the Main and Five-Quarter coals (Figure 20)

In the south part of the district, around Pontop, Annfield Plain and Burnhope, the separation between these coals is as little as 0.35 m, the coals forming a composite seam up to 2.38 m in thickness. East of an approximate line from Tanfield to Stanley, and thence south-eastwards, this interval thickens rapidly and ranges up to an exceptional 27 m, although the more general separation is between 14 and 18 m. With local exceptions, the strata are largely argillaceous, with only subordinate sandstones.

Strata overlying the Main Coal are exposed [2392 4979] on the south side of the Cong Burn, north-east of Edmondsley, where the following section has been recorded:

	Thickness m
Sandstone, thin-bedded	2.43
Shale	1.21
Shale, sandy	3.65
Shale	0.91
Main Coal	1.52

A sandstone midway in the interval is locally present in the Marley Hill and Kibblesworth area, and also in the western outskirts of Gateshead. In the Bensham Road Borehole [2482 6223], Gateshead, these strata are nearly 19 m thick, and in the upper part include a coal 0.07 m thick overlain by grey shale containing mussels. North of the Tyne, in the north-eastern part of the district, a higher proportion of sandstone is commonly present. In the western part of Newcastle, a bed of fireclay, the so-called 'Lister Horizon', has been worked in a limited area south of the site of the crematorium and the A69 road. In Westerhope No. 4 Borehole [2036 6646], in the north-west part of the city, undetermined mussels were collected at a similar horizon to those recorded in the Bensham Road Borehole. In Heathery Lane No. 11 Borehole*, in the extreme south-east of the adjacent Morpeth district, mussels were recorded at two levels, notably in 4.39 m of mudstone in the roof of the Top Main, and in a bed 1.55 m below the Bottom Five-Quarter Coal.

Five-Quarter Coal (Figure 20)

In much of the southern part of the district, the Five-Quarter Coal forms a single seam, of variable quality, ranging in thickness between 0.60 and 2.70 m. In the western part of the outcrop, notably between Pontop and Annfield Plain, the Five-Quarter Coal lies close above the Main Coal. In the south-east, the coal becomes split into Top and Bottom Five-Quarter coals separated by up to 9 or 10 m of strata. Both coals tend to deteriorate as a result of splitting. The Bottom Five-Quarter Coal ranges from as little as 0.02 up to 0.99 m in thickness, whereas the Top Five-Quarter Coal ranges between 0.48 and 0.71 m. Farther east, in the Sunderland district, the Top and Bottom Five-Quarter coals reunite to form a single seam up to 1.25 m thick. North of the River Tyne, in the north-eastern part of the district, both Top and Bottom Five-Quarter coals are generally present and are separated by up to 12 m of strata. The Bottom Five-Quarter ranges up to an exceptional 1.29 m in thickness, although in some shafts it was not recorded, and the Top Five-Quarter forms a seam in up to three leaves, with many dirt partings.

The strata between the Top and Bottom Five-Quarter coals are largely argillaceous, but with some sandstone in the middle or upper part of the interval north of the River Tyne. In Westerhope No. 4 Borehole [2036 6646], in the north-western part of Newcastle, mussels including *Anthracosia* sp. were recorded at three separate levels in 11 m of strata between the Top and Bottom Five-Quarter coals; a thin coal underlies the uppermost mussel band.

The Five-Quarter Coal has been widely worked in the south of the district, but workings become more patchy farther east, in the Gateshead area, and the coal has not been worked around and to north of Springwell. North of the River Tyne the seam is generally of poor quality, but has been worked in limited areas.

Strata between the Five-Quarter and High Main coals (Figure 20)

This interval, which includes the Metal Coal, ranges between 15 and 36.5 m in thickness, the thicker sections being present towards the west, and largely where the underlying beds, between the Five-Quarter and Main coals, are thinnest. The strata are mainly argillaceous although one or two thin sandstones are generally present. Atypically, at South Tanfield

Colliery Shaft [1793 5218] the whole sequence, over 27 m thick, is recorded as sandstone. Along the eastern margin of the district, and including the western outskirts of Gateshead, the proportion of sandstone increases significantly as recorded in the Isabella [2730 5966] and Fanny [2673 6018] pits at Sheriff Hill. Up to three impersistent coals have been recorded in this part of the district. Towards the east and south-east, one of these, which lies roughly midway between the Five-Quarter and High Main coals, thickens sufficiently to have been worked locally in Gateshead, and to the south at Ravensworth Colliery. This seam, thought to be equivalent to the Metal Coal of the Tynemouth and Sunderland districts, is generally less than 0.6 m but locally thickens to 0.9 m.

In the north-eastern part of the district, north of the River Tyne and including Newcastle, the interval is generally thinner, in the range 15 to 22 m. The strata are also largely argillaceous, but generally include a sandstone near the base. Mussels, including *Anthracosia* sp., have been recorded in the roof measures of the Top Five-Quarter in Westerhope No. 4 Borehole [2036 6646]. A thin coal locally recorded 5 to 9 m below the High Main Coal is thought to be the equivalent of the Metal Coal. It locally thickens, and boreholes in the Gallowgate area of Newcastle suggest that there were ancient workings in the seam.

High Main Coal (Figure 20)

In the south of the district this coal forms a consistently thick seam ranging between 1.27 and 2.59 m in thickness. A dirt parting about 0.22 m thick (exceptionally up to 0.73 m) is generally present in the middle or lower part of the coal. Locally the seam is split by a number of very thin dirt bands or beds of inferior coal. The seam is also subject to local impoverishment or washout, as for example in a south- and south-east-trending belt around Pontop Pike and Harelaw. In the south-east the High Main Coal locally splits into two leaves, a lower leaf, up to 2.0 m thick, separated by thin argillaceous measures from an upper leaf, up to 0.60 m in thickness. The High Main Coal has been extensively worked across the south and east parts of the district.

North of the River Tyne, in the Newcastle area, the High Main Coal ranges between 1.95 and 2.41 m thick (Figure 20). The coal has been worked very widely, in some places almost to the point of total extraction. It is probable that in parts of Newcastle, notably in the Town Moor, Castle Leazes, Fenham and Benwell areas, and also in parts of Gateshead, the High Main Coal was one of the first seams in the district to be dug at outcrop, perhaps as early as the 12th or 13th century.

Strata between the High Main Coal and the base of the High Main Marine Band (Figure 20)

These strata range between 7 and 24 m in thickness. In most of the south and east of the district the interval is occupied by a thick bed of sandstone, the High Main Post. This bed is rarely exposed, although outcrops are widespread, including areas west of Annfield Plain, around Stanley, near West Pelton, in north and north-west Gateshead, and on the higher ground overlooking the Tyne and Team valleys. The High Main Post mainly comprises cross-bedded fine- and medium-grained sandstone with some coarse-grained beds. More locally it is medium- to coarse-grained and pebbly, filling erosive-based channels with a south to south-easterly trend which cut down into, and wash out, the underlying High Main Coal. In some areas, as for example around Annfield Plain and Stanley and Beamish, sandstone extends up to the High Main Marine Band position, but argillaceous measures, ranging up to 6 or 7 m in thickness, more generally succeed the High Main Post. These commonly include a seatearth overlain by a thin coal, the High Main Marine Band Coal, which is relatively persistent in parts of the district. For example, in the neighbourhood of Edmondsley, this coal locally attains a thickness of 0.60 m, and was worked opencast [212 482] north of Broom Hill. In the Isabella Pit [2730 5966], in the Sheriff Hill area of Gateshead, the High Main Marine Band Coal was 0.60 m thick.

North of the River Tyne, including the Newcastle area, the High Main Post, up to 20 or 25 m thick, also dominates this interval. The sandstone forms a large subcircular crop in the north-west part of the city, and also underlies most of the Town Moor and West Jesmond. North-west of the Ninety Fathom Fault, however, notably in the Denton and Westerhope areas of the city, the interval thins significantly, sandstone is subordinate or absent, and the High Main Marine Band is only a few metres above the High Main Coal. In Westerhope No. 4 Borehole [2036 6646] the strata, 6.85 m thick, are wholly argillaceous. '*Estheria*', with fish scales below, is present in a bed some 2.28 m above the High Main Coal.

High Main (Maltby) Marine Band (Figure 21)

The High Main Marine Band has been recorded in Westerhope No. 4 Borehole [2036 6646], where 0.10 m of dark grey shale containing *Lingula* sp. was recorded. Several boreholes and shafts that have penetrated this bed in the district have proved the characteristic lithology of dark grey and black shale, but the marine fauna was absent or unrecognised. Locally the marine band is cut out by sandstone, which may coalesce with the underlying High Main Post.

Strata between the High Main Marine Band and the Ryhope Little Coal (Figure 21)

These strata, which include the High Main Shell Bed and the Moorland Coal, show considerable variation in thickness, the interval thickening to the east and north-east. In the Stanley area the strata range between 12 and 25 m thick and are mainly argillaceous, with a thin sandstone generally present in the lower part and another immediately underlying the Ryhope Little Coal. A coal locally termed Crow No. 2 Coal, ranging in thickness between 0.10 and 1.06 m, is also widely present in the lower part of the interval. This seam is probably too low in the sequence to equate with the Moorland Coal of the Tynemouth district.

On the eastern margin of the district, and in the higher parts of Gateshead, including Sheriff Hill, Beacon Lough and towards Wrekenton, these strata are thicker, ranging between 24 and 33 m, and include a high proportion of sandstone (Figure 21). At the Isabella Pit [2730 5966], Sheriff Hill, five sandstones are present, the thickest lying at the base and top of the interval. Although coal was not recorded in this neighbourhood, a thin seam up to 0.45 m thick, correlated with the Moorland Coal, occurs in the middle to upper part of the sequence beneath parts of central Gateshead, its crop trending north-north-east.

North of the River Tyne, in the Newcastle area, these strata are not well known. The full sequence was, however, proved in the Westerhope No. 4 Borehole [2036 6646], north of the Ninety Fathom Fault (Figure 21). Here the strata, 34 m thick, are predominantly argillaceous but two thin sandstones occur, the lower being the thicker. Two coals, the lower 0.27 m thick and the upper 0.33 m thick, are present in the lower and middle part of the sequence, the upper one being correlated with the Moorland Coal. Faunas have been collected at three levels. The lowest of these, the High Main Shell Bed, is in the roof to the High Main Marine Band and consists of grey shale nearly 7 m thick containing *Naiadites* sp. and *Spirorbis*. The next shell bed, in the roof of the overlying coal, consists of 1.52 m of grey mudstone containing *Anthracosia* sp. The third shell bed is in the roof of the Moorland Coal, and comprises 0.96 m of black shale with sporadic *Anthracosia* sp. and *Naiadites* sp. A coal thought to be the Moorland Coal, and an overlying thin sandstone, crop out [259 658] in Jesmond Dene. There are no plans of workings in the Moorland Coal, but it is possible that some extraction may have taken place north of the Ninety Fathom Fault near Kenton Bar.

Ryhope Little Coal (Figure 21)

This seam shows much variability, with sudden changes of thickness, and is commonly split into two leaves. In the area around Annfield Plain and Stanley, the coal generally ranges between 0.58 and 0.70 m in thickness, attaining an exceptional 1.36 m in 'A' Pit at South Tanfield Colliery [1793 5218], where the seam was described as 'coal and black stone'. However, west of Harelaw and Annfield Plain, and near the seam's western crop, it is split into two thin leaves, the Top and Bottom Ryhope Little coals, separated by up to 9 m of argillaceous strata. South of Quaking Houses, near Stanley, the seam is also split, the Top Ryhope Little Coal, up to 1.37 m thick, being separated by up to 4 m of strata from the Bottom Ryhope Little Coal, itself split into two thin leaves. A small area of Top Ryhope Little Coal was worked opencast east of New Acres Farm [1922 5037]. East of Craghead, near White Hill [2328 5109], an outlier of Bottom Ryhope Little Coal, here a single seam up to 0.51 m thick, also has been worked opencast. Hereabouts, a thin Top Ryhope Little Coal up to 0.73 m thick has been recorded, ranging to 12 m above the Bottom leaf. Areas of Bottom Ryhope Little Coal have also been worked opencast on the southern margin of the district, south-west of Edmondsley, where the general thickness was between 0.45 and 0.91 m.

On the eastern margin of the district, in the Sheriff Hill area of Gateshead, the Ryhope Little Coal is present as a single seam 0.58 to 0.68 m thick. It splits into two leaves towards the north-east and the Tyne Valley, but these reunite again some distance north of the river. No workings are present in these coals hereabouts, but a possible lower leaf of the Ryhope Little Coal, 0.7 to 0.9 m thick, was worked nearby at Portobello in the Sunderland district.

In the eastern part of Newcastle the coal generally contains many thin argillaceous bands, and is widely split into Top and Bottom coals by mudstone. Exceptionally, in the South Gosforth and Jesmond areas, the coals unite to form a single seam 1 m, very locally 2 m, thick. The coal was worked from South Gosforth Colliery and several shafts in Jesmond Dene. North of the Ninety Fathom Fault, the few data available suggest that the coal is not split, and ranges between 0.30 and 0.48 m thick.

Strata between the Ryhope Little and Ryhope Five-Quarter coals (Figure 21)

These strata range between 8 and 20 m in thickness, the interval being generally thinner in the north-east, and thicker in the east and south. Around West Stanley, argillaceous strata and sandstones are present in roughly equal proportions. Two sandstones, one near the base, the other occupying a broadly intermediate position, are commonly present, separated by an impersistent thin coal or seatearth. In the southern part of Gateshead, these strata consist mainly of sandstone or sandy 'mixed' beds, but north-east towards the River Tyne the sandstone appears to thin out. There is a sandstone at the same level in the Jesmond Dene area of Newcastle, where it forms the roof to the Ryhope Little Coal. This sandstone is visible in Jesmond Dene [2574 6724], where some 6 m of fine- to medium-grained sandstone crop out in the stream bed, and on the east bank of the Dene. The sandstone passes up into beds of alternating siltstone and sandstone underlying the Ryhope Five-Quarter Coal. North-west of the Ninety Fathom Fault, these strata consist of between 7 and 8 m of argillaceous beds which tend to be sandy towards the top.

Ryhope Five-Quarter Coal (Figure 21)

In the Stanley area the Ryhope Five-Quarter Coal generally forms a single seam ranging between 0.38 and 0.69 m thick. The coal has been worked at a number of opencast sites south of Stanley, and south-west of both Craghead and Edmondsley, typically with a thickness of between 0.36 and 0.53 m. In the Gateshead area the coal is variable, and either forms a single seam up to 0.60 m thick, or is split into two leaves up to 0.35 and 0.51 m thick separated by up to 3 m of argillaceous strata. North of the River Tyne, in the eastern part of Newcastle, the coal typically contains dirt bands and is generally in two leaves, 0.66 and 0.25 m, thick separated by 2 or 3 m of argillaceous measures. What appears to be the lower leaf of the seam was worked locally around Gosforth, where the coal ranged up to a recorded maximum of 0.96 m thick. North of the Ninety Fathom Fault the seam is up to 0.60 m thick and contains many dirt partings.

Strata between the Ryhope Five-Quarter Coal and Kirkby's Marine Band (Figures 21 and 22)

These are the highest recorded strata around Stanley, and also south-west of Craghead, ranging up to about 35 m in thickness. A sandstone up to 10 m thick commonly overlies the Ryhope Five-Quarter Coal. It was formerly exposed at Morrowedge Quarry [1910 5033], south of Quaking Houses, where over 4.2 m of medium-grained sandstone were separated by a few metres of argillaceous strata from the roof of the Ryhope Five-Quarter Coal. The succeeding strata, poorly documented, are largely argillaceous and contain two impersistent coals. The lower of these is thin; the upper, locally termed the Crow, is up to 0.38 m thick and lies some 25 m above the Ryhope Five-Quarter Coal. The coal is overlain by a thin impersistent sandstone.

In the Gateshead area the Ryhope Five-Quarter Coal is overlain by, or separated by a few metres from, a sandstone up to 27 m thick; this is commonly split into two beds, the upper of which is discontinuous. Where the sandstones are separate, the lower bed is overlain locally by an impersistent coal or a seatearth, which possibly equates with the lower thin coal in the Stanley and Craghead area. An impersistent coal near the top of the interval may be equivalent to the Crow Coal. North of the Tyne, in the Jesmond Dene area of Newcastle, the interval is up to 27 m thick and includes one or two sandstones and three thin impersistent coals. The uppermost of these is thought to underlie the position of Kirkby's Marine Band (Figure 22). North of the Ninety Fathom Fault, in the Blakelaw area of Newcastle, the sequence is predominantly argillaceous and contains two thin coals.

Kirkby's (Haughton) Marine Band (Figure 22)

In this district Kirkby's Marine Band comprises alternating marine and nonmarine phases. It is only exposed [2574 6724] on the east side of Jesmond Dene, Newcastle, the sequence being mudstone with *Lingula* sp. and fish debris 0.15 m, siltstone and mudstone with mussels 1.0 m, on mudstone with

Lingula sp. A detailed collection has not been made from this locality. Just to the east, in the Sunderland district, it has been proved in boreholes at Byker and is also exposed on the north bank [2816 6329 to 2840 6312] of the River Tyne at St. Anthony's, from where fucoids, *Lingula mytilloides, Planolites ophthalmoides* and *Lioestheria vinti* were collected (Smith, 1994).

Strata overlying Kirkby's Marine Band (Figure 22)

The presence of higher beds in the district, including the Hylton Marine Band, the Seventy Fathom Post, the Burradon Coal, the Ryhope Marine Band and the Grindstone Post, has been largely inferred from boreholes and other subsurface information, together with a restricted amount of surface data. These higher strata occur more widely to the north-east and east of the district, and have been described in the sheet memoirs for the Tynemouth and Sunderland districts (Land, 1974; Smith, 1994).

Over 50 m of strata occupy the interval between Kirkby's Marine Band and the Ryhope Marine Band (Figure 22). In the eastern part of Newcastle, the strata overlying Kirkby's Marine Band mainly consist of sandstones or beds of a largely sandy character. Neither the Hylton (Sutton) Marine Band nor the thin Burradon Coal have been recorded in the district, but they may locally be present near the middle of this interval. Both are known to be cut out over wide areas in adjacent districts by a sandstone, the Seventy Fathom Post. This sandstone is present in the north-east corner of the district, and was exposed [2827 6879] some 2 km beyond the district boundary at Benton Junction (Land, 1974, p.102). The presence of the Ryhope (Aegiranum) Marine Band has been inferred in a small down-faulted area [2718 6127] south-west of High Felling, Gateshead. The nearest exposure of this bed is also at Benton Junction [2826 6888] (Land, 1974, p.104). An overlying thick sandstone, the Grindstone Post, is separated from the Ryhope Marine Band by a few metres of argillaceous strata. This sandstone was formerly quarried extensively for grindstone, building stone and road material, and together with the underlying Seventy Fathom Post forms much of the high ground in the south part of Gateshead and in eastern Newcastle.

Little is known of these strata in the part of the district that lies north of the Ninety Fathom Fault. The full sequence was, however, proved in Heathery Lane Borehole No. 11*, some 2 km north of the district boundary (Figure 22). A thick bed of sandstone a few metres above Kirkby's Marine Band was formerly worked at both Blakelaw [2122 6683] and Kenton [2205 6753] quarries, the latter just north of the district. In the first of these, more than 7.62 m of white, yellow and brown, flaggy to laminated, fine- to medium-grained sandstone were exposed.

FOUR
Structure

The structure of the concealed Lower Carboniferous and older rocks, and the structural controls on Carboniferous sedimentation, were reviewed in Chapter 2. In this chapter, the structure of the Upper Carboniferous rocks is considered in more detail.

TECTONIC HISTORY

Syndepositional fault activity probably continued during Upper Carboniferous times (Fielding, 1984a; Collier, 1989), but on a much smaller scale than in the earlier Carboniferous (Kimbell et al., 1989) (Figure 4). Late Namurian and younger syndepositional fault displacements were small, so they cannot be detected on seismic reflection profiles (Chadwick et al., 1993a, 1995). Upper Carboniferous sedimentation in the district was in response to broad regional subsidence, and the differences in stratigraphy and thickness that can be detected between the Northumberland Trough and the Alston Block result as much from differential compaction as from contemporary displacement of the Ninety Fathom–Stublick Fault System. The youngest Carboniferous rocks preserved are high Middle Coal Measures of early Bolsovian age, but by analogy with adjacent districts it is probable that deposition continued in the district until Westphalian D times.

As a result of broadly east–west compression, brought about by the end-Carboniferous Variscan earth movements, the Carboniferous rocks of the district were uplifted, faulted and gently folded (Jones et al., 1980; Dunham, 1990; Chadwick et al., 1995) (Figures 23 and 24). Some older faults, most notably the Ninety Fathom and Stublick faults, were reactivated. The Northumberland Trough was inverted at this time, and basin margin folds formed in response to reversed oblique-slip displacement of the Ninety Fathom Fault. A gentle doming and an easterly tilt were imparted to the Alston Block, and significant thicknesses of Upper Carboniferous rocks were eroded away. The Whin Sill and its associated dykes were intruded at a late stage in these movements, following the inversion of the Northumberland Trough and the doming and tilting of the Alston Block.

The subsequent tectonic history of the district is conjectural because of the lack of any preserved strata younger than Bolsovian. However, it seems probable from regional considerations that extensional faulting and related basin subsidence took place sporadically from Permian to early Cretaceous times (Chadwick et al., 1993b). It is likely that Permo-Triassic and early Jurassic rocks once covered the district (Holliday, 1993). In mid- to late Cretaceous times, active extension probably ceased, and regional subsidence allowed the Chalk to be deposited over the whole of the district.

Another deformational phase of basin inversion and regional uplift, resulting from north–south compression, began in Palaeogene (early Tertiary) times. This led to the reactivation of earlier faults as well as to renewed faulting and minor folding in the district, locally accompanied by tholeiitic dyke injection (Chapter 5). Uplift of the Pennines, in response to the subsidence of the North Sea, imparted a further easterly tilt to the rocks of the district (Figures 5 and 24), and led to the removal of all post-Carboniferous cover rocks and to further erosion of Upper Carboniferous strata.

FAULTS

As indicated in Chapter 2, the **Ninety Fathom–Stublick Fault System** was a major influence on Carboniferous sedimentation. It is the most important and fundamental structure in the Carboniferous rocks of the district, and formed by reactivation of a major basement shear zone. Other major faults in the district are the Muggleswick, Healeyfield, Tantobie, Felling and Heworth faults (Figure 23). Syndepositional displacement on these cannot be wholly discounted, but was likely to have been small. It is not known whether they are reactivated basement structures, but the Tantobie and related faults could be related to the western and southern margins of the Rowlands Gill cupola of the Weardale Granite.

The **Ninety Fathom Fault** has a persistent, if slightly sinuous, east-north-easterly course across the district. The fault consistently throws down to the north, the throw increasing slowly and erratically towards the east-north-east. The fault dies out to the west near Acton Fell, in the Hexham district; at the western margin of the Newcastle district the throw is barely 10 m; between Labourn's Fell and Ryton the displacement ranges between 50 and 100 m; in the western suburbs of Newcastle the throw is 200 to 210 m; and it is about 250 m at the northern margin of the district. To the north-east, the fault can readily be traced across the south-eastern part of the Morpeth district, the southern part of the Tynemouth district, where the throw ranges between 137 and 366 m (Land, 1974), and for a few kilometres offshore (Chadwick and Holliday, 1991). The fault is of simple, normal character with a hade of up to 45°; on the north (downthrow) side of the fault, the dip of strata towards the fault locally increases rapidly over a limited distance, but there is usually little or no disturbance on the south (upthrow) side.

North-west of the Ninety Fathom Fault is a series of parallel or subparallel faults in a belt up to 8 km wide (Figure 23). At Corbridge and Riding Mill two of these faults, both of which trend north-east, form the eastern

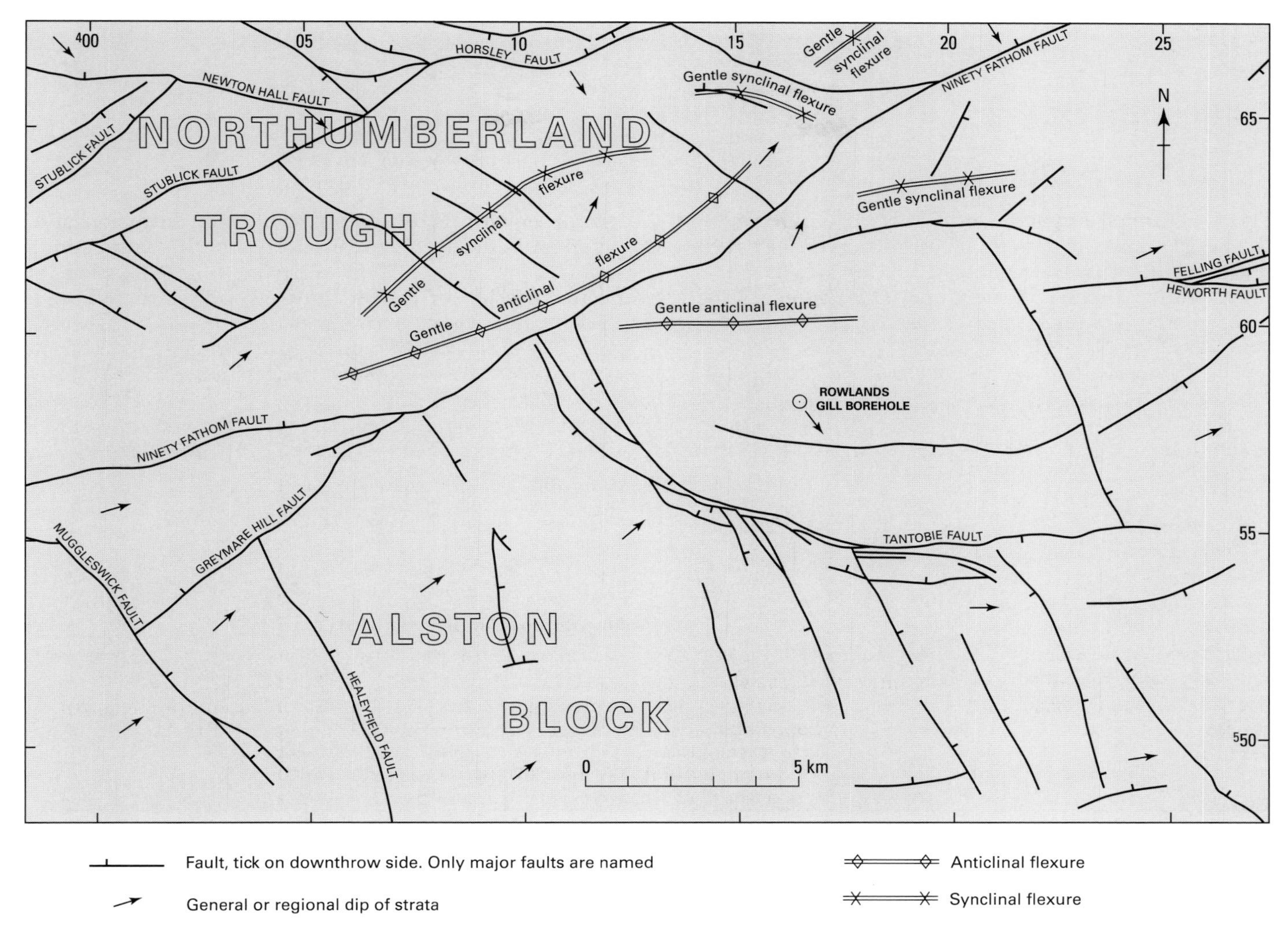

Figure 23 Surface position of main structural elements of the district.

continuation of the **Stublick Fault**, which is the westward, *en échelon* continuation of the Ninety Fathom Fault. The Stublick Fault System is a major feature of the Hexham and Brampton districts to the west. In this district, the southern of the two faults, locally termed the **Horsley Fault**, with a maximum displacement of 155 m, throws Coal Measures down on its northern side. This extension to the coalfield, which has been extensively mined in opencast workings to the north and west of Horsley and north of Heddon-on-the-Wall, was not mapped by the Primary surveyors. Near the eastern limit of this system, Coal Measures are faulted down into an east–west-trending graben with Stainmore Group rocks to both north and south. Several other faults are associated with the Stublick and Ninety Fathom faults, but in most cases their throws are comparatively small.

In the south-west of the district, the **Muggleswick Fault** trends north-west along the Derwent Reservoir from Muggleswick, where it appears to be split into two components. Faults encountered in the Derwent–Wear Tunnel (Appendix 3) at chainages 714 and 1177 m are thought to belong to this system; the second of these has a throw estimated at over 50 m. The fault was also proved during the site investigations for the Derwent Dam. Beneath the Derwent Reservoir the fault truncates the **Greymare Hill Fault**. Farther to the north-west, at Bale Hill, where the throw is estimated to have increased to 125 m, the fault abruptly changes direction to west-north-west (Figure 23), and the throw rapidly diminishes. The fault probably dies out altogether a short distance into the adjacent Hexham district.

The north-west-trending **Healeyfield Fault** can be traced from the southern boundary of the district, south-west of Castleside, to near Greymare Hill where it is truncated by the Greymare Hill Fault (Figure 23). A zone of intense disturbance at chainages 3416 to 3436 m in the Letch House–Airy Holm–Derwent Tunnel (Appendix 3) is thought to form part of this fault. The throw, up to 85 m in the south, decreases progressively to the north. The Healeyfield Fault, part of a more extensive system in the Wolsingham district to the south, is associated with vein deposits of lead ore at Healeyfield itself.

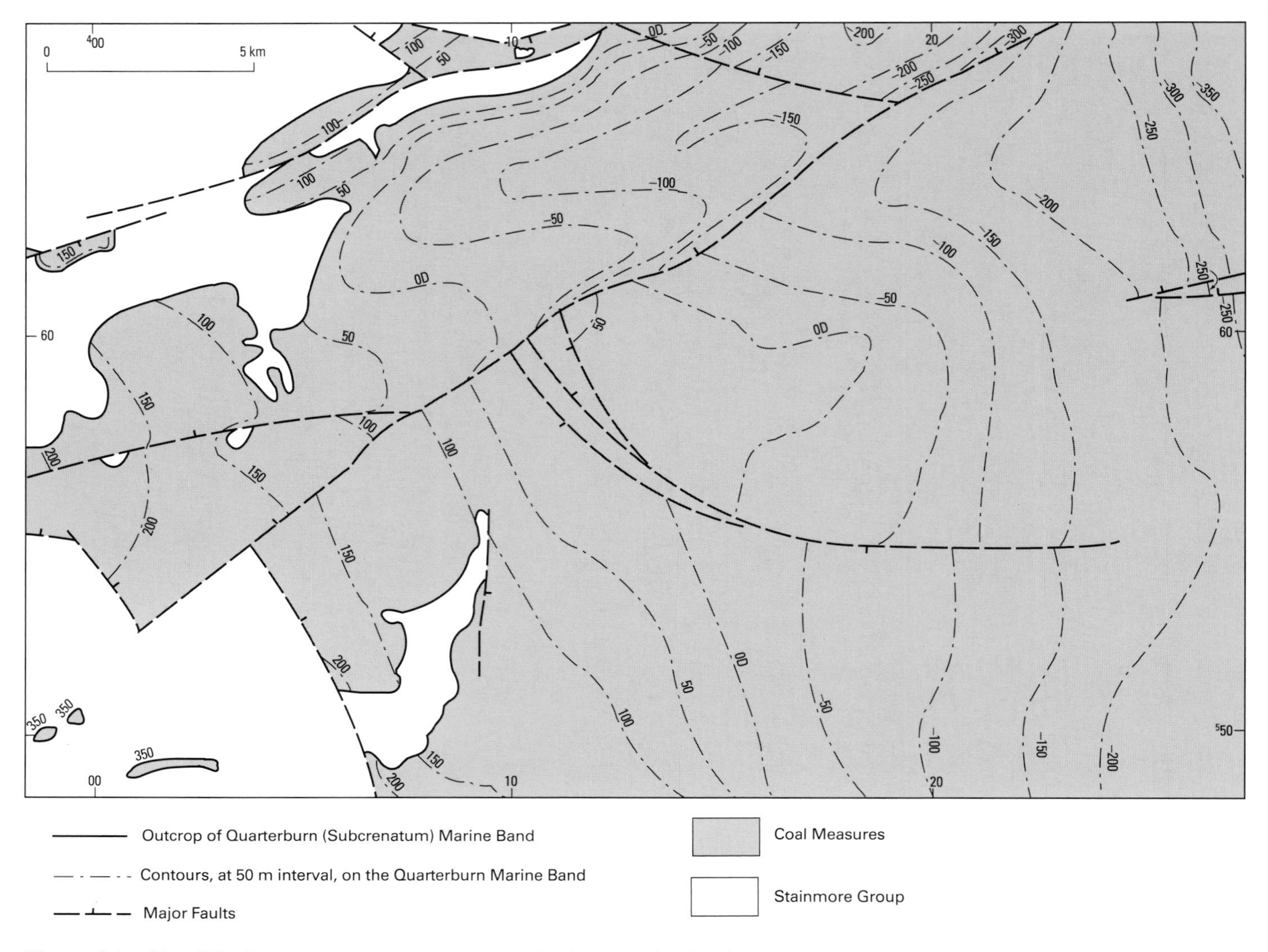

Figure 24 Simplified structure contour map on the base of the Coal Measures in the district (after Lawrence, 1991). Data largely based on extrapolation from coal seam plans.

A complex of faults, trending between south-east and east, extends in an arcuate belt from the Ninety Fathom Fault north-west of Chopwell, across the Derwent Valley east of Hamsterley, thence eastwards towards Tantobie, south of Tanfield, and dies out in a series of minor disturbances between Stanley and Kibblesworth (Figures 23 and 24). Over most of its length this fault complex consists of two or three parallel or subparallel fractures, including linking members. Between Hamsterley Park and Tantobie it forms a zone of severely dislocated strata up to 300 m wide, in which many coals were largely unworkable or remain unworked. In this zone the strata locally dip very steeply to the north, and near Tantobie they approach vertical. The detailed structure hereabouts is highly complex and its true nature is not fully understood. The northern, and longest, member of this fault complex, the **Tantobie Fault**, has a downthrow west of some 60 m in the Chopwell area, and up to about 200 m, 1 km north-west of Tantobie. East of Tantobie the throw of the fault rapidly diminishes and becomes dissipated in a number of subsidiary subparallel fractures trending south-east.

Near the eastern margin of the district the **Felling** and **Heworth faults**, better known in the Sunderland district, unite in the western part of Gateshead to form a single fault with a maximum downthrow of 17 m to the north (Figure 23). This fault dies out to the west in the vicinity of Whickham.

Small normal faults are widespread throughout the district and are well documented on the numerous coal seam mine plans. They have little or no effect on the overall structure of the district. Three main fault trends are present, notably north-north-west, between east-north-east and north-east, and broadly east to west, the last being the least common. Most of these faults have throws of less than 10 m and hades broadly ranging between 20° and 25°. Many of the smaller faults affect only one seam or group of seams, but locally caused considerable disturbance to mining activities. Although minor faults are everywhere common, clusters of such faults are more evident in certain areas, notably in the Tyne Valley, at and to the north-east of Wylam, in the Gateshead and Low Fell areas, and in the south-eastern part of the district, more particularly between Burnhope and Waldridge. There is no reason to presume that the overall density of minor faulting is less outside the worked coalfield area, but the relative absence of underground data hinders their detection. However, the greater proportion of more competent rocks in the lower part of the sequence (Stainmore Group and lower part of Lower Coal Measures) may have resulted in a lesser density of faulting. The overall density of faulting encountered in the Kielder tunnels (Appendix 3) was low, and most of the displacements were small. In the Letch House–Airy Holm–Derwent Tunnel, both the Greymare Hill Fault at chainage 1663 m and the Unthank Fault at 2289 m appear to have reversed components.

Steep dips are commonly recorded in strata immediately adjacent to faults, but do not necessarily indicate a throw of high magnitude. Many small faults encountered in underground workings are associated with great stratal disturbance out of all proportion to the observed displacement. The degree of dislocation appears to be more related to the relative competence of beds. The effects of such a relatively small fault are visible [0390 5621] in Backworth Letch near Greymare Hill, where horizontal strata are separated from beds dipping 40° to the south-west. Steeper dips were associated with only a minority of the faults in the Kielder tunnels (Appendix 3).

A detailed study of **joints** has not been carried out in this district, but measurements in the Edmundbyers and Hedley on the Hill areas demonstrate two sets, trending approximately north or east, within the ranges north-north-west to north-north-east and east-north-east to east-south-east (Edmonds and Davies, 1982). Joint measurements were recorded during site investigations for the Kielder tunnels (Carter and Mills, 1976; Coats et al., 1977) and during the tunnel excavation (Appendix 3). A broad scatter of trends was apparent, especially in the Letch House–Airy Holm–Derwent Tunnel. In the northern half of the tunnel there was a main set at 065°, with subsidiary sets at 145° to 165° and 125°; in the southern half the sets tended to range from 090° to 120°, 150° to 170° and 040° to 060° A much more consistent pattern was present in the Derwent–Wear Tunnel, with a main direction between 050° and 060° and subsidiary sets at 140° to 160° and 020° to 030°. Most joints are vertical or steeply subvertical.

FOLDS

Variscan and Tertiary (Palaeogene/Neogene) uplift resulted in a regional dip of up to 5° (generally between 1° and 3°) towards the east and north-east in the Upper Carboniferous rocks of the district (Figures 5b and 24). Except where the strata are folded, and in the immediate vicinity of some faults, there is little variation from this general range. Only minor folding and flexuring are present in the Kielder tunnels (Appendix 3).

A broad, basin-margin inversion anticline occurs in the hangingwall-block of the Ninety Fathom Fault along much of its length, with its axis subparallel to the trend of the fault (Figures 5 and 24). A complementary synclinal flexure is present to the north, extending from Painshawfield in the west to Ryton in the Tyne Valley in the east, and farther north-east to between Walbottle and Westerhope; this gentle fold has a slightly steeper northern limb. A very shallow and broad anticlinal flexure is present on the south side of the Ninety Fathom Fault east of Chopwell; this flexure dies out entirely in the Whickham area. Minor rolls, flexures and gentle doming are relatively common throughout the coalfield area, although rarely visible at the surface. However, an example of such a dome is visible in the Stocksfield Burn [052 590] south-east of Hindley, where Stainmore Group strata at the level of the Whitehouse Limestone and above are revealed by just such a flexure.

FIVE

Igneous rocks

The igneous rocks consist of intrusive basalts and dolerites of the Whin Sill suite, thought to be present at depth throughout the district, and two Palaeogene ('Tertiary') dykes present, though no longer visible, in the north-east (Figure 25).

THE WHIN SILL

The Carboniferous rocks of the Northumberland Trough and the Alston Block are intruded by a set of quartz-dolerite sills, collectively known as the Whin Sill, and related dykes (Randall, 1980a; Francis, 1982; Dunham and Strasser-King, 1982; Dunham, 1990). The Whin Sill suite was intruded in late Carboniferous to early Permian times, towards the end of the Variscan deformation. The Whin Sill is exposed in the Northern Pennines and Northumberland, between Bamburgh and Teesdale, and concealed extensions, to the east of its outcrop, have been proved by several deep boreholes.

The sill neither crops out nor has been penetrated by boreholes within the district. However, it is believed to be present everywhere at depth, having been proved, 38.53 m thick, at a depth of 543.79 m in the Throckley Borehole*, just beyond the northern boundary. The sill can also be traced locally within the district by means of seismic reflection profiles (Kimbell et al., 1989). It is not intruded everywhere at the same level within the Carboniferous rocks, and more than one sill may be present locally. In the Throckley Borehole the sill is intruded in the upper part of the interval between the Great and Little limestones. From this and other regional information (Randall, 1980a; Dunham, 1990) it can be inferred that in northern, central and eastern parts of the district the sill probably is intruded at levels low in the Stainmore Group or near the top of the Upper Alston Group. Towards the west and south-west of the district, the sill appears to be intruded into lower levels within the Upper Alston Group.

The metamorphic effects of the sill on the Carboniferous country rocks are locally marked. In particular, the organic matter is significantly affected for some considerable stratigraphical thickness away from the sill (cf. Ridd et al., 1970; Jones and Creaney, 1977; Creaney, 1980; see also discussion of hydrocarbon prospectivity in Chapter 7). The core logging of the Throckley Borehole detected the metamorphic effects of the sill in strata to about 30 m above the upper contact. However, subsequent study has shown that the organic matter is thermally altered to more than 200 m above the sill. For example, spores become

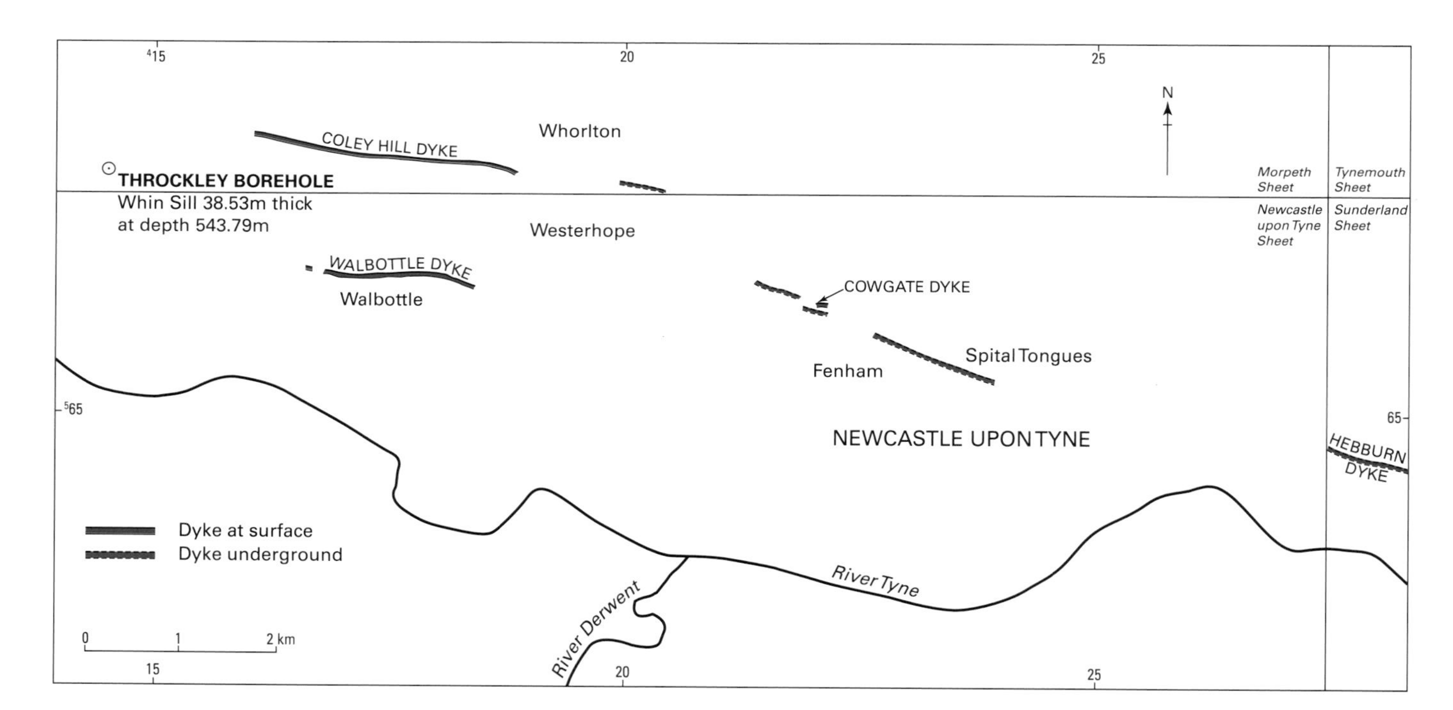

Figure 25 Distribution of Palaeogene ('Tertiary') dykes in the north-eastern part of the district and adjacent areas.

increasingly blackened with depth and none were obtained below 400.51 m (Owens, 1972).

A thin, north-easterly trending basaltic dyke was formerly visible in old lead workings near Thornbrough [012 653] (Smith, 1923). From its trend, this dyke is judged to belong to the Whin Sill suite, rather than to the Palaeogene intrusions.

PALAEOGENE ('TERTIARY') DYKES

West- to west-north-west-trending basic dykes, no longer visible at the surface, are present in the north-eastern part of the district (Figure 25). They are closely related to dykes of similar composition and the same general trend elsewhere in north-eastern England, and form part of a dyke swarm emanating from the Tertiary Igneous Centre of Mull in western Scotland. The dykes have been described by Teall (1884), Lebour (1878, 1886), Holmes and Harwood (1929), Tomkeieff (1953) and Randall (1980b). Within the district the dykes are less than 15 m wide, generally vertical or steeply subvertical, and in a laterally discontinuous *en échelon* arrangement along their length. The metamorphic effects of the dykes are generally restricted in extent, but are more noticeable in coals and in other organic matter (Marshall, 1936; Raistrick and Marshall, 1939; Jones and Creaney, 1977). Locally, coal has been reduced to cinders more than 50 m from a dyke. The contacts are commonly brecciated with considerable intermingling of cindered coal and altered basalt.

The dykes are formed of basalt and dolerite of the type commonly referred to as tholeiite (Holmes and Harwood, 1929). These are olivine-free or olivine-poor plagioclase-augite rocks, in which the feldspar and pyroxene are generally in ophitic relationship, and have an insertal texture with a glassy mesostasis, commonly devitrified or microcrystalline. Pyroxene and olivine are typically more abundant than feldspar. Anorthite phenocrysts are commonly present. Several different varieties of tholeiite have been recognised based on petrological character.

The Walbottle (Newburn) Dyke has been traced over a distance of some 2.1 km between Walbottle in the west and West Denton in the east (Figure 25). The overall trend of the dyke, although slightly sinuous, is about 280°. The dyke was formerly exposed [1631 6655] and quarried for a short distance along its length on the east side of Walbottle Dene, but unfortunately no specimens are available. The width of the dyke is probably less than 15 m, and the workings extended possibly over 70 m; It is uncertain whether this dyke is present at or near the surface throughout its length. It seems likely that it is discontinuous with an apparent gap of some 450 m a little to the east of old workings near Walbottle Dene. However, it was apparently continuous where recorded underground in the workings of the Harvey Coal.

At the former 'Fenham Red Brick and Tile Works' [223 661] at Cowgate, in the north-western part of Newcastle upon Tyne, Tomkeieff (1953, p.91) noted an exposure of a tholeiitic dyke in the northern part of the shale pit. The total width of the dyke, at least 1.82 m, could not be established as only its southern margin was exposed. The trend of the dyke was apparently east–west, and the bulk of the rock was greatly decomposed. The surface extent of the dyke is not known, but it would appear to be directly related to a dyke at least 350 m in length, trending 290°, which was encountered in the workings of the Durham Low Main Coal and is offset some distance to the south-west of the surface position. Dykes encountered in underground workings in two other places nearby may also be continuous with the Cowgate Dyke. Some distance to the north-west, a dyke trending 282° was encountered in Brockwell Coal workings over a length of at least 450 m. To the south-east, a dyke trending 297° was noted in coal workings for a distance of at least 1350 m from north-east of Fenham to the Castle Leazes and Spital Tongues area of the city.

It would appear that the Cowgate Dyke and adjacent dykes encountered underground are part of the same intrusion (Figure 25). Tomkeieff (1953, p.91) has pointed out that the Cowgate Dyke is an *en échelon* continuation of the Coley Hill Dyke, which trends between 280° and 290° for a distance of over 4.5 km in the Morpeth district (Figure 25). An underground occurrence, apparently of this dyke, in workings of the Harvey Coal stops short of the northern boundary of the district by a matter of a few metres [2072 6751] (Figure 25). According to Richardson (1983), the Cowgate Dyke can be traced farther to the east-south-east in underground workings as far as Walker in the Sunderland district, and may be continuous with the Hebburn Dyke of that district (Figure 25). Although there is strong evidence that these dykes belong to the same intrusion, there are nevertheless significant petrographic differences between the various occurrences. For example, Tomkeieff (1953) has pointed out that the Cowgate rock is of Brunton tholeiite-type and that the Coley Hill Dyke is of Acklington type. The Hebburn Dyke is generally of Cleveland type, but also shows affinities with the Acklington type (Teall, 1884; Holmes and Harwood, 1929; Randall, 1980b). The relationship of the Walbottle Dyke to the Cowgate and associated intrusions is not known. It is broadly parallel to the Coley Hill Dyke, at a distance of between 1 and 1.5 km to the south.

SIX

Quaternary

Quaternary (Drift) sediments are widespread throughout the district and largely conceal the underlying Carboniferous rocks. They include a wide range of superficial deposits formed during the advance and retreat of the Pleistocene ice sheets, and also those that have accumulated since the disappearance of the ice. These last include material being deposited today as part of the continuing geological process. The district was probably subjected to several periods of glaciation during the Pleistocene, but the glacial deposits preserved all belong to the last (Late Devensian, or Weichselian) glaciation. Sediments of earlier glaciations and interglacial periods have been removed and recycled. Generally the Quaternary deposits range up to 10 m in thickness, but values well in excess of 50 m, and locally up to 90 m, have been recorded in the buried valleys, more especially in those associated with the rivers Tyne, Derwent and Team. Boulder clay is by far the most widespread deposit and extensive spreads of sand and gravel also are present. Relatively thick deposits of laminated clay occur locally, commonly forming the main infills of buried valleys. Although drift deposits are present over much of the district, they are generally very poorly exposed, and most of the information relating to them has been derived from temporary exposures and boreholes. Some of the higher ground and steeper slopes, and some incised river and isolated stream sections, have been mapped as drift-free. However, bedrock is rarely seen at the surface because of a residual mantle of clay, weathered rock and soliflucted material.

Among the more important accounts relating to the Quaternary of the district and adjacent areas include those by Howse (1864), Dwerryhouse (1902), Woolacott (1905, 1921), Smythe (1908, 1912), Herdman (1909), Merrick (1910, 1915), Raistrick (1931), Anderson (1940), Peel (1949, 1956), Sissons (1958, 1960), Beaumont (1970), Cuming (1970), Francis (1970), Taylor et al. (1971), Thabet (1973), Sladen (1979), Lunn (1980), Eyles and Sladen (1981), Giles (1981), Lovell (1981), Edmonds and Davies (1982), Mills (1982), Cox (1983), Richardson (1983) and Allen and Rose (1986). Quaternary deposits in adjacent districts have been described by Smith and Francis (1967), Land (1974), Frost and Holliday (1980), Jackson et al. (1985), Lawrence and Jackson (1986, 1990) and Smith (1994). Beaumont (1968) summarised the history of Quaternary research in northern England and gave an extensive list of references relating to the district and a much wider region.

PREGLACIAL SURFACE

A long period of erosion was initiated in Palaeogene times during which the Alston Block was uplifted and tilted, and around 1500 to 2000 m of Mesozoic strata, and lesser thicknesses of Upper Carboniferous rocks, were removed (Holliday, 1993). The detritus was transported eastwards into the subsiding North Sea Basin. In common with many adjacent areas, the main geographical features of the district were established by the end of the Neogene Period, and a pattern of consequent and subsequent drainage, broadly similar to that of the present day, was established (Merrick, 1915; Trotter, 1929). This pattern was subsequently modified by glacial and postglacial processes, depositional and erosional, during the Quaternary Period.

Therefore, at the outset of the Late Devensian glaciation, the main elements of the present-day topography had already been established, albeit with a slightly bolder relief. On this surface, a widespread veneer of unconsolidated and weathered debris probably was present, including sediments formed during earlier glacial and interglacial periods. Gravel deposits probably were present within the main river valleys. The approximate lines of the main pre-Late Devensian valleys, of which the Tyne, Derwent and Team are the most important, are shown in Figure 26. The Late Devensian ice spread over this surface, engulfing the whole district. Interpretation of the Quaternary sequence, and of the modification of the landscape that resulted from glaciation, is considered following a general description of the deposits.

GLACIAL DEPOSITS

With the exception of buried valley deposits, the main types of glacial sediment are described here in terms of their lithology rather than as stratigraphical or generic units. Where relevant, reference is made to associated landforms. On the published Drift Edition of the 1:50 000 map of the district, various minor occurrences of glacial deposits were not differentiated from the boulder clay. Such deposits include thin beds and lenses of clay, silt, sand and gravel whose occurrence appears to be restricted or thin, and whose relationship to the surrounding deposits is obscure.

Boulder clay (till)

Boulder clay is the most widespread of the Quaternary deposits and covers between 50 and 60 per cent of the district. It rests directly on the underlying Carboniferous rocks and in many places it is the main or only drift deposit present. Outside the confines of buried channels the boulder clay has a maximum thickness of 30 m; over much of the district it rarely exceeds 10 m.

In general the boulder clay is a tough, overconsolidated, dark brown, greyish brown or dull yellowish

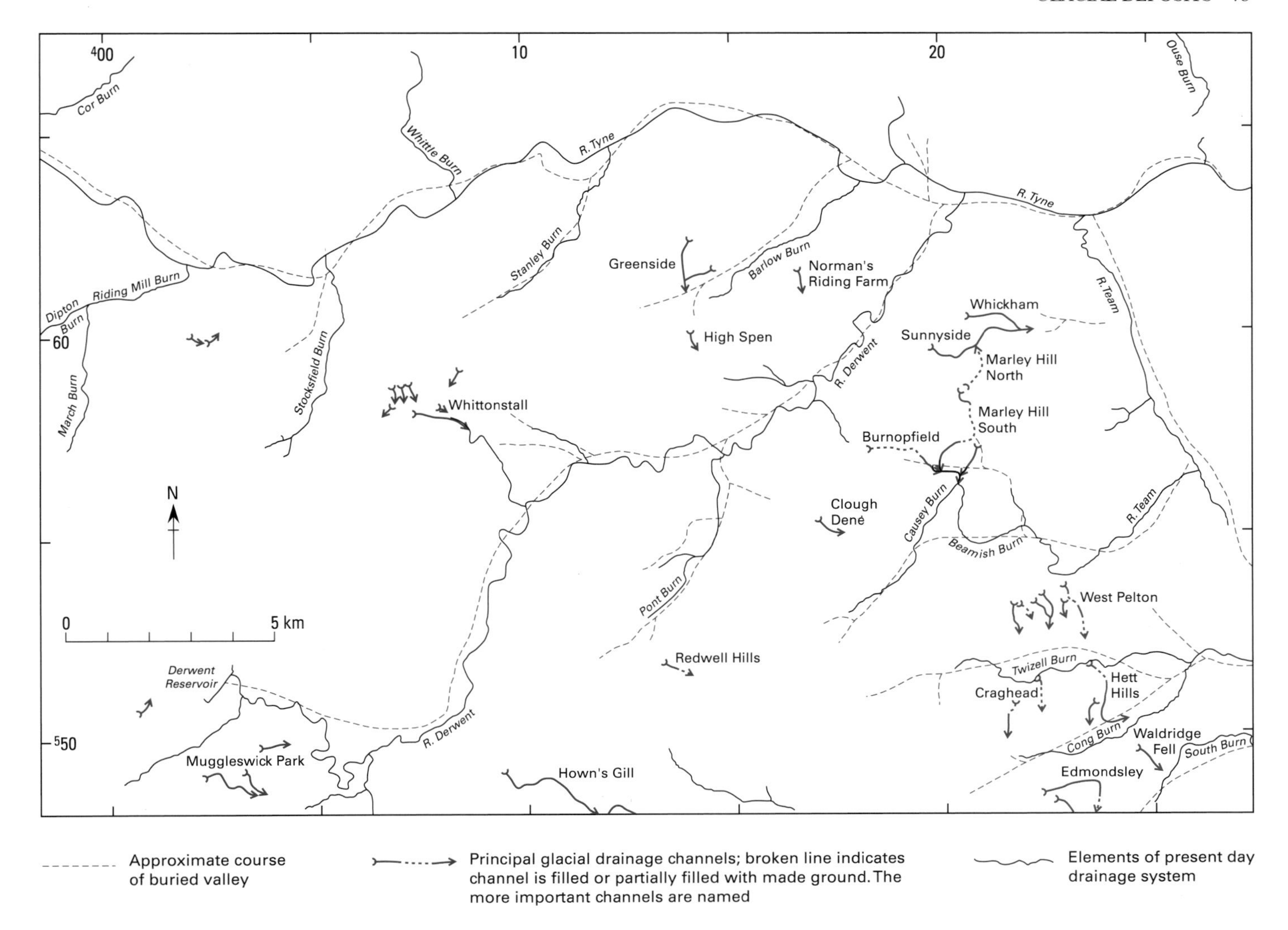

Figure 26 Distribution of main buried valleys, glacial drainage channels and present-day drainage system.

brown, silty, locally sandy clay, which generally contains abundant clasts of assorted bedrock lithologies. Illite and kaolinite are the dominant clay minerals in the clay matrix. The clasts are unsorted and range from subangular to subrounded chips, pebbles and cobbles to rarer boulders or even larger sizes. The larger-sized material generally occurs near the base of the deposit. Some clasts are scratched and highly polished.

The colour and composition of the main boulder clay sheet shows some local variation, but the limits of these changes are not easily defined. On the higher ground, and more especially in the vicinity of the major sandstone outcrops, the boulder clay is commonly a dull yellow or yellowish grey and more sandy deposit, with numerous fragments of angular to subangular degraded sandstone. Although these deposits occur widely they are usually very restricted in extent; they are for example locally present on the high ground between Prudhoe, Chopwell and Blaydon, between Consett and Stanley, and in the far south-west of the district, more especially south of the Derwent Valley.

Locally the upper 1 to 2 m of boulder clay is highly weathered, being grey, brown or slightly mottled and more sandy, with fewer and smaller clasts. These deposits are regarded as a weathering or otherwise modified product of the boulder clay, and have certain resemblances to the Upper Clays described below.

Most of the clasts are of relatively local derivation and mainly comprise Carboniferous sandstones and limestones, although fragments of ironstone, siltstone, mudstone and coal are also present. Significantly less abundant are fragments of quartz-dolerite of the Whin Sill suite, as are clasts derived from much farther west, including south-west Scotland, the northern part of the Lake District and the Vale of Eden. Such far-travelled debris includes granites from several Lake District and Scottish sources, the Borrowdale Volcanic Group, Lower Palaeozoic greywackes, and Triassic St Bees Sandstone.

Carboniferous rocks of distant origin are probably also present, but are generally not readily distinguished from more local rocks. A general eastward or east-south-eastward ice movement, modified by local relief effects, is implied by the distribution of the clasts and by measurements of their orientations (Dwerryhouse, 1902; Raistrick, 1931, Taylor et al., 1971; Lunn, 1980). Small amounts of debris from the Cheviot Hills and south-eastern Scotland have been found in the extreme north-eastern part of the district, but the main ice stream from that direction appears to have been deflected to the east of the district, nearer to the coast, by the eastward-flowing ice.

Sporadic occurrences of glacially transported rafts or erratic blocks, several metres thick and extending over several hundreds of square metres, have been recorded in north-east England. Some notable examples of glacially transported rafts of Coal Measures occur in the adjacent Wolsingham district (Cuming, 1970; Mills, 1974). An erratic raft of fractured and disturbed, partially dolomitised Great Limestone, at least 3.75 m thick and extending over an area of about 800 m by 200 m, occurs [005 685] just beyond the northern boundary of the district, some 3 km north-north-east of Corbridge. This raft has been transported at least 12 km from the nearest outcrop of the limestone. A similar area of poorly exposed limestone, 400 m by 200 m, occurs within the district at Whittington Hill [042 656] and is apparently a raft of Great Limestone which has been moved a greater distance from the outcrop. Several isolated and poorly exposed occurrences of limestone and sandstone in the north-west of the district were assumed to be in situ during the mapping, but a raft origin for at least some of these cannot be discounted, and they are probably much more common than has hitherto been recognised. One example of a glacially transported erratic has been recognised near the Sycamores [086 483], at Castleside on the southern margin of the district, where a large raft of Lower Coal Measures, including a coal believed to be the Brockwell Coal, occurs at the crest of the hill and appears to have been moved bodily forward to the east for a short distance, probably between 250 and 300 m.

Small mounds of what appear to be partially transported masses of morainic material, consisting of very stony boulder clay but also including some sand and gravel, are present [0070 5120] under the sandstone escarpment of Berry Bank and between there and Pow Hill Country Park. The mounds also incorporate slipped masses of sandstone derived from the escarpment. Similar, but smaller, mounds are also present [0390 4990] on the west side of the road to the west of Muggleswick.

Deposits similar to the predominant boulder clay type occur commonly in, or on the flanks of, the major buried channels (see below). These clays differ, however, in that the fabric tends to be more silty and less well ordered, and the clasts are generally smaller and less numerous than in the main boulder clay. They are thought to occur largely as lenses, wedges or rafts within a mélange of other glacial deposits in the channels.

In general the boulder clay forms a good geotechnical foundation material, which commonly can stand unsupported in excavations for limited periods. However, this is not always the case and for a number of reasons difficulties may be encountered during construction or excavation. These include the occurrence of less competent material both within and below the boulder clay itself, the lower strength of the top layer where substantially degraded or weathered, and also the presence of larger erratics within the clay. The related stony clays occurring on the flanks of, or within, buried valleys are generally substantially weaker and less overconsolidated, and support for these is almost always required.

The boulder clay of the district is the equivalent of the Durham Lower Boulder Clay of the adjacent areas to the east and south-east (Smith and Francis, 1967; Smith, 1994). It is largely a ground moraine or lodgement till deposited by a moving ice sheet. The less stony clays on the flanks of, or within, buried valleys probably formed as a result of remobilisation, flow and mass downslope movement of waterlogged lodgement till or ground moraine, resulting in the main from periglacial solifluction processes.

Glacial sand and gravel

Glacial and glaciofluvial sand and gravel form extensive spreads in some parts of the district. They are found on the flanks of the Tyne Valley from Corbridge to Winlaton, on the northern side of the Derwent Valley north of Newlands, and around Rowlands Gill. Irregular areas of sand and gravel are present also in the south-east of the district, between Sunniside and Chester-le-Street. Smaller areas and patches of sand and gravel occur elsewhere, for example north of Corbridge, and in the south-west near Edmundbyers and Muggleswick in the Derwent Valley above Consett. The surface expression of these deposits, more especially the larger spreads, commonly gives rise to an irregular, hummocky, well-drained topography, as for example west of Highfield near Rowlands Gill. Such hummocks are rarely more than 15 m high. The sand and gravel generally rests on the boulder clay, commonly with a spring line at the base. Locally, lenses and channels of sand and gravel occur within the boulder clay or, as in many buried valleys, are interbedded with other superficial deposits.

The nature and composition of the deposits is varied and complex. They range from fine- to medium-grained, well-sorted, evenly bedded sands, coloured pale brown, dull grey or dull yellow, to poorly sorted medium- and coarse-grained angular to subrounded gravels with a variable sand, silt and clay matrix. They are derived largely from Carboniferous sandstones, but fragments of limestone, mudstone, ironstone and coal are also present, the coal commonly in the form of small grains and chips. Farther-travelled material, including igneous debris, though not common, is also present. The direction of transport was generally towards the east or east-south-east. Sands, often of considerable thickness, also occur within the confines of buried valleys, where they commonly contain lenses and wedges of laminated clay and silt, fine and medium gravel, and, near the flanks, boulder clay. These sands tends to be more uniform and level bedded, commonly with abundant pebbles and small coal fragments, particularly near the valley centre lines.

The sand and gravel deposits of the Tyne Valley and its flanks, between Hexham in the west and the outskirts of Newcastle upon Tyne and Gateshead in the east, were investigated as part of a mineral assessment programme (Giles, 1981; Lovell, 1981). This investigation included a detailed programme of boreholes, a synthesis of pre-existing information, and a quantitative and qualitative summary of results.

The great volume of sand and gravel present is consistent with deposition having taken place dominantly in a subglacial environment, and can be attributed to deposition from seasonal streams and sheets of water under wasting or stagnant ice. However, some of the deposits, in particular those intimately associated with thick laminated clays, may have been laid down in ephemeral ice-impounded proglacial lakes. The deposit between Hollings Hills, Blackhall and Newlands (see below) may be an example. Other sand and gravel occurrences, particularly those most intimately associated and interbedded with boulder clay, are probably englacial in origin.

In the Tyne Valley above Ovingham and Eltringham, recorded thicknesses of sand and gravel (excluding alluvium and river terrace deposits) range between 2.3 and 18.0 m, with an estimated mean of 6.3 m. At a point [0130 6358] near the northern margin of a working at Styford Lodge, 6.4 + m of coarse gravel containing subangular to rounded sandstone, together with some fragments of limestone, ironstone and far travelled volcanic rocks, were recorded. The thickest sequence known in this general area was proved in a borehole [0426 6090] near Shilford East Wood, where 18.0 m of pebbly sand and sandy gravel were recorded overlying pebbly and sandy clay. Sand and gravel are currently being worked at Thornbrough [015 630] and Merry Shield [063 616]. Good quality sand, albeit locally clayey, is present at both localities. However, lenses of clay, gravel and boulders are also present. This is particularly true of Merry Shield where some very large boulders have been encountered.

A large irregularly shaped belt of locally hummocky sand and gravel is present on the south side of the Tyne Valley between the Eltringham–West Wylam area in the west and Winlaton in the east, including Crawcrook, Ryton and Greenside. Recorded thicknesses vary widely in this area, ranging from as little as 2 m to a maximum of 25 m. In the more western part of this area, more notably between Eltringham and the Stanley Burn, the sands and gravels are overlain by patchy overburden including clay and silt. In a borehole [1109 6357] at Eastwoods Farm near West Wylam, 6.6 m of silty clay, sandy clay and pebbly sand were recorded overlying 18.4 m of clayey sand with some gravel at the top. In most of the eastern part of the area, east of Stanley Burn, overburden to the sands and gravels is thinner and usually restricted to a sandy soil. The maximum recorded thickness was formerly visible [1574 6255] at the western end of workings at Mosspool Hill, east-north-east of Greenside, where 2.1 m of medium- and coarse-grained sand, with gravel, overlay 22.9 m of largely fine-grained clayey quartz sand with lithic grains including coal. Extensive areas of sand and gravel have been worked between Greenside and Stargate, a distance of some 3.5 km; the maximum width of working was up to 750 m. The deposit is currently being worked at Crawcrook [128 637] and Blaydon [155 622].

A large spread of sand and gravel is present on the north side of the River Derwent above and below Rowlands Gill, with a lobe south of the river opposite Lintzford. The form and nature of the deposit is very similar to that described on the south side of the Tyne Valley. Moundy and hummocky topography, locally with kettle holes, is present in the area south and south-east of Hooker gate, south of Highfield, and thence extending into the upper part of Lintzford Wood [143 578]. Only small old pits and diggings are present in this spread, and the deposit appears to consist largely of sand with isolated and irregular beds and lenses of gravel. A well-developed esker, nearly 600 m long, is present in Chopwell Wood north-east of Forest Lodge [1370 5802]. The form of the esker is largely concealed by woodland, but is locally visible [between 1370 5818 and 1420 5800] as a well-defined sharp ridge up to 8 m in height. In the west, the crest is almost level, but traced eastwards it falls by some 30 m, and the orientation of the ridge changes from east and east-south-east to almost south. The ridge then widens and loses its identity in a south-facing slope, largely composed of sand and gravel, in the southern part of Chopwell Wood. The esker is composed of sand and small rounded gravel, and flanked by boulder clay at its western end.

Another less extensive deposit of sand and gravel is present farther west, in the vicinity of Newlands (Allen and Rose, 1986). Anderson (1940) considered that this flat-topped deposit formed a steep delta front, some 46 m in height, facing the present River Derwent. A similar example was quoted near Barlow Burn, 1.5 km east of Greenside. South of Hollings Hill [097 574], boreholes recorded sand and gravel thickening southwards, the deposit resting on boulder clay. Broad Oak Quarry [098 567], worked for sand and gravel, locally exposes over 10 m of sand and pebbly sand, commonly cemented and in places cross-bedded. A detailed description and study of the sediments exposed in this quarry has been given by Allen and Rose (1986). Hollings Hill Pit [096 575], north-west of Broad Oak Farm, was opened in 1988. The deposit consists largely of clean, medium- to coarse-grained cross-bedded sand, but includes some silty sand layers, gravelly beds and lenses. The stone content mainly consists of sandstone, but some limestone and dolerite are also present and granite has been recorded. Rafts of boulder clay locally overlie the sand. The deposit is used both for tile manufacture and as an aggregate.

In the south-western part of the district, related patches of sand and gravel are present on either side of the Derwent Valley above Consett. These deposits typically consist of poorly sorted medium- to coarse-grained sands and gravels with a variable silt and clay content. They are commonly strongly cross-bedded and locally contain discrete beds of laminated clay and silt. At Edmundbyers, old pits [0115 5015] occur in a large isolated deposit of sand and gravel. Deposits south of the Derwent Dam contain a small disused pit which exposed up to 5.0 m of cross-bedded sand and gravel. A borehole [0467 5004] in an irregular mound at Muggleswick proved at least 8 m of sand and gravel. A mound of clayey sand and gravel is also present [0390 4990] just west of Muggleswick; it incorporates masses of sandstone slipped from the nearby escarpment to the south. In the vicinity of the south portal [044 516] of the Letch House-Airy Holm–Derwent Tunnel, 3 to 4 m of clayey gravel and cross-bedded sand and silt, interbedded with grey and brown boulder clay, were exposed during construction.

Small spreads of sand and gravel, chiefly sand, are present near Stanley and Annfield Plain. They rise to over 244 m above OD and in places are up to 12 m thick. These spreads contain only small isolated workings, one of which at Riding Hills [172 523] exposed up to 12 m of sand.

Extensive spreads of sand and gravel are present in the south-east of the district, in an irregular belt trending broadly north-west to south-east from Sunniside to West Pelton, Waldridge and Kibblesworth. The thickest deposits occur between Causey

[205 564] and Pockerley Hills [227 549], and farther south-east near Grange Villa [232 521]. Near Causey, a prominent ridge, Oxpasture Hill [200 554], aligned north-north-east to south-south-west, consists of a mass of sand and gravel resting on boulder clay and flanked on its western side by laminated clays formerly worked at the East Tanfield clay pits. The ridge also lies athwart the buried valley of the Beamish Burn (Figure 26). Most of the deposits of sand and gravel in the south-eastern area are thought to consist essentially of two elements, an upper sequence mainly comprising medium- and coarse-grained angular to subangular gravel up to 3 m thick, and a lower sequence consisting primarily of well-sorted, locally cross-bedded sand up to 30 m in thickness, which contains beds and lenses of gravel, silt and clay. All the deposits rest on till.

These spreads continue south-eastwards into adjoining districts. Some 300 m south of the district [2643 4773], between 3 and 4 km south-south-west of Chester-le-Street, disturbed, faulted and folded gravels and coaly sands have been recorded above a thin, but apparently continuous, till sheet. Fine-grained sand and silt underlying the till has been sheared during emplacement. Frost wedges were also noted.

Laminated clays and silts

Laminated clays and silts are widespread and relatively thick, locally in excess of 40 m, in the buried valleys of the rivers Tyne, Derwent and Team. They are generally dark brown to greyish brown and purple-brown, the colour range probably resulting from variations in the sand and silt content. They range from exceptionally homogeneous to even-bedded, with films, laminae and fine beds of micaceous silt and mud. Clasts, generally of pebble grade, are commonly present, and boulders and larger erratics, though rare, have been recorded. The greatest thicknesses of laminated clay have been proved along the central axis of the buried valley of the River Team where, in up to 60 m of drift, laminated clay is thought to form the major component. Laminated clays and silts generally overlie boulder clay, but can also be present within deposits of sand and gravel, although the extent and thickness of such occurrences are commonly small.

Laminated clays and silts constitute a major geotechnical problem, because their relatively high water content renders them plastic. The shear strength of such deposits ranges from low to moderate, and they are generally weak or very weak, especially when moist. Under vertical load they are prone to strong compression and ductile flow, and foundations need to be specially designed to overcome such weaknesses. The deposits have a low angle of rest, and excavations require close support; tunnelling and excavating in these beds is particularly hazardous.

Laminated clays and silts are thought to have been deposited in glacial lakes, in essentially low-energy environments with sporadic incursions of coarser material. Deposits of both subglacial and wholly or partially ice-dammed subaerial lakes are probably present. The occurrence of scattered boulders, or rafts of boulder clay, within the laminated clays and silts could have resulted from subglacial melt-out of englacial rafts, or from the melting of floating masses of stone-laden ice. Where they appear at surface, as in the Team Valley, they are shown on the Drift edition of the published 1:50 000 map as Glacial Lake Deposits.

The following records of laminated clays at the surface in the Team Valley indicate the range of thicknesses and associated deposits that may occur from near the mouth of the Team Valley for a distance of some 6 km to the south-south-east.

Dunstonhaugh No. 1 Borehole [2267 6264]: laminated clay with beds of sand and boulder clay, 43.9+ m;

Dunston No. 3 Borehole [2314 6188]: laminated clay with beds of silt and boulder clay, 20.1 m, on sand and gravel;

Team Valley Trading Estate Borehole [2452 6053]: laminated clay with thick beds of boulder clay and silt, 54.3 m, on rockhead;

Lamesley Church Borehole [2523 5789]: clay 41.45 m, on 'boulders' 1.22 m, on ironstone fragment 0.23 m, on blue clay with stones and coal fragments 2.13 m;

Birtley No. 4 Borehole [2657 5583]: blue and reddish brown clay 13.18 m, on clay with pebbles 1.45 m, on bluish brown clay 1.06 m, on fine sand with clay partings near top 23.32+ m.

Laminated clays and silts also occur widely in the lower parts of the Tyne Valley in the eastern part of the district, downstream of Swalwell. They are likewise present in the lower reaches of the Derwent Valley, and in the south-east of the district in some of the subsidiary channels leading into the Team and Wear valleys. In the higher reaches of the Derwent Valley, between Eddy's Bridge and the Derwent Dam, laminated clays are present within a mélange of boulder clay, gravel and sand. Boreholes sunk prior to the construction of the Derwent Dam recorded a high proportion of laminated clay within the drift sequence of the buried valley of the River Derwent, north-north-east of Edmundbyers (Figure 29). Laminated clay and silt interbedded with boulder clay has been recorded in boreholes east and north-east of Tanfield Lea. Near the line of the Causey Burn, laminated clay was formerly worked at East Tanfield clay pits [194 548] on the western margin of a large expanse of sand and gravel.

Isolated deposits occur at the surface elsewhere, notably in an abandoned brickworks [NY 986 653] now occupied by the A69 Corbridge bypass road; near Houghton [120 669], south of the Roman Wall; and near Peepy [037 627] on the north side of the Tyne Valley north-west of Bywell, where laminated clays were seen in a landslip [028 622] and proved in nearby site exploration boreholes for the diversion of the A68 road.

Upper Clays

Upper Clays or Upper Stony Clays of south Northumberland have been mapped in the extreme north-east of the district, and are shown incorrectly as Head on the published Drift edition of the 1:50 000 map of the district. Cox (1983) considered that these clays were widely present on the lower ground along the east side of the Team Valley, more especially perhaps north of Birtley. Similar clays may be more widespread than indicated within the east and north-east of the district, but they have not been positively identified elsewhere. The Upper Clays consist of grey, brown, reddish brown and purplish red, blocky, silty clays with a low stone content, and form a surface mantle generally between 1 and 2 m thick. Thin beds of sand may occur locally within the clay. The scattered small pebbles and cobbles are similar to those of the boulder clay. Closely spaced subvertical polygonal joints are common at the top of these clays. They rest on the Durham Lower Boulder Clay (Smith, 1994).

The Upper Clays resemble the Superficial clays described by Smith (1994) in the Sunderland district to the east, where they include both Pelaw Clay and overlying Prismatic Clay. However, the two units have not been separately delineated in the present district on account of the lack of good exposures. On balance, the Upper Clays seen here would appear to more closely relate to the Prismatic Clay. Smith (1994) has tentatively suggested that these deposits may have originated by the redistribution of earlier deposits, as a result of periglacial processes and by downhill creep and incorporation of hillwash.

The Upper Clays are the weakest of the stony clays, more especially in the lower part of the deposits where they can be soft and plastic if in contact with water-bearing beds. They can be very unstable in excavations.

Buried valley deposits

The presence of drift-filled linear depressions, or buried valleys, cut into rockhead has long been recognised within the district (e.g. Wood and Boyd, 1864). The distribution of the main buried valleys is shown in Figure 26, and rockhead contours in the buried valleys of the rivers Tyne and Team around Gateshead and Newcastle are shown in Figure 27. The maximum known thickness of superficial deposits in the district, 90.83 m, is recorded in a borehole [0558 5015] near the presumed centre line of the buried valley of the River Derwent between Crooked Oak and Wallish Walls. Geophysical investigations of the buried valleys of the district have been carried out by Cornwell and Johnson (1976), Andrew and Lee (1977) and Wilson (1980).

The buried valley deposits comprise an interlensing complex of boulder clay, sand and gravel, and laminated silt and clay. The proportions of each of these components within the confines of a buried valley commonly varies very abruptly. On the published Drift edition of the 1:50 000 map of the district, these deposits are not distinguished as a separate unit but are shown by the appropriate surface lithology. A schematic section through the glacial (and later) deposits of one buried valley within the district is shown in Figure 28. In this example, the central part of the buried valley fill comprises laminated silt and clay which pass both laterally and vertically into sands containing a varying proportion of gravel. On the flanks of the valleys, stony clays, probably mass flow deposits of redistributed lodgement till, overlie the rockhead surface. At any one locality, a borehole may prove irregular beds, wedges and lenses of a variety of lithologies, apparently at random, forming the whole sequence or interspersed through a dominant lithology. In practice there is considerable departure from this idealised sequence, and much variation has been recorded in the Team Valley. On the south side of the Tyne Valley, between Derwenthaugh and the northern limit of the Team Valley, the deposits consist largely of laminated clay and silt of very variable thickness. By contrast, in the higher reaches of the Derwent Valley a more complex distribution of drift is present. Boreholes sunk prior to the construction of the Derwent Dam, across the buried valley of the River Derwent north-north-east of Edmundbyers (Figure 29), illustrate this point. Similarly, deposits contained within the buried valleys of smaller streams and tributaries are probably dominated by a more chaotic and variable pattern of lithologies.

The buried valley deposits broadly equate with those of the Tyne–Wear Complex described by Smith (1994) for the Sunderland district to the east. In general they overlie the Durham Lower Boulder Clay, but locally rest directly on rockhead.

The rockhead surface beneath the buried valleys of the lower reaches of the Tyne and Team valleys are generally graded to a base level deeper than 30 m below OD. However, borehole and mining evidence suggests that the valleys are locally overdeepened, and have humped profiles. In both the Tyne and Team valleys, near their confluence, rockhead is deeper than 46 m below OD (Figure 27). Several broad, closed, elongate hollows in the rockhead surface are evident in the Team Valley, and a distinct col is present at about 15 m below OD in the Lady Park area. Farther to the south-south-east, the base of the buried Team Valley deepens progressively and is more than 30 m below OD north-north-east of Chester-le-Street, in the immediately adjacent part of the Sunderland district. There is some evidence for assuming the presence of similar hollows between 15 m and 30 m below OD along the Tyne Valley on the extreme eastern margin of the district. Geophysical evidence (Andrew and Lee, 1977) suggests the existence of several closed hollows on rockhead surface in the Tyne Valley between Wylam and Ryton, falling to as low as 40 m below OD near Wylam, where the buried valley appears to take a wide swing to the south of the present course of the river.

There is little evidence regarding base levels in the main buried valleys near the western margin of the district. In the Tyne Valley near Stocksfield, where the present river level is about 15 m above OD, a figure of at least 11 m below OD has been recorded; in the Derwent Valley between the Derwent Dam and Consett, where the present river level is about 130 m above OD, a figure of not more than 115 m above OD has been proved for the base of the buried valley near Wallish Walls. Neither record is necessarily on the centre line of the respective buried valley. The margins of the preglacial valleys tend to have rather shallower sides than their modern counterparts, whereas the longitudinal base level profiles are broadly similar to the modern ones or are slightly more pronounced. Although there are large areas for which no information is available, most of the smaller present-day streams appear to have a buried counterpart, the lines of which broadly correspond to the present system. In a small number of cases there is a wider divergence from the present-day course over limited areas (Figure 26).

Due to the variability of lithology within buried valley deposits, considerable problems are commonly encountered in the construction of suitable foundations. This is

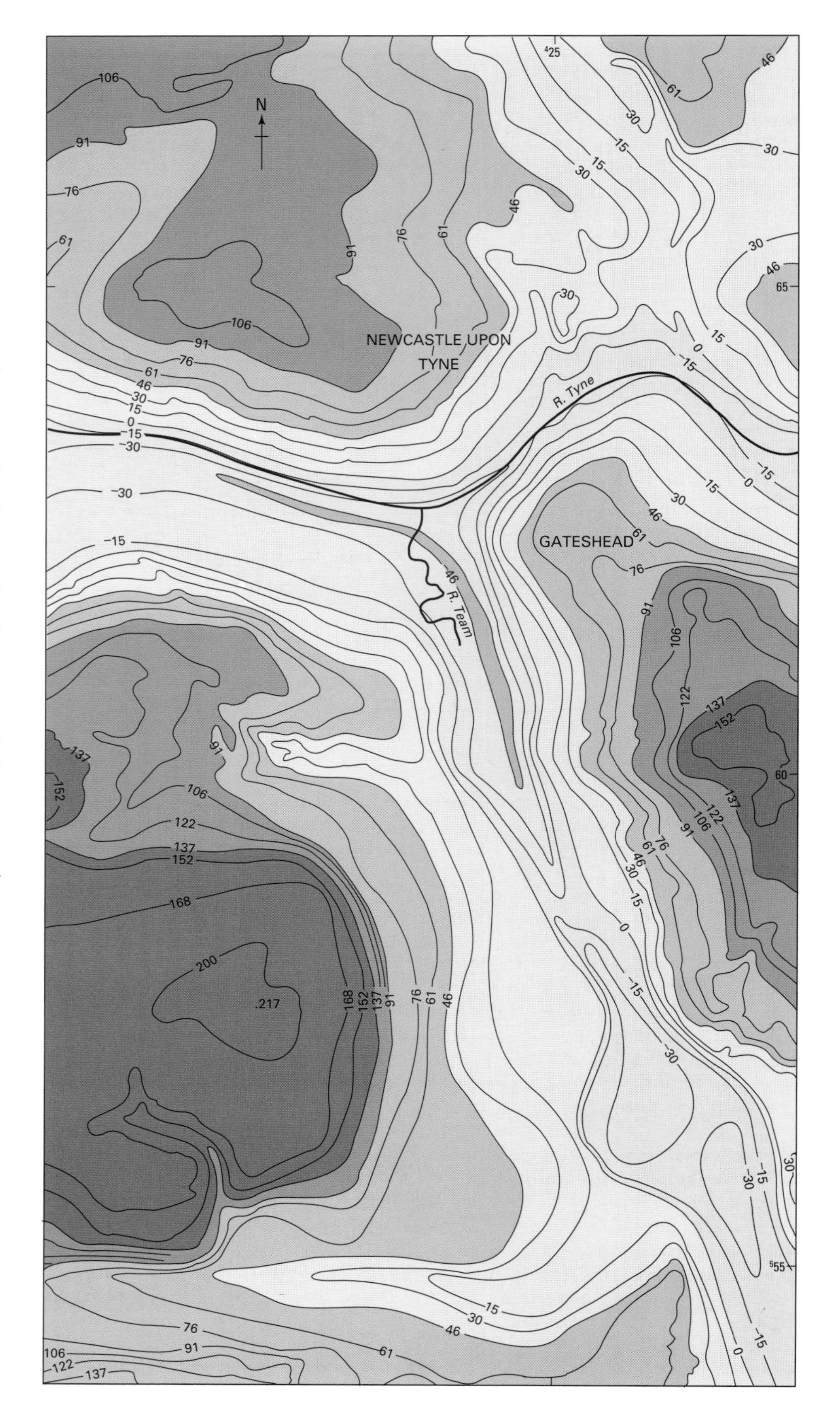

Figure 27 Rockhead contours in the north-eastern part of the district, showing buried valleys of the Rivers Tyne and Team. Modified from Cox (1983) and Richardson (1983). Contours in metres; figures preceded by − are below Ordnance Datum. Note that the present valleys are filled with deposits up to levels between the 0 and + 15 m contours.

particularly the case where laminated clays and silts are present, because these are plastic and subject to flow when wet (see above).

Without exception, the buried valleys of the district appear to be pre-Late Devensian valleys that were infilled and locally overdeepened during the glacial period, with some local late glacial and postglacial modification. Similar Pleistocene buried valley or channel-like incisions, locally overdeepened, have been observed in many parts of the world. Their origin is uncertain, but of the alternatives considered by Ehlers and Wingfield (1991) only overdeepening of pre-existing valleys by glacial scour (e.g. Gripp, 1975) or erosion by subglacial meltwater (e.g. Wingfield, 1990) would appear to be applicable to the examples in the district. The age of the sediments infilling the valleys is also uncertain. The dominantly glaciolacustrine laminated clays and silts are probably contemporaneous with deglaciation, but some of the older deposits may have formed during glaciation.

GLACIAL DRAINAGE CHANNELS

Channels cut into bedrock or drift deposits by glacial meltwaters are numerous in parts of the district, and the more important of these are shown and named in Figure 26. Some of the channels are occupied by minor streams of the present-day drainage, but the great majority have been abandoned and are rela-

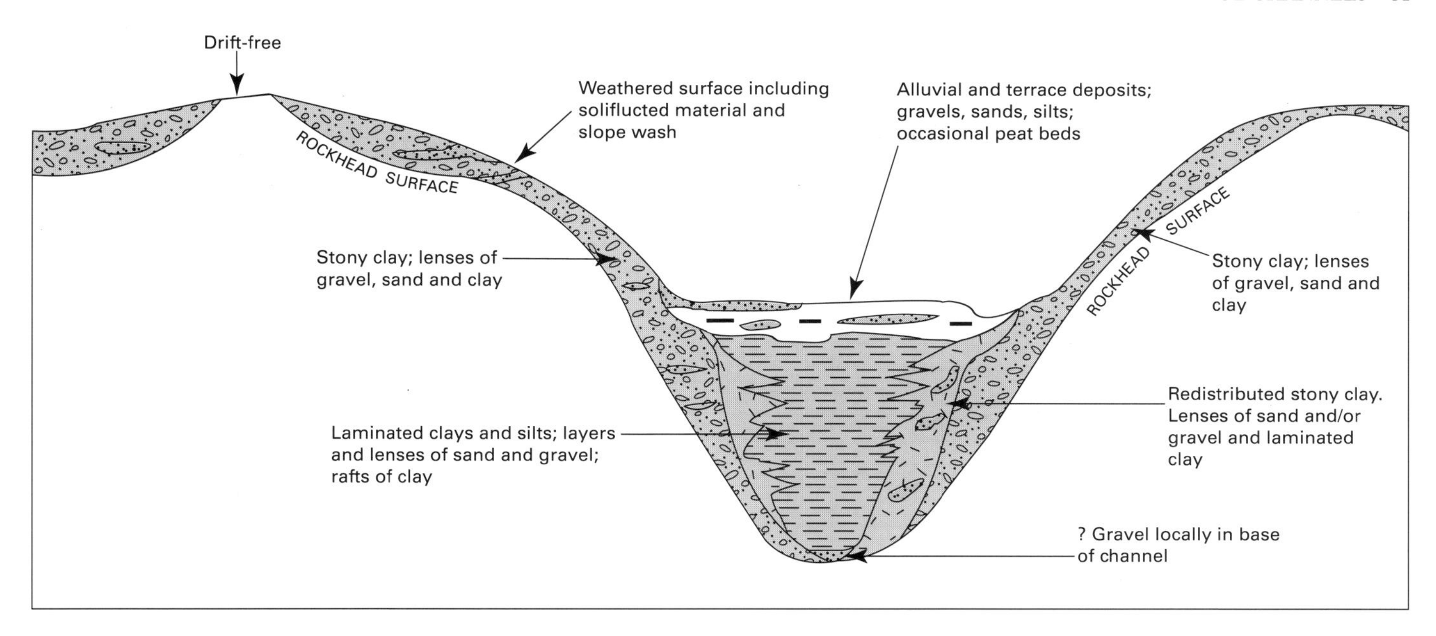

Figure 28 Schematic section across a typical buried valley, showing range and nature of glacial and postglacial deposits (modified after Richardson, 1983, fig. 10).

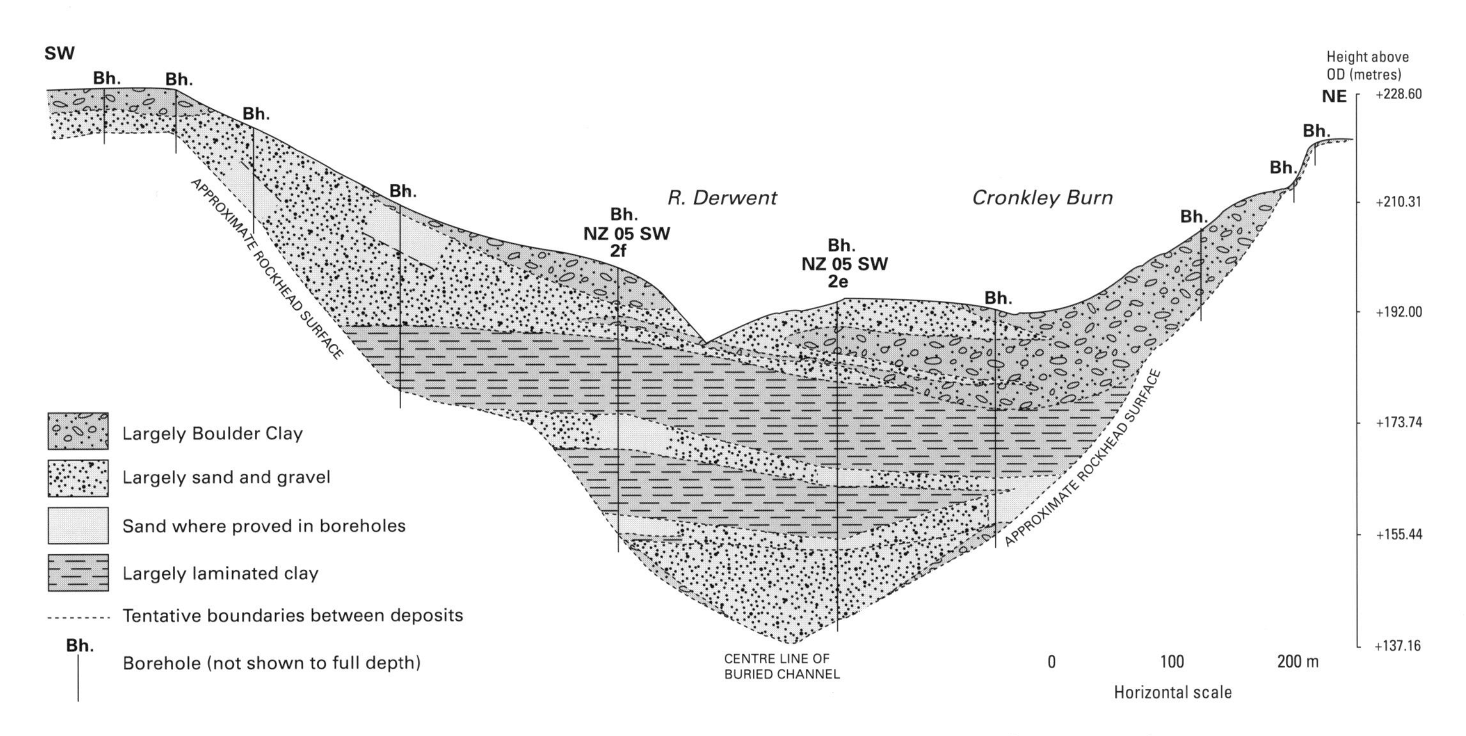

Figure 29 Buried valley deposits of the River Derwent north-north-east of Edmundbyers. Based on boreholes sunk prior to construction of the Derwent Dam and interpretation by J H Hull (unpublished notes, 1961).

tively dry. Typically, such channels are steep-sided and flat-bottomed, and some have a 'humped' longitudinal profile. Channels range in length up to 5 km or more, though most are significantly shorter than this. The channels were believed by early workers to be subaerial in origin but, in the light of later investigations (Peel, 1949, 1956), most are now believed to have been cut subglacially by meltwater which commonly was flowing under high pressure and hydrostatic head (Sissons, 1958). Some of the channels were studied by Herdman (1909), Anderson (1940) and Allen and Rose (1986), and Francis et al. (1970) made brief reference to two channels in the south-east of the district. A number of channels unfortunately can no longer be seen, or at least are only partly visible, having been infilled by made ground.

The valley of Dipton Burn, upstream [NY 996 608] from around its confluence with March Burn (Figure 26), and continuing westwards into the Hexham district, has many of the characteristic features of a glacial drainage channel (Peel, 1949). The channel forms a deep, narrow, steep-sided, flat-bottomed valley, and differs from the examples described below only in presently carrying a sizeable stream. Between the downstream end of the glacial channel and the confluence of the burn with the River Tyne, there is a spread of sand and gravel, perhaps deposited from subglacial meltwaters flowing down the Dipton Burn channel from the present-day valley of the Devil's Water in the adjacent Hexham district.

In more central parts of the district, the Whittonstall and related channels [081 579] (Figure 26) also appear to be related genetically to delta-like spreads of sand and gravel in the Derwent Valley. They have received considerable attention, forming part of the studies of Herdman (1909) and Anderson (1940), who interpreted the channels as essentially subaerial in origin. More recently, Allen and Rose (1986) have reinterpreted them as subglacial features, carrying south-easterly flowing meltwater which deposited a sand and gravel delta in a proglacial lake situated in the present Derwent Valley.

Among the more scenic, and indeed dramatic, examples of meltwater channels in the district is Hown's Gill [095 491], south of Consett, near the southern margin of the district (Plate 1). The channel is bridged by a spectacular viaduct, designed by Thomas Bouch and opened in 1858. The viaduct formerly carried the railway to Weardale and is now a preserved industrial monument. The trend of the channel is broadly east-south-east (Figure 26), but with a slightly sinuous course. It is over 5 km in length and continues beyond the southern margin of the district. It is up to 46 m deep and has a minimum width of 250 m. The channel intakes at about 190 m above OD, and its outlet is at about 137 m. The channel is cut throughout in bedrock, in the measures beneath the Brockwell Coal, and includes a high proportion of sandstone at a position close to the Marshall Green Coal. The channel is cut into the interfluve between the valley of the River Derwent and that of the River Browney, a tributary of the River Wear. The channel was probably largely subglacial in origin and, to judge from its size, it carried very large volumes of meltwater in glacial times from the upper reaches of the Derwent Valley into the Browney drainage system.

A complex system of glacial drainage channels is present between Sheep Hill and north of Causey, some 2.5 km to the east-south-east. They lie on, and to the east of, the interfluve between the Derwent and Team drainage systems (Figure 26). A long sinuous channel, cut in drift from a point [182 571] near Sheep Hill, is joined on its north side near Marley Hill [207 579] and Hedley West House [209 566] by two shorter subsidiary channels. The two subsidiary channels cut across a buried channel of a former tributary of the preglacial Beamish Burn. This channel system is separated by a gap from a series of north- and east-flowing channels which extend from north of Marley Hill towards Sunniside and Whickham Fell (Figure 26). Unfortunately a considerable length of the Burnopfield Channel, between Sheep Hill and the upper reaches of Bobgins Burn [196 566], has been filled with made ground. Similarly, much of the northern end of the Marley Hill South Channel is also infilled, largely by colliery waste and reclamation from the former Marley Hill Colliery. Although the channels in Whickham and between Whickham and Sunniside are visible, a substantial part of their length is in built-up ground. A considerable length of the upper part of the Sunniside Channel near Long Hill [198 595] is well exposed, however.

In the south-east of the district a major channel is present near Tribley Farm, east of Craghead, from north-west of Hett Hills [238 514] to the Humble Burn [243 503], a tributary of the Cong Burn west of Waldridge (Figure 26). A small section at the northern end is partly obscured by made ground and a waste tip. The southern end is marked by an abrupt change of direction from nearly south to east. In the extreme south-east of the district, a major channel over 3 km in length has been mapped south-east [224 485] of Holmside, west of Edmondsley, and extends into the adjacent district north-west of Sacriston [241 470]. Both margins of this channel are largely in bedrock but drift deposits are present along parts of the centre line of the channel. At a point 1.5 km from its entrance, the channel changes course by almost 90° to a southerly direction, and cuts through the Cong Burn and River Browney interfluve. The lowest part of the channel is concealed by made ground.

Other channels within the district include examples at Redwell Hills [135 517], east of Leadgate; at High Spen [140 600] (Lebour, 1906); south [138 613] of Greenside, where two interconnected channels are present, one elongated north to south, the other west to east; near [167 615] Norman's Riding Farm, south-west of Winlaton; at Clough Dene [172 552], north of Tantobie; at West Pelton [233 530], where only very limited parts of a channel, some 1.5 km in length, are now visible; north [217 525] of Twizell Hall, near Stanley; north [226 528] of Eden Hill Farm, between Beamish and West Pelton, where twin channels largely cut into bedrock merge into a short broader channel south of Beamish; south Waldridge Fell [249 493]; and at White House Farm [216 504], south of Craghead, the uppermost part being obscured. In the south-west of the district, several shallow peat-bottomed channels at Muggleswick Park [032 493], south-east of Edmundbyers, are thought to have formed as ice-margin channels (Figure 26).

POSTGLACIAL AND RECENT DEPOSITS

River terrace deposits

Well-developed river terrace systems are largely confined to the western part of the Tyne Valley between Corbridge and Ovingham, and the Derwent Valley between the Derwent Reservoir and Hamsterley. Small poorly developed terraces occur in association with some of the tributary streams. The deposits consist of gravels, sands, silts and silty clays, and commonly exhibit much lateral variation. The higher terraces generally contain the coarser material.

In the Tyne Valley, terrace surfaces in the western part of the district range between about 18 and 27 m above OD. The few boreholes that have been drilled in these terraces record up to a maximum of 7.3 m of sand and gravel, underlying a thin stony soil (Giles, 1981; Lovell, 1981). A typical sequence was recorded in a borehole [0330 6159] on the terrace north-west of Low Shilford:

	Thickness m
Soil	0.3
Gravel, coarse, with subangular to subrounded sandstone fragments, and including some ironstone and volcanic debris; and sand, medium-grained, subangular to subrounded, quartzitic; coal fragments	2.7

A broadly similar sequence of sand and gravel, 7.3 m thick, was recorded in a borehole [0552 6189] near Stocksfield Hall, beneath 2.0 m of very clayey sand.

In the Derwent Valley, at and below Derwent Bridge [032 513], and around Allensford [077 503] and Low Waskerley [086 533], remnants of a postglacial terrace system consist largely of irregularly bedded silts, sands and gravels, the latter locally very coarse and containing boulders. Small terrace deposits along Burnhope Burn [NY 992 483 to NZ 012 495] in the extreme south-west are particularly coarse.

Alluvium

A wide, almost continuous belt of river alluvium is associated with the River Tyne, more especially between Wylam and the confluence with the Team Valley on the western outskirts of Gateshead. A similar, though significantly narrower belt is generally present along the Derwent Valley, mostly below Ebchester. The construction of the Kielder Reservoir in the North Tyne Valley, and the Derwent Reservoir, have largely mitigated the effects of flooding over many parts of these valleys that were previously regarded as floodplain. Alluvium partly overlies the laminated clays of the Team Valley, between the River Tyne and Chester-le-Street. Narrow discontinuous belts of alluvium are associated with many of the smaller streams in the district. The alluvium consists of gravels, sands, silts and silty clays, commonly in thin beds, and lateral variation is characteristic. Thin beds of peat and organic clays within the alluvium have been recorded, more particularly in the Tyne Valley near its confluence with the River Team. In the lower parts of the Tyne, Derwent and Team valleys the alluvium is locally covered by substantial amounts of made ground. Currently, the only pit working alluvial sand and gravel, albeit on an occasional basis, is at Farnley Haughs [005 634] between Corbridge and Riding Mill.

The following sequence was recorded in a borehole [1787 6421] at Newburn Haughs:

	Thickness m
Made ground	1.4
Silt, sandy, olive-grey	2.0
Sand, clayey	2.0
Silt, as above	3.0
Gravel, coarse, and sand, medium-grained; sporadic peat layers	8.6
Sand, clayey	8.0+

Peat and shelly layers were recorded in 5.2 m of alluvial silt and clay in a borehole [1932 6301] at Blaydon.

Alluvium occurs in a number of small hollows formerly occupied by shallow lakes, more especially in the west and north-west of the district. The largest area [025 665] occurs west of Newton Fell House. Small areas are also present in the south-east, between Beamish and Chester Moor. Narrow deposits of high-level peaty alluvium are present on the floors of glacial drainage channels near Muggleswick Park [026 492].

Alluvial fan deposits

Several very small alluvial fans or cones have been mapped in the area of high ground in the south-west of the district, particularly in the vicinity of Berry Bank [007 512] and in the valley of the Burnhope Burn. The fans occur at the bases of steep sandstone escarpments where small streams descend from higher ground. They consist of coarse, largely angular and subangular gravel and sand.

Peat

Hill peat, rarely exceeding 1.5 m in thickness, overlies bedrock or boulder clay and mantles several relatively small, scattered areas of high ground in the vicinity of Edmundbyers, in the south-west of the district. In the same general area, small peaty hollows are present, and tracts of peat-covered alluvium occur along the floors of some glacial drainage channels.

Isolated hollows with a thin peat fill, probably overlying lacustrine alluvium consisting of silt and clay, occur in the north and west of the district, notably near Rudchester [119 670] and at Fotherley Moss [016 575], where an area of peat approximately 500 m by 300 m is present. The thickness of peat in these hollows is unknown but is unlikely to exceed 3 m. In the south-east of the district, peat is locally present in some small hollows, and examples were seen east of Twizell Hall [at 2160 5184], and north of Edmondsley [at 2400 5025]. In both localities the peat overlies lacustrine alluvium.

The presence of thin peat layers within the alluvium in the lower parts of the Tyne valley has already been mentioned.

Landslip

Some important areas of landslip have been mapped on steep banks of deeply incised streams, and in places where slopes formed by boulder clay and/or poorly consolidated drift deposits have been undercut and oversteepened by river action.

A few relatively small areas of slipped boulder clay, gravel, sand and laminated clay are present in the Tyne Valley between the western boundary of the district and the Ryton area, the most extensive being on the north side of the river south of Ovington.

Landslip has been proved in site investigations along the margins of the Tyne and Team buried valleys, even on low-angled slopes, where laminated clay and silt occur at or near the surface. These landslips commonly have no natural expression in the landscape and no attempt has been made to map them.

Extensive areas of landslip are present in the Derwent Valley, between Eddy's Bridge in the west and Consett in the east. The largest of these is on the north side of the river between Crooked Oak [055 498] and Allensford [077 502], where slipped masses of stony clay with irregular bands and lenses of gravel, sand, silt and laminated clay have been observed. These landslips extend for a distance of some 2.5 km and have a height of between 30 and 40 m. A similar, though less-extensive mass of slipped clay and other materials is present on the south side of the river between Allensford and Consett. Crags on the south side of the River Derwent, immediately west of Allensford [0770 5024], show signs of landslipping and disturbance in cross-bedded sandstones largely referred to the 'First Grit'. In the same general area, masses of slipped clay and unstable ground are present in the lower reaches of Coalgate Burn and Horsleyhope Burn, both of which join the River Derwent east of Combfield House [059 491]. Masses of slipped clay also occur in drift along the crest of the deeply incised meander of the River Derwent south-east of Muggleswick, and more especially along the south bank. In the south-west of the district, small areas of slipped sandstone have come down from the sandstone escarpments north of Berry Bank and Muggleswick Park [0070 5120 and 0390 4990].

In the south-east of the district small landslips are present on the margins of many streams, of which the Cong Burn above Waldridge, and the steeper flanks of the Twizell and Beamish burns, provide good examples.

Head, hillwash and related deposits

Deposits of surface material, largely clay but including also some coarser debris, are common in many parts of the district, more especially on the steeper slopes and higher ground. They include soliflucted material deposited under periglacial conditions following the retreat of the ice sheet, together with more recent hillwash. These deposits do not normally exceed 1 to 2 m in thickness, and have not been mapped. An area shown as Head in the north-eastern part of the district on the published Drift edition of the 1:50 000 map was incorrectly identified; it is in fact Upper Clay.

Made ground

Made ground occurs widely in the district, more especially in the centre and east where mining and industrial activity have been longest concentrated. The deposits are of very variable composition, including domestic waste, industrial and chemical waste, colliery and quarry spoil, overburden piled up from excavations, and surface material redistributed during landscaping. In particular, in the more built-up areas of the Tyne Valley, a wide range of material has been tipped, and the surface is commonly redistributed. Made ground has been shown generally where it exceeds 2 to 3 m in thickness; where thinner, it has been mapped only where it can be clearly defined. Spreads of urban rubble generally are not shown but are widely present, 1 to 2 m thick, under older and redeveloped areas.

Domestic and industrial waste has commonly been tipped into valleys and quarries, and may infill or form a rim around them; the full extent and thickness of such deposits is in many cases poorly documented. In urban areas especially, careful study of old records and maps, supplemented by site exploration, is vital if engineering works or redevelopment are envisaged in places where the existence of such fill is known or suspected.

Dearman et al. (1977a, fig. 7) discussed engineering geological mapping in the Tyne and Wear conurbation, and described the infilling of Pandon Dene [254 645]. Up until the early years of the last century this was a natural valley with a stream entering the River Tyne [2554 6394] downstream of the Tyne Bridge. The stream was culverted and the valley progressively filled by a variety of waste, much of it general refuse, during the last century. The deposits are up to 24 m thick, and the dene has presented particular foundation problems ever since. Except near its downstream end, hardly any surface trace remains of the dene, although recent redevelopment work behind the new Law Courts on the Quayside has exposed its position. Other deeply incised valleys (denes) on the north side of the Tyne have been subjected, in whole or in part, to such infill by made ground. They include, from west to east, the upper and lower reaches [2030 6550 and 1962 6456] of the Denton Burn on the west side of Newcastle; the Skinner Burn which had its origins in the Waterloo Street and Bath Lane area of the city, ran just west of Forth Banks and entered the River Tyne [2474 6343] some distance below the King Edward Railway Bridge; a small dene, the Swirle [2578 6405], which extended back towards City Road and Gibson Street; and much of the lower reaches of the Ouse Burn [2613 6473] more particularly north of Shields Road.

Away from the city, a number of former glacial drainage channels (Figure 26) have also been fully or partially infilled by waste and spoil, including the Burnopfield Channel east of Sheep Hill; part of the Marley Hill system; the West Pelton Channel east and south of Beamish; and the southern end of the Edmondsley Channel. Waste-infilled abandoned quarries and clay pits present similar construction problems to those encountered in infilled valleys.

Wedges of made ground, much of it relatively old, are present locally on both banks of the Tyne in the north-east of the district, in the lower part of the Derwent Valley and also along parts of the Team. These deposits are commonly up to 5 m thick. Chemical waste is widespread on the south bank of the Tyne at Prudhoe [095 641], and between the north-eastern part of Gateshead and Pelaw in the adjacent Sunderland district.

Little remains of the characteristic cone or hump-shaped colliery spoil tips which formerly dominated the landscape in the coalfield area. During the last twenty years, a policy of landscaping these spoil heaps and tips has been adopted. In most cases, the spoil has been regraded, landscaped and sown over in situ. Part of the contents of some tips have been used to fill holes, depression and valleys, or alternatively to raise levels and build embankments. Among the more dramatic examples of regrading and landscaping is that of the great area of waste and slag formerly occupied by the Consett steelworks (Plate 8). Little or no trace now remains of the great industrial complex that occupied this site for well over 100 years, and which formed such a familiar and dominant feature of the landscape for many miles around.

Restored and backfilled opencast sites constitute another type of made ground, which are widely present in many areas. Reinstated opencast coal workings form a special case in that they are infilled with rock waste from the site itself, the topsoil being returned to the surface on completion. In nearly all cases, the former depth of excavation and the nature of the infill have been well documented.

The very varied nature and quality of made ground, its extent and depth, present a wide range of engineering problems in construction and development, whether it

Plate 8 Consett steelworks [095 506] (now demolished). Made ground is encroaching on to the natural landscape of boulder clay and Lower Coal Measures. L1766.

lies on drift deposits or bedrock. Some areas of made or restored ground have been built on, but many others have been returned to agriculture or alternatively turned over to some type of amenity use.

INTERPRETATION OF QUATERNARY DEPOSITS

Pre-Late Devensian events

The main elements of the present-day drainage system were established in Neogene (late Tertiary) times (Merrick, 1915; Trotter, 1929). In detail the drainage pattern was much modified in the Quaternary by glacial processes, and the pattern of buried valleys (Figure 26) was in existence by early Devensian times. No Quaternary deposits older than Late Devensian have been recorded but some may be preserved in the buried valleys. The Team buried valley, which links the Tyne and Wear valleys, marks the only example of a major divergence between the locations of preglacial and postglacial river systems. Its origin remains an enigma. The suggestion that the River Wear formerly flowed north along this valley to join the River Tyne, before glacial diversion, has received considerable support (e.g. Woolacott, 1905; Raistrick, 1931). However, it is equally likely that the River Tyne formerly flowed south to join the old River Wear system near Chester-le-Street, because downstream of the Team Valley confluence the Tyne Valley is relatively narrow and restricted. Another possible explanation is that the Team Valley may have been formed originally by two minor consequent streams, which flowed into the rivers Tyne and Wear. Whatever its origin, the Team Valley probably has been considerably further modified by glacial and postglacial events.

Late Devensian glaciation

The Late Devensian ice sheet which overrode the district emanated from two source areas, south-western Scotland and the northern part of the Lake District (Lunn, 1980; Boulton et al., 1985). The two ice masses, of which the former is thought to have been the larger, coalesced and streamed eastwards and east-south-eastwards across the northern part of the Northern Pennines and through the Tyne Gap, covering a wide area of north-east England, including the whole of the present district, in a thick blanket of ice. The upper surface of the ice sheet probably lay at between 800 and 950 m above OD. Another stream of ice, flowing southwards from south-eastern Scotland and the Cheviot Hills, impinged on the coastal area to the east of the district (Smith, 1994). The exact western limit of this ice, and the extent to which it extended inland, is uncertain. Areas immediately to the south-west, in the adjacent Alston and Wolsingham districts, were probably more under the influence of relatively local ice emanating from the high Northern Pennines, including the watersheds of the rivers South Tyne, Wear and Tees. Boulder clay (till) was the most widespread deposit formed during this phase of the glaciation. Basal sands and gravels formed under or in advance of the ice may be locally present. The major valley lines, notably the Tyne and to a lesser extent the Derwent, would probably be the first to be affected by lobes of advancing ice. It seems probable that the whole district was covered by till or closely related englacial deposits as the advancing ice choked the pre-glacial drainage system.

It is probable that both the Tyne and the Team valleys were overdeepened, and perhaps widened, at this time, as a result of direct glacial scouring or from the erosive power of sediment-loaded subglacial meltwater. Closed hollows with rockhead levels deeper than 45 m below OD were formed (Figure 27). Early lobes of western ice moving down the confines of the Tyne Valley may have divided into two streams on meeting the rock obstruction of the high ground now occupied by the north-western suburbs of Gateshead, so that one lobe of ice was diverted down the present lower Tyne Valley and another down the Team Valley. Later, following more general inundation by western ice, such a marked divergence in flow direction at lower levels in the ice sheet could have caused immense strain within the ice sheet, leading to overdeepening as a result of enhanced glacial scour. Some of the older part of the buried valley infill may date from this time.

The southern, Team Valley, lobe of this ice probably continued south towards Chester-le-Street. Cox (1983) has drawn attention to the presence of several apparently closed, elongate hollows cut into rockhead in the floor of the valley in this area (Figure 27). He also has pointed out that the lower courses of the easterly-flowing tributary buried valleys, notably those of the proto-Team and proto-Twizell, have been truncated and apparently left hanging at a point near where they join the main valley. Cox considered that the origin of the buried valley may, at least in part, be explained by the erosive action of highly pressured meltwater flowing beneath thick ice. In the opinion of the senior author this can only account for further overdeepening of the preglacial profile rather than inception of the whole system.

Deglaciation

Deglaciation can be considered almost wholly in terms of wasting of stagnant and dead ice rather than actual westward 'retreat'. Wasting of ice was largely effected by the thinning and gradual disappearance of ice from the higher ground, with a general progressive reduction in the surface level of the ice in lower-lying areas. In the final stages of deglaciation, masses of dead ice probably remained in the main valleys, particularly those of the Tyne, Derwent and Team. The gradual decay of ice from the surface would also have been supplemented by internal wastage and basal undermelting. Meltwater released in this way would have contributed significantly to the cutting of glacial drainage channels, many of which would have carried water under high hydrostatic pressure.

Downwasting of ice suggests steadily shrinking ice margins as levels progressively lowered, leading to the formation of ice-dammed lakes. Dwerryhouse (1902) and many later workers (e.g. Raistrick, 1931) proposed a system of subaerial ice-dammed lakes connected by glacial drainage channels. Related sand and gravel spreads were interpreted as lake delta deposits. Abandonment of higher lakes, and the establishment of new lakes at lower levels, were thought to have occurred in response to the progressive contraction of ice margins, as the height of the ice surface was reduced.

Anderson (1940), following Dwerryhouse (1902) and Herdman (1909), developed this concept further with reference to the north-western part of County Durham, and to the Derwent Valley in particular. Glacial meltwater channels (Figure 26) are relatively common in this area, and were related by Anderson to a series of ephemeral lakes at progressively lower ice- margin levels during deglaciation. Associated spreads of sand and gravel were largely interpreted as glacial lake deposits. A steep frontal bank [102 562] north of the River Derwent, between Newlands and Blackhall Mill, was regarded as the leading edge of an extensive deposit of sand and gravel, forming a delta laid down in 'Lake Derwent' when the level of the lake had fallen to about 126 m above OD. A similar example was quoted by Anderson north [156 621] of the Barlow Burn, east of Greenside. The occurrence of extensive spreads of sand and gravel in the north-central and south-eastern parts of the district was attributed to the period of maximum melting, when the volume and carrying power of meltwaters were at their greatest. It was considered that meltwaters were borne away to the east and south-east, during most of the deglaciation, via a complex sequence of overflow channels, initially into the Browney drainage system of the Wolsingham district and subsequently into the Cong and Chester Burns, both of which ultimately drained to the Wear. Not until a relatively late stage were meltwaters directed into the Team system.

Although there are many aspects of Anderson's (1940) study which are plausible both in terms of the morphology and the nature of some of the deposits, it leaves a number of unanswered questions. It implies the existence of large areas of water over lengthy periods of time, that lake levels fell to prescribed heights as new channels were cut, that all ice margins were virtually watertight, and that there was no way of escape other than by the glacial drainage channels. Moreover, there is little or no evidence that any lake margin deposits formed during temporary stands, although such evidence may have been removed by subsequent periglacial processes.

More recent studies in north-eastern England (e.g. Sissons, 1958; Clapperton, 1970; Burgess and Holliday, 1979; Arthurton and Wadge, 1981; Allen and Rose, 1986) have supported the view that meltwater channels, together with constructional features such as eskers and other stratified drift deposits, are the result of englacial or subglacial stream action in a wasting ice sheet, with water commonly under considerable pressure and hydrostatic head (Carruthers, 1953; Boulton, 1972). Lunn (1980) drew attention to the possibility that subglacial or englacial meltwater drainage could be strongly guided by the tectonic structure of the ice rather than by structures in the underlying bedrock or by rockhead relief. Allen and Rose (1986) have reinvestigated part of the Derwent Valley area studied by Herdman (1909) and Anderson

(1940), in particular the Whittonstall glacial drainage channel system and the related sand and gravel deposits near Ebchester and exposed in Broad Oak Quarry [098 567]. They have interpreted the channels as subglacial in origin, but regard the associated sediments as proglacial lake deposits.

Deposition of sediment beneath the ice may be associated with channel erosion. It is likely to occur where the transporting capacity of meltwater is reduced by subglacial obstructions, such as irregularities in the lower surface of the glacier or in rockhead topography, particularly in the later stages of the wasting process. Deposition beneath ice may account for the moundy topography characteristic of many water-laid glacial sediments.

Richardson (1983) and Cox (1983) attributed the formation of most of the sand and gravel deposits in the areas they studied (1:25 000 squares NZ 16 and NZ 25) to seasonal streams within stagnant melting ice. They suggested that, during progressive ice wastage, the large volumes of impounded water and sheet streams laid down sediments irregularly over a wide area. Their view is largely supported here, with the proviso that some deposits may owe their origin to proglacial rather than sub- or englacial phenomena. Ephemeral lakes may have developed at various levels as outlets became blocked by temporary obstructions, whether of ice or superficial deposits. These conclusions are probably applicable to the whole district, including the deposits of the Derwent Valley discussed above. Subglacial drainage probably dominated during the early part of deglaciation, but temporary proglacial lakes were likely to have formed in the later stages. During the course of deglaciation, some lakes may have evolved from being totally subglacial, through a stage with a significant floating ice cover, to a largely ice-free proglacial lake.

In the later stages of deglaciation, when the higher ground was becoming free of ice, the subglacial drainage pattern was strongly influenced by the preglacial valley system. However, during earlier phases of deglaciation the older valleys seem to have been partially or totally choked by ice, and meltwater was obliged to find new outlets. Flowing under pressure and hydrostatic head beneath the ice, meltwaters found new pathways, moving uphill and crossing pre-existing interfluves, and cutting glacial drainage channels in the process. For example, meltwaters from the upper reaches of the Derwent Valley were transported via the Hown's Gill channel into the Wear drainage. Another major drainage system carried meltwater from tributary valleys of the Tyne through the Whittonstall channels into the Derwent drainage, and thence by way of the Burnopfield channel into the Team catchment. During this phase, most deposition of sediment from meltwaters was sub- or englacial in origin. As deglaciation progressed and ice in the upland areas thinned and melted away, the higher levels of the subglacial drainage system were abandoned. At this time, ephemeral proglacial lakes, dammed by ice or other constrictions, were able to form. As the ice further wasted, the level of such lakes progressively fell.

In the closing phases of deglaciation a major glacial lake, Lake Wear, was thought by Smith (1994) to have occupied a large area to the east of the district. This lake was believed to be part of a series of ice-dammed water bodies formed as a result of inhibition of meltwater movement at the boundary, near to the present coastline, between stagnant, westerly derived ice and still-active ice flowing from the Cheviot Hills and south-eastern Scotland. The level of the lake seems to have fluctuated considerably with time, as outlets were opened or closed, and details of its history are still uncertain. On its western side, Lake Wear was thought by Smith (1994) to have extended into the present district, in the Team Valley and the lower parts of the Tyne Valley. The upper level of the lake within the district is difficult to determine, however. It is unlikely to have exceeded about 50 m above OD, a level which would have been sufficient to produce westerly directed fingers of the lake some considerable distance up the Tyne Valley, perhaps almost as far as Riding Mill and Corbridge, and also in the lower part of the Derwent Valley. Considerable thicknesses of the laminated clay deposits, including some of the infill of the Team and Tyne buried valleys, are inferred to have had their origin in this lake. Rafts of boulder clay and other glacial deposits within the clays may have been transported in by floating ice, or may represent the occasional incursion of coarser water-borne sediment. Deposits resulting from the mass flow of material during periglacial conditions surround the lake margins. Laminated clays and overlying sand and gravel in the Bywell area may also be deposits of Lake Wear, but the elevation of the laminated clays near Corbridge suggests that they formed in an older water body whose level was at about 60 m above OD.

Postglacial events

The final stages of deglaciation saw the complete decay of ice, the drainage of any remaining ice-dammed lakes, and the gradual amelioration of the climate from periglacial and permafrost conditions. This period of slow amelioration saw the evolution of the drainage from a dominantly subglacial and lacustrine system to the present-day subaerial drainage pattern, which corresponds in general outline to its preglacial counterpart. Much of the present-day drainage is cut into drift deposits, including those of the supposed Lake Wear and other glacial lakes. More locally, the rivers and streams are cut into bedrock, particularly where they have been most diverted from their former courses, leading to the development of postglacial rock gorges. Among the more spectacular of these in the district is the incised meander [052 497 to 060 494] of the River Derwent south-east of Muggleswick, which is nearly 50 m deep, precipitous in places, and locally as little as 250 m in width. Other examples include that of the Causey Burn [197 554 to 203 561] from some 1.5 km north-east of Tanfield Lea to near Causey; that west [228 542] of High Urpeth farther downstream [232 540] in the same valley, where it is named the Beamish Burn; that of the Twizell Burn [215 513 to 227 515] below Craghead; part of the Cong Burn [236 496 to 246 502] above Waldridge; Aydon Dene [000 662] cut by the Cor Burn near Corbridge; and Whittle

Dene [072 651] cut by Whittle Burn north-west of Ovingham (Figure 26). It is probable that many of these rock gorges were initiated as subglacial channels during the late stages of ice wasting. Whatever its origin, the Team Valley was largely abandoned as a major drainage route at this time, apart from present-day drainage from the Stanley area.

Since the end of the ice age, sea-levels have fluctuated considerably, as a result of the glaci-eustatic and glaci-isostatic effects of the melting and unloading of ice. The local magnitude and timing of these changes are not certainly known. It is probable that from the end of the ice age sea-levels gradually rose from a level significantly lower than that at present to one somewhat higher. More recently, sea-level has gradually fallen to the current level. The effects of these base level changes on the river systems of the district, and on their deposits are not known. During this time, the rivers have laid down the sediments of the river terraces and the present-day alluvium, although the pattern of the latter has been significantly modified by river regulation schemes, and other human activities.

Localised hill peat formed in postglacial times. Beds of peat have been recorded within the alluvium, or as small patches within depressions on the surface of glacial drift. These probably mark the sites of former shallow lakes and ponds, some of which may have been initiated in late glacial times.

SEVEN

Economic geology

COAL

Until comparatively recently coal formed the main mineral product of the district. This part of north-east England is one of the oldest coal mining areas in the country, the earliest recorded workings having taken place along the banks of the River Tyne in the 12th and 13th centuries (Galloway, 1882; Forster, 1930; Smailes, 1935; Middlebrook, 1950; Atkinson, 1979; Land, 1974). Progress in the winning of coal over the next three or four centuries was slow. Initially the coal was probably dug or quarried at outcrop, while later efforts saw the development of a system of bell pits, in some places accompanied by limited gallery working. However, by the 17th century, sufficient technical advances had been made to enable coal to be mined in many parts of the district (Hair, 1844; Galloway, 1882; Nef, 1932). Pits, rarely more than 30–40 m deep, were sunk throughout the district, generally down to the first thick coal, and in areas where the superficial deposits were comparatively thin or absent. The pits were grouped into collieries, examples being Benwell, Fenham and Walbottle north of the Tyne, and Beamish, Ellison Main, Felling and Team south of the Tyne. Transport methods were significantly improved by the development of the wooden waggonway.

The greatest impetus to the further development of the coal industry came, however, with the advent of steam power, not only for use in machinery at the mines themselves but also for improved methods of transport both on land and at sea. The arrival of mechanisation transformed the industry and developments were more rapid than at any time during the preceding five centuries. Improved methods of ventilation, pumping and drainage allowed deeper shafts to be sunk, so enabling several coals to be worked from one shaft or group of shafts. Mining communities sprang up over many parts of the district, and townships such as Stanley, Annfield Plain, Burnopfield, Tanfield Lea, Wylam, Prudhoe and Consett largely owe their existence to this phase of the coal industry's development. The zenith of the coal industry was reached in the closing years of the 19th century and the early years of the present century. With the steady dwindling of reserves, allied to the difficulties and expense of mechanising many of the pits, the pace of closure began to accelerate in the years after the second war. By the 1970s few working pits remained open and with the closure of Marley Hill Colliery in 1983 deep mining in this district finally came to an end.

Opencast working, initiated during the second world war, became important in some parts of the district. Most of the early sites were comparatively small and were located mainly on the interfluves. They rarely worked more than two or three seams. More recently, however, larger sites working several seams have been exploited in the north and west of the district, particularly on the fringes of the main coalfield area. These have included sites at Horsley and Heddon-on-the-Wall in Northumberland, and Greymare Hill, Chapman's Well, near South Moor, and Whittonstall in County Durham.

Many of the coals are of relatively high rank (in the medium- to low-volatile bituminous category), especially in the central and western parts of the district. They were used as prime coking coals, reputed to be among the best in the world, and were particularly suitable for the manufacture of foundry coke. They were also used as general purpose coals and as high-quality coals for steam raising. In general the rank of the coals increases with depth, and also in a westerly direction towards the central part of the Alston Block (Trotter, 1954; National Coal Board, 1965; Jones and Magraw, 1980). This rank pattern has been related to the thermal effects of burial and the high heatflow associated with the concealed Weardale Granite (Bott, 1967; Creaney, 1980). Relatively high ranks in the central part of the district occur above the Rowlands Gill cupola of the granite. Except where influenced by Tertiary dyke intrusion, present-day coal ranks appear to have been attained during late Carboniferous burial, modified at depth by the intrusion of the Whin Sill (Ridd et al., 1970; Creaney, 1980). Post-Carboniferous burial generally was insufficient to bring about any further increase in coal rank within the district.

Prospects for future coal working are on a relatively small scale. Some scope remains for small drift mines and opencast workings to exploit areas of unworked coal, and to remove remnants of coal from previously worked deep mined areas, but the potential is limited. In respect of opencasting, social and environmental factors may play an increasing role in inhibiting further substantial development.

HYDROCARBONS

The hydrocarbon prospectivity of the Carboniferous rocks of north-eastern England has been reviewed by Scott and Colter (1987), Fraser et al. (1990) and Chadwick et al. (1995). The last-named authors concluded that the chances of finding economic oil and gas accumulations was low, in view of the lack of significant proven source potential below the Coal Measures, the overall poor reservoir qualities of the Carboniferous limestones and sandstones, and the generally unfavourable burial and thermal history.

Despite such pessimistic assessments, a significant area, including the northern part of the district, has been licensed and investigated for hydrocarbons in the past. That part of the district which includes the Ninety Fathom-Stublick Fault System and the southern margin of the Northumberland Trough probably attracted interest because of the possible presence of structurally closed inversion anticlines at the basin margin, into which hydrocarbons generated within the basin and migrating towards the Alston Block may have been trapped. In the course of hydrocarbon exploration of this part of the district, several seismic reflection surveys were carried out but no boreholes were drilled.

The organic matter within the Carboniferous rocks of the district is likely to have been severely affected by the high regional heatflow (Chapter 2; see also Geothermal Energy, below) and, more particularly, by the intrusion of the Whin Sill (Chapter 5). The sill was intruded in late Carboniferous or early Permian times, generally at levels close to the base of the Stainmore Group. Comparison with boreholes in adjacent districts (Ridd et al. 1970; Jones and Creaney, 1977; Creaney, 1980), suggests that all Carboniferous rocks within the district, with the exception of the Coal Measures and the upper part of the Stainmore Group, have been overmature for oil and gas generation since the intrusion of the sill. Any hydrocarbons that formed before, or at, the time of maximum burial, in late Westphalian times, are likely to have escaped before the Variscan deformation and the formation of trapping structures; alternatively, they are likely to have been destroyed at the time of Whin Sill injection. Because a high organic rank was attained in late Carboniferous to early Permian times, there is little prospect that renewed generation of economic quantities of hydrocarbons will have occurred in the area since that time, beneath the relatively modest Mesozoic cover (cf. Holliday, 1993). Overall, therefore, hydrocarbon prospects in the district are not good.

COAL-BED METHANE

The hydrocarbons considered in the previous section were expelled from their source rocks and subsequently migrated into trapping structures or escaped. However, significant quantities of methane, which formed during coalification, commonly remain trapped within coal seams. Although the volumes are generally much less than those of more conventional sources of natural gas, methane extracted directly from beds of coal ('coal-bed methane') may prove to be an important future source of energy. The coal-bed methane potential of the coalfields of north-eastern England is generally thought to be low, because of the great extent of former coal workings and the high proportion of coal already extracted. Mean methane quantities in the region are also relatively low, at around 1.3 m^3/tonne, compared with values significantly in excess of 3.0 m^3/tonne in many other coalfields (Creedy, 1991; Ayers et al., 1993). This applies even in areas such as the central and western parts of the district where coal ranks are relatively high.

SANDSTONE

Sandstone for use as a building stone, as a roofing stone, and also to a lesser extent as an aggregate, was formerly quarried widely throughout the district. Working was always on a relatively small scale, and reserves in general have hardly been touched. The many beds that have been worked for good stone include some of the sandstones in the higher part of the Stainmore Group and the lower part of the Lower Coal Measures, and higher in the sequence the Durham Low Main, High Main, Seventy Fathom and Grindstone posts of the Middle Coal Measures. The last-named formerly was used in the manufacture of grindstones, the bulk of production having come from quarries just east of the present district near Wrekenton, including Eighton Banks and Springwell quarries. Among the larger sandstone quarries formerly worked are those near Castleside [077 495], at Hown's Gill [097 489] near Consett (Plate 6), Heddon-on-the-Wall Quarry [130 668] where the sandstone worked was used in the construction of Newcastle Central Station, and at Blakelaw [212 668] in the north-western suburbs of Newcastle upon Tyne. Old small quarries, now largely infilled, are particularly numerous in the built up areas. There are many such old workings, for example in the High Main Post in the western part of Newcastle upon Tyne. Bearl Quarry [053 641] near Ovington is still in use.

Small quantities of ganister for refractory purposes were obtained from quarries at Dean Howl [066 484], near Healeyfield, and at Castleside [077 495]. Significant quantities of sand for moulding purposes have been obtained from High Acton Quarry, just to the west of the district, off the Slaley–Blanchland road, and smaller amounts have been obtained from the quarries at Castleside. Deeply rotted coarse feldspathic sandstones are widely present in north-western County Durham, and have provided a source for first-class moulding sand (Carruthers and Anderson, 1943).

LIMESTONE

In the north-western part of the district, the Lower Felltop (Corbridge) Limestone near Corbridge was quarried until recently, largely as a source of lime. The largest of these quarries, Deadridge Quarry (Plate 3) [NY 996 655], north of Corbridge, also worked the underlying fireclay. Elsewhere in the same general area are numerous old workings for lime in the Belsay Dene, Lower Felltop, Upper Felltop (Thornbrough) and Grindstone (Newton) limestones.

IRONSTONE

The presence of ironstone in certain parts of the district has been known for a considerable time. In the early eighteenth century, a family of Toledo sword makers from Germany settled near Shotley Bridge to continue their trade, and it is believed that ironstone was obtained at outcrop from the Shotley Low Quarter area north of Shotley Bridge, from near Delves south-east of Consett, and on the site [097 506] of the former steel making complex at Consett (the 'German Bands' Ironstone). The existence of clay ironstones, albeit thin, together with the presence of abundant supplies of coal, was undoubtedly the reason for the establishment of the Consett Iron Company in 1839 (Anon, 1858; 1954). Clay ironstone ribs, and beds containing ironstone nodules, were mined from three levels in the Coal Measures of the Consett area and the country immediately to the east. Local ironstone mining effectively ceased in the last century with the acquisition of iron ores from Cleveland, Cumberland and Lancashire and, since 1872, from Spain. All iron and steel manufacture has now ceased within the district, and there is no prospect of renewed local mining of ironstone.

FIRECLAY

Fireclay for the manufacture of sanitary ware and refractory goods has been obtained from surface and underground workings in both the Stainmore Group and the Coal Measures, chiefly the latter. Fireclay is commonly associated with coal seams. Workings were generally of limited extent, and the fireclay was commonly worked as an adjunct to the extraction of other materials such as coal or limestone. The largest underground workings are thought to have been in the Scotswood area of Newcastle upon Tyne, where fireclay was worked in conjunction with the Plessey or Ruler Coal. Some distance farther north, the so-called Lister Fireclay, between the Main and Five-Quarter coals, was worked in the western outskirts of Newcastle near the cemetery and crematorium. The fireclay beneath the Lower Felltop (Corbridge) Limestone has been extensively worked in the area immediately to the north of Corbridge, associated with the quarrying of the limestone (Plate 3), but also in its own right from underground mine workings. A small area was being worked at the time of survey. Significant deposits of fireclay probably remain within the district, but in current circumstances they are unlikely to be worked.

DOLERITE

Dolerite is commonly used as a source of aggregate but has little economic potential in the district because of its restricted area of occurrence. The Palaeogene ('Tertiary') dolerite dyke at Walbottle (Chapter 5) has been quarried to a very limited extent, but much of the outcrop is now built over.

MUDSTONE AND SHALE

Shale and mudstone, for use in brick manufacture, have been worked at Throckley (West) Quarry [152 674], north-west of Throckley. More extensive workings were located north of Throckley in the adjacent Morpeth district. Shale and mudstone near the level of the Hutton Coal have been extracted from Phoenix (Crawcrook) Quarry [135 628], south of Crawcrook. Mudstones are abundant in the district (Chapter 3), but it is unlikely that in the foreseeable future they would be regarded as a source material in view of extensive supplies available from elsewhere.

BRICK CLAY

Deposits of Quaternary laminated clay and silt (Chapter 6) are present in several parts of the district and are particularly extensive in the Team Valley. Here they have been worked as a source of brick clay for many years, and the area contains numerous old clay pits, the majority of them backfilled. The main workings are located at Kibblesworth Pit [254 564, now closed] south-east of Kibblesworth, and at Union Brickwork's Pit [264 561], Birtley, a little farther to the south-east on the west side of the main Newcastle to London railway line. The deposits are relatively thick, extensive and easily worked, and reserves are sufficient for several years' production.

There were important workings of laminated clay just to the north-west of Corbridge, and there are also numerous small clay pits, now long since abandoned and infilled, in many parts of the district. Many of these workings, which were mainly for supply to the local market, also used the till for manufacture of bricks and tiles. Amongst the larger of these clay pits were those of East Tanfield [194 548] north-east of Tanfield Lea, where over 7 m of laminated clay were formerly worked, with boulder clay, for brick manufacture.

SAND AND GRAVEL

Extensive deposits of glacial and postglacial sand and gravel (Chapter 6) occur in many parts of the district, but more especially along the upper parts of the Tyne Valley, on the interfluve between the Tyne and Derwent valleys east of an approximate line from Wylam to west of Rowlands Gill, and in an extensive but irregular belt in the east and south-east extending from Sunniside to west of Chester-le-Street. The sand and gravel resources of the Tyne Valley between Hexham and Blaydon have been assessed as part of a mineral assessment programme (Giles, 1981; Lovell, 1981). Considerable quantities of sand and gravel have been worked at many localities in the Tyne and Derwent valleys, and also in the south-east of the district between Craghead and Chester-le-Street. At the time of writing (1994), the following pits were active: Farnley Haughs [005 634], Thornbrough [015 630], Merry Shield [063 616], Broad Oak (Allen and Rose, 1986) [098 567], Hollings Hill [096 575] (sand used in tile making), Crawcrook [128 637] and Blaydon [155 622]. Large quantities of sand and gravel have been removed from the extensive deposit on the north side of Barlow Burn, but the longer term potential in this area is becoming restricted, due to housing and other developments. Many present and former workings have been used as landfill sites under carefully monitored conditions.

The sand and gravel resources of the remainder of the district are comparatively undeveloped. The nature of the deposits, both on the north side of the River Derwent in the neighbourhood of Rowlands Gill, and again in the south-east of the district between Sunniside and Chester-le-Street, are in many respects similar in nature, form and thickness to those in the Tyne Valley described by Giles (1981). Extensive reserves are present, although a detailed assessment would be required to fully evaluate these deposits because of their complex nature and variable content. The thickest deposits are thought to occur west and north-west of Highfield near Rowlands Gill and in a north-west-trending belt west of Chester-le-Street. In some areas, more especially perhaps in the Derwent Valley near Rowlands Gill and Lintzford, and also west of Chester-le-Street, development and amenity considerations might constrain major exploitation.

Farther west up the Derwent Valley, small pits containing sand and gravel, which was commonly clayey, were formerly worked for local use at Edmundbyers and Muggleswick.

METALLIFEROUS VEIN DEPOSITS

The south-west of the district lies within the north-eastern part of the Northern Pennine Orefield. The mineral deposits of this area have been described in detail by Dunham (1990, pp.221, 227–229, 273–276). Some additional occurrences, in the north-west of the district, were described by Smith (1923, pp.19–21). Most of the veins in both areas were of very limited extent and supported only small-scale workings, commonly of considerable antiquity. They are summarised below. In addition, minor veining in which calcite, ankerite and barite are the most common minerals, is widespread in the district. Such veins rarely exceed 0.13 m in width.

Swandale Head Vein, Edmundbyers Common; barytes, trending 068°; shallow opencast workings [NY 9927 4967].

College Edge Vein, south-east of Edmundbyers Common; barytes, trending east-north-east [NY 9940 4910].

Pedams Oak Vein, north side of Burnhope Burn; lead ore, trending c. 325° [NY 991 489].

Swandale Vein, Burnhope and Harehope burns; lead ore, trending c. 325°; worked from three separate shafts, notably Harehope Gill Mine [0097 4846], Burnhope Mine [0061 4884] and Swandale Mine [0040 4912]; vein continues into adjacent Wolsingham district.

Silvertongue Veins, Derwent Valley south-east of Muggleswick; lead ore, trending c. 345°; two, or possibly three, short veins [057 491].

Healeyfield Vein, Dean Howl, west of Castleside; lead ore, trending 345°–350°, but sinuous and branching; main adit level [0689 4866], Whitwell Shaft [0687 4863]; extensive vein workings also extending south from Dean Howl into adjacent Wolsingham district.

Some 10 km to the east, witherite was discovered in the workings of the Morrison Pit [1734 5109], South Moor Colliery, Annfield Plain, when a lengthy drift to the south-south-east was made from the Hutton Engine Plane. The workings, almost wholly within the Wolsingham district, just extend into this district [at 1822 4804] south-west of Burnhope. The witherite vein is thought to form a branch of the Great Spar 'Dyke' of the Wolsingham district.

Craghead Deposits [c. 186 485 and c. 205 494], west of Holmside Hall, between Craghead and Burnhope, and also south-east of Craghead; witherite; two small witherite deposits on east-west trending faults.

Tanfield Moor Deposit, Tantobie; a small deposit of witherite, contaminated with calcite, was encountered in a drift driven through a north-north-west trending fault some 400 m at 350° from Willey Pit [1699 5438].

Thornbrough Mine [012 653], east-north-east of Corbridge; lead ore in a mineralised fault trending north-east and locally containing a basalt dyke; according to Smith (1923) some minor occurrences of lead ore nearby were also investigated by a shaft.

Future prospects for development of the known metalliferous deposits in this district would appear to be poor. The mineralisation of the Alston Block is concentrated in the rocks of the Alston Group and lower part of the Stainmore Group, and its distribution is closely related to the form and location of the Weardale Granite (Dunham, 1990). Thus, it is probable that there are vein deposits in the concealed Carboniferous rocks of the district, particularly in the vicinity of the Rowlands Gill cupola of the Weardale Granite. However, their delineation and characterisation would require a major investigation and considerable expenditure without any guarantee of success.

HYDROLOGY AND GROUNDWATER

The annual precipitation in the district ranges from as little as 700 mm, in the lower-lying areas of the north-east and south-east, to about 890 mm on the higher ground in the extreme south-west. In the coalfield area the natural pattern of groundwater drainage has been extensively modified by mining, particularly from the late 18th century onwards as improved mining techniques led to a great expansion of deep mining. Connections were commonly made between underground workings, taking water from one colliery to another, and certain collieries acted in effect as drainage levels from which water was pumped to the surface. As a result of prolonged pumping of mine workings the water table was substantially lowered, but with the closure of mines the general water table has inevitably begun to rise again. The extensive system of underground workings will continue to have a profound effect on the local hydrogeology and geochemistry of minewaters. Younger (1993, 1994), Younger and Sherwood (1993), Younger and Bradley (in press), and Sherwood and Younger (in press), have discussed some of these effects and the problems they present, more particularly in relation to the Durham coalfield.

Modification to the local, essentially near-surface, groundwater flow can be expected, even if only temporarily, in the vicinity of opencast sites, and around other excavations connected with major civil engineering enterprises.

The north-west and south-west parts of the district, occupied principally by strata of the Stainmore Group and the lower part of the Lower Coal Measures, have been subject to relatively little deep mining activity. The natural groundwater pattern has therefore been much less disturbed by human activities in these areas, and the groundwater hydrology is probably considerably less complex. Although the sandstones and thicker limestones in these areas have low intergranular porosity and permeability, they generally contain a system of interconnected joints and fractures and, within a local context, commonly constitute good aquifers. Together they form a multi-layer aquifer system, each bed having a perched water table. A similar situation probably existed within the coalfield prior to the main phase of deep coal extraction. Overall, flow of water is towards the north-east and south-east, but around the margins of the worked coalfield, or around any local workings, some modification of the pattern of flow might be anticipated.

Within the drift sequence, localised perched water tables are probably present within glacial sands and gravels; however, the presence of these is probably more important in causing instability in excavations than in providing a water supply.

The bulk of the district is now provided with mains water, although some supplies, largely agricultural, are met by spring water or by wells, especially in the higher and more remote areas. The south-west part of the district includes the major impounding Derwent Reservoir on the upper reaches of the River Derwent near Edmundbyers. The district also includes part of the Kielder tunnels, a component of the transfer works of the Kielder Water Scheme (Burston and Coats, 1975; Coats and Ruffle, 1982; Carter and Mills, 1976; Coats et al., 1982; Davies et al., 1981), the purpose of which is the southward transfer of water from the Tyne drainage system to that of the River Wear and River Tees. The elements of the transfer works in this district include a pumping station on the south bank of the River Tyne near Riding Mill, a pumping main from this station to the northern flank of Greymare Hill, the Letch House–Airy Holm–Derwent Tunnel, including a headpond reservoir at Airy Holm Farm, a cut-and-cover main crossing the Derwent valley near Eddy's Bridge, and between 2 and 3 km of the Derwent–Wear Tunnel (Appendix 3).

The nature of the groundwater supply and flow is of particular importance in deciding the location and content of landfill sites. Observation and metering holes are of vital importance, as is consideration of the porosity and structure of bedrock strata and superficial deposits adjacent to such sites.

Hydrogeological information is available from several organisations. The National Water Well Record Collection is maintained by the British Geological Survey, and a wealth of hydrochemical and physical data is held in the archives of the appropriate local authorities.

GEOTHERMAL ENERGY

Two main methods of geothermal energy extraction from the concealed rocks of the district have been proposed (Downing and Gray, 1986). These are the low enthalpy system and the hot dry rock (HDR) system. Low enthalpy systems attempt to extract hot groundwater directly from deep aquifers. This method of extraction has been successfully applied in Southampton and in the Paris Basin in France. In a hot dry rock system, cold water from the surface is pumped down a borehole into hot, fractured, impermeable rock at depth. There the water is warmed, and when it returns to the surface its newly acquired heat can be used for space heating or, if the temperature exceeds 200°C, electricity generation. Although a number of pilot projects have made encouraging progress, nowhere has economic application of this method been achieved.

The potential for the successful establishment of a low enthalpy system in the Carboniferous rocks of north-eastern England, including the present district, is thought to be low (Holliday, 1986; Chadwick et al., 1995). Within the district, suitable temperatures (c.75°C) are likely to occur at depths of about 2 km because of the relatively high geothermal gradients resulting from the high regional heat flow. However, the general lack of adequate aquifers at the appropriate depths makes any economic exploitation unlikely.

The high heat production and associated high heat flow anomaly of the Weardale Granite have made it a major target for potential HDR systems. A detailed review of the evidence has been compiled by Evans et al. (1988). The Rowlands Gill Borehole* was drilled and two seismic reflection surveys were carried out as part of that study. Evans et al. (1988) concluded that the most promising location in the region occurs within the district, in the Rowlands Gill area, about 10 km south-west of Newcastle. This area is underlain, at relatively shallow depths, by a cupola projecting from the Weardale Granite to the sub-Carboniferous unconformity (Chapter 2; Figure 6). Evans et al. (1988) predicted that temperatures around 230°C would be met at a depth of 7 km below this area. However, much more detailed investigations, including deep drilling into the granite, together with major advances in the necessary exploitative technology, will be necessary before any attempt to make economic use of HDR geothermal energy in the district can be contemplated.

GEOLOGICAL CONSTRAINTS ON PLANNING AND DEVELOPMENT

In terms of longer-term planning and environmental considerations, the mineral resources of part of the district have been the subject of a study by the former Tyne and Wear County Council (Anon, 1984a; 1984b).

Although mining and quarrying of mineral products are now much reduced in the district, major civil engineering and infrastructural developments are likely to continue in the future. Examples of such schemes in the recent past include the Tyneside Metro System, the Tyneside Interceptor Sewer, the Newcastle Western Bypass and part of the tunnel system and related works of the Kielder Water Scheme. All such developments demand a focused and well- planned site investigation programme, and the above projects were no exception in this respect. Major civil engineering works in much of this district are faced by two particularly serious geological constraints, notably the widespread occurrence of old underground coal workings, and the local presence of relatively thick deposits of laminated clays and silts. Neither of those factors need in any way inhibit development, but decisions can be made with regard to planning and development only after a particularly careful approach to the planning, execution and interpretation of site investigations.

Dearman et al. (1977a, 1977b), Sladen (1979), Eyles and Sladen (1981), Cox (1983), Richardson (1983), Jackson et al. (1985) and Lawrence and Jackson (1986) have outlined and discussed these and other geotechnical factors in relation to various parts of this district and adjacent areas. Although a comprehensive discussion of these factors is beyond the scope of this book, some of the main issues are summarised below.

Old coal workings present many related problems in site investigation and design, including the presence or absence of underground workings, the depth of workings, the nature and method of mining, the type of support, the presence of methane gas, the presence or absence of shafts and adits, the extent of dislocation of strata above old workings (Plate 7), subsidence factors, the nature of superficial cover, the dislocation to groundwater, and the extent of backfilled opencast workings. Around 20 named coal seams have been extensively mined, and in addition several other thin coals have been worked at or near outcrop. Near-surface coal seams with little or no drift cover have been worked since the 13th century, and more extensive and deeper workings began in the 17th and 18th centuries. For the vast majority of these early workings little or no adequate information exists, and few plans have been preserved which adequately record the extent or depth of mining. Even up to the early years of this century, many mine plans did not provide adequate levels or suitable geological data. Although workings in more modern times have been comprehensively surveyed and documented, it is vital that, where possible, original documentation and archive material is consulted, even though such data may have been abstracted on to compilation plans by the Coal Authority (and its predecessors British Coal and the NCB). It is also essential to remember that ancient, unrecorded, workings and shafts may occur anywhere in the district where coal is present near the surface (Dearman et al., 1977b). Thus, although the mine shafts plotted on maps prepared by both the British Geological Survey and the Coal Authority are as comprehensive as possible, they cannot be regarded as definitive and the presence of unrecorded shafts cannot be ruled out. To illustrate the problem, at least 140 pits were recorded in the mid-18th century as being present in the Denton area of Newcastle alone.

Laminated silts and clays, part of the Drift sequence (Chapter 6), have a high compressibility, and are generally weak below the water table or when waterlogged. Under vertical load they are prone to strong compression and ductile flow. They tend to be unstable, even on gentle slopes, and disturbance can result in landslipping. In excavations, heave is a common feature, and excavations and tunnels cut in such deposits require close support. Heavy structures require pile foundations through the laminated clay into underlying more competent material. Thus, the presence of these deposits can prove extremely hazardous if unrecognised, and any locality where they are thought or suspected to be present requires particularly detailed site investigation. Such deposits are widely present in some areas, more particularly in the buried valleys.

Other potential constraints which require investigation include the nature and form of the rockhead surface, the presence and infill of buried valleys, the possibility of landslip, and the presence, extent and nature of made ground. Made ground infilling old quarries, pits, and former stream valleys can prove particularly hazardous where adequate documentation does not exist and the nature of the fill is unknown. As an example, Dearman et al. (1977a) described the infilling of Pandon Dene (p.84) in Newcastle during the last century, and reviewed the many and varied sources of data which it was necessary to peruse in order to obtain the required data. Waste tipping into landfill sites generally is now carried out under carefully controlled conditions, but the infiltration of toxic substances into permeable strata, and also the presence of gas-generating materials, are factors which require serious consideration in the location of such sites. The geotechnical properties of the waste materials likely to be encountered is a further factor which should influence all studies for engineering developments.

In summary, the necessity for adequate site investigation is paramount. Although published documents and unpublished British Geological Survey and the Coal Authority archive material will serve as a basis for site investigation, the information in them cannot in any way pre-empt the necessity for a site-specific study. This must particularly be the case in a district such as this, where man's hand has had so great an influence on the environment, both underground and at the surface.

REFERENCES

Many of the references listed below are held in the library of the British Geological Survey at Keyworth, Nottingham. Copies of the references can be purchased subject to the current copyright legislation.

ALLEN, P, and ROSE, J. 1986. A glacial meltwater drainage system between Whittonstall and Ebchester, Northumberland. 69–88 in *Quaternary river landforms and sediments in the northern Pennines*. MACKLIN, M G, and ROSE, J (editors). (British Geomorphological Research Group and Quaternary Research Association.)

ANDERSON, W. 1940. Buried-valleys and late-glacial drainage systems in northwest Durham. *Proceedings of the Geologists' Association*, Vol. 51, 274–281.

ANDREW, E H, and LEE, M K. 1977. Geophysical surveys for buried valleys in North East England. II. Newcastle upon Tyne. *Report of the Applied Geophysics Unit, Institute of Geological Sciences*, 37/2.

ANON. 1858. History and progress of the Consett Iron Works. *The Newcastle Chronicle.* June 1858.

ANON. 1954. Consett Iron Company Limited; a technical survey. History of the Company. *Iron and Coal Trades Review*, Vol. 41, 9–16.

ANON. 1984a. *Tyne and Wear County Minerals Local Plan: Background to the proposals.* Tyne and Wear County Council.

ANON. 1984b. *Tyne and Wear County Minerals Local Plan: Draft proposals.* Tyne and Wear County Council.

ARMSTRONG, G, and PRICE, R H. 1954. The Coal Measures of north-east Durham. *Transactions of the Institution of Mining Engineers*, Vol. 113, 973–997 and Vol. 114, 83–86 and 111–114.

ARTHURTON, R S, and WADGE, A J. 1981. Geology of the country around Penrith. *Memoir of the Geological Survey of Great Britain*, Sheet 24 (England and Wales).

ATKINSON, F. 1979. *The Great Northern Coalfield, 1700–1900* (3rd edition). (Newcastle upon Tyne: Frank Graham.)

ATKINSON, F. 1989. *Victorian Britain: the North East.* (Newton Abbot: David and Charles.)

AYERS, W B, TISDALE, R M, LITZINGER, L A, and STEIDL, P F. 1993. Coalbed methane potential of Carboniferous strata in Great Britain. 1–14 in *Proceedings of the 1993 International Coalbed Methane Symposium*, Birmingham, Alabama. (Tuscaloosa, Albama: University of Alabama.)

BEAMISH, D. 1986. Deep crustal geoelectric structure beneath the Northumberland Basin. *Geophysical Journal of the Royal Astronomical Society*, Vol. 84, 619–640.

BEAMISH, D, and SMYTH, D K. 1986. Geophysical images of the deep crust: the Iapetus Suture. *Journal of the Geological Society of London*, Vol. 143, 489–497.

BEAUMONT, P. 1968. A history of glacial research in northern England from 1860 to the present day. *University of Durham Occasional Paper*, No. 9, 21pp.

BEAUMONT, P. 1970. Geomorphology. 25–45 in *Durham County and City with Teesside.* DEWDNEY, J C (editor). (Newcastle upon Tyne: Hindson Reid Jordison, British Association for the Advancement of Science, Durham.)

BOLTON, E. 1926. Fossil flora of the Northumberland and Durham Coalfield. *Transactions of the Natural History Society of Northumberland*, Vol. 6 (new series), 167–181.

BORINGS and SINKINGS. 1878–1910. *An account of the strata of Northumberland and Durham as proved by borings and sinkings.* 7 vols in 4 books. (Newcastle upon Tyne: Council of the North of England Institute of Mining and Mechanical Engineers.)

BOTT, M H P. 1967. Geophysical investigations of the northern Pennine basement rocks. *Proceedings of the Yorkshire Geological Society*, Vol. 36, 139–168.

BOTT, M H P, and MASSON SMITH, D. 1957. The geological interpretation of a gravity survey of the Alston Block and Durham Coalfield. *Quarterly Journal of the Geological Society of London*, Vol. 113, 93–117.

BOULTON, G S. 1972. Modern Arctic glaciers as depositional models for former ice sheets. *Journal of the Geological Society of London*, Vol. 128, 361–393.

BOULTON, G S, SMITH, G D, JONES, A S, and NEWSOME, J. 1985. Glacial geology and glaciology of the last mid-latitude ice sheets. *Journal of the Geological Society of London*, Vol. 142, 447–474.

BROWN, M W. 1888. A further attempt for the correlation of the coal seams of the Carboniferous Formation of the north of England, with some notes upon the probable duration of the coalfield. *Transactions of the North of England Institute of Mining and Mechanical Engineers*, Vol. 37, 3–25.

BUDDLE, J. 1831a. Synopsis of the several seams of coal in the Newcastle district. *Transactions of the Natural History Society of Northumberland*, Vol. 1, 215–224.

BUDDLE, J. 1831b. Reference to the sections of the strata in the Newcastle coalfield. *Transactions of the Natural History Society of Northumberland*, Vol. 1, 225–240.

BURGESS, I C, and HOLLIDAY, D W. 1979. Geology of the country around Brough-under-Stainmore. *Memoir of the Geological Survey of Great Britain*, Sheet 31 (England and Wales).

BURSTON, U T, and COATS, D J. 1975. Water resources in Northumberland with particular reference to the Kielder Water Scheme. *Journal of the Institution of Water Engineers*, Vol. 79, 226–251.

CALVER, M A. 1968a. Distribution of Westphalian marine faunas in northern England and elsewhere. *Proceedings of the Yorkshire Geological Society*, Vol. 37, 1–72.

CALVER, M A. 1968b. Coal Measures inverterbrate faunas. 144–177 in *Coal and coal-bearing strata.* MURCHISON, D G, and WESTOLL, T S (editors). (Edinburgh: Oliver and Boyd.)

CALVER, M A. 1969. Westphalian of Britain. *Compte Rendu 6e Congrès International Stratigraphie Géologie Carbonifère, Sheffield 1967*, Vol. 1, 233–254.

CARRUTHERS, R G. 1953. *Glacial drifts and the undermelt theory.* (Newcastle upon Tyne: Harold Hill & Son Ltd.)

CARRUTHERS, R G, and ANDERSON, W. 1943. Some refractory materials in north-eastern England. *Wartime Pamphlet of the Geological Survey of Great Britain*, No. 31.

CARRUTHERS, R G, DUNHAM, K C, HEDLEY, W P, HICKLING, G, HOLMES, A, HOPKINS, W, MOCKLER, G S, RAISTRICK, A, TOMKEIEFF, S I, and TRECHMAN, C T. 1931. Contributions to the geology of Northumberland and Durham. *Proceedings of the Geologists' Association*, Vol. 42, 217–296.

CARTER, P G, and MILLS, D A C. 1976. Engineering geology investigations for the Kielder tunnels. *Quarterly Journal of Engineering Geology*, Vol. 9, 125–141.

CHADWICK, R A, and HOLLIDAY, D W. 1991. Deep crustal structure and Carboniferous basin development within the Iapetus Convergence Zone, northern England. *Journal of the Geological Society of London*, Vol. 148, 41–53.

CHADWICK, R A, EVANS, D J, and HOLLIDAY, D W. 1993b. The Maryport Fault: the post-Caledonian tectonic history of southern Britain in microcosm. *Journal of the Geological Society of London*, Vol. 150, 247–250.

CHADWICK, R A, HOLLIDAY, D W, HOLLOWAY, S, and HULBERT, A G. 1993a. The evolution and hydrocarbon potential of the Northumberland–Solway Basin. 717–726 in *Geology of northwest Europe: Proceedings of the 4th Conference*. PARKER, J R (editor). (Geological Society of London.)

CHADWICK, R A, HOLLIDAY, D W, HOLLOWAY, S, and HULBERT, A G. 1995. The structure and evolution of the Northumberland-Solway Basin and adjacent areas. *Subsurface Memoir of the Geological Survey of Great Britain.*

CHAPMAN, A J, RICKARDS, R B, and GRAYSON, R F. 1993. The Carboniferous dendroid graptolites of Britain and Ireland. *Proceedings of the Yorkshire Geological Society*, Vol. 49, 295–319.

CLAPPERTON, C. 1970. The evidence for a Cheviot ice cap. *Transactions of the Institute of British Geographers*, Vol. 50, 115–127.

COATS, D J, and RUFFLE, N J. 1982. The Kielder Water Scheme. *Proceedings of the Institution of Civil Engineers*, Vol. 72, 135–147.

COATS, D J, BERRY, N S M, and BANKS, D J. 1982. The Kielder Transfer Works. *Proceedings of the Institution of Civil Engineers*, Vol. 72, 177–208.

COATS, D J, CARTER, P G, and SMITH, I M. 1977. Inclined drilling for the Kielder tunnels. *Quarterly Journal of Engineering Geology*, Vol. 10, 195–205.

COLLIER, R E L. 1989. Tectonic evolution of the Northumberland Basin; the effects of renewed extension upon an inverted extensional basin. *Journal of the Geological Society of London*, Vol. 146, 981–989.

COOPER, A H, MILLWARD, D, JOHNSON, E J, and SOPER, N J. 1993. The early Palaeozoic evolution of northwest England. *Geological Magazine,* Vol. 130, 711–724.

CORNWELL, J D, and JOHNSON, C E. 1976. Geophysical investigations of buried river valleys. I. The upper Derwent valley. *Report of the Applied Geophysics Unit, Institute of Geological Sciences*, 37/2.

COX, F C. 1983. *Geological notes and local details for 1:10 000 sheets NZ 25 NW, NE, SW and SE (Kibblesworth, Birtley, Craghead and Chester-le-Street).* (Newcastle upon Tyne: Institute of Geological Sciences.)

CREANY, S. 1980. Petrographic texture and vitrinite reflectance variation on the Alston Block, north-east England. *Proceedings of the Yorkshire Geological Society*, Vol. 42, 553–580.

CREEDY, D P. 1991. An introduction to geological aspects of methane occurrence and control in British deep coal mines. *Quarterly Journal of Engineering Geology*. Vol. 24, 209–220.

CUMING, J S. 1970. Rockhead relief of south-east Northumberland and lower Tyne valley. Unpublished PhD thesis, University of Newcastle upon Tyne.

DAVIES, T P, CARTER, P G, MILLS, D A C, and WEST, G. 1981. Kielder tunnels-predicted and actual geology. *Report of the Transport and Road Research Laboratory*, No. 676, 1–46.

DAY, J B W. 1970. Geology of the country around Bewcastle. *Memoir of the Geological Survey of Great Britain*, Sheet 12 (England and Wales).

DEARMAN, W R, MONEY, M S, COFFEY, J R, SCOTT, P, and WHEELER, M. 1977a. Engineering geological mapping of the Tyne and Wear conurbation, north-east England. *Quarterly Journal of Engineering Geology*, Vol. 10, 145–168.

DEARMAN, W R, BAYNES, F J, and PEARSON, R. 1977b. Geophysical detection of disused mineshafts in the Newcastle upon Tyne area, north-east England. *Quarterly Journal of Engineering Geology*, Vol. 10, 257–269.

DOWNING, R A, and GRAY, D A (editors). 1986. *Geothermal energy — the potential in the United Kingdom.* (London: HMSO for British Geological Survey.)

DUNHAM, A C, and STRASSER-KING, V E H. 1982. Late Carboniferous intrusions of northern Britain. 277–283 in *Igneous rocks of the British Isles.* SUTHERLAND, D S (editor). (New York: John Wiley & Son.)

DUNHAM, K C. 1948. Geology of the Northern Pennine Orefield, Vol. 1. Tyne to Stainmore (1st edition). *Memoir of the Geological Survey of Great Britain.*

DUNHAM, K C. 1950. Lower Carboniferous sedimentation in the northern Pennines (England). *Report of the 18th Session of the International Geological Congress, Great Britain, 1948*, Vol. 4, 46–63.

DUNHAM, K C. 1990. Geology of the Northern Pennine Orefield, Vol. 1. Tyne to Stainmore (2nd edition). *Memoir of the Geological Survey of Great Britain .*

DUNHAM K C, DUNHAM, A C, HODGE, B L, and JOHNSON, G A L. 1965. Granite beneath Viséan sediments with mineralisation at Rookhope, northern Pennines. *Quarterly Journal of the Geological Society of London*, Vol. 121, 383–417.

DUNHAM, K C, and WILSON, A A. 1985. Geology of the Northern Pennine Orefield, Vol. 2. Stainmore to Craven. *Memoir of the Geological Survey of Great Britain.*

DWERRYHOUSE, A R. 1902. Glaciation of Teesdale, Weardale, the Tyne valley and their tributaries. *Quarterly Journal of the Geological Society of London*, Vol. 58, 572–608.

EDMONDS, E A, and DAVIES, J R. 1982. *Geological notes and local details for 1:10 000 sheets NZ 05 NW, NE, SW and SE (Edmondbyers and Hedley on the Hill).* (Keyworth: Institute of Geological Sciences.)

EHLERS, J, and WINGFIELD, R T R. 1991. The extension of the Late Weichselian/Late Devensian ice sheets in the North Sea Basin. *Journal of Quaternary Science*, Vol. 6, 313–326.

ELLIOTT, T. 1974. Abandonment facies of high-constructive lobate deltas, with an example from the Yoredale Series. *Proceedings of the Geologists' Association*, Vol. 85, 359–365.

ELLIOTT, T. 1975. The sedimentary history of a delta lobe from a Yoredale (Carboniferous) cyclothem. *Proceedings of the Yorkshire Geological Society*, Vol. 40, 505–536.

ELLIOTT, T. 1976. Sedimentary sequences from the Upper Limestone Group of Northumberland. *Scottish Journal of Geology*, Vol. 12, 115–124.

EVANS, C J, KIMBELL, G S, and ROLLIN, K E. 1988. *Hot dry rock potential in urban areas. Investigation of the geothermal potential of the UK.* (Keyworth, Nottingham: British Geological Survey).

Eyles, V A, and Sladen, J A. 1981. Stratigraphy and geotechnical properties of weathered lodgement till in Northumberland, England. *Quarterly Journal of Engineering Geology,* Vol. 14, 129–141.

Fairbairn, R A. 1980. The Great Limestone (Namurian) of south Northumberland. *Proceedings of the Yorkshire Geological Society,* Vol. 43, 159–167.

Fielding, C R. 1982. Sedimentology and stratigraphy of the Durham Coal Measures, and comparisons with other British coalfields. Unpublished PhD thesis, University of Durham.

Fielding, C R. 1984a. A coal depositional model for the Durham Coal Measures of NE England. *Journal of the Geological Society of London,* Vol. 141, 919–931.

Fielding, C R. 1984b. Upper delta plain, lacustrine and fluvio-lacustrine facies from the Westphalian of the Durham Coalfield, NE England. *Sedimentology,* Vol. 31, 547–567.

Fielding, C R. 1984c. 'S' or 'Z' shaped coal seam splits in the Coal Measures of County Durham. *Proceedings of the Yorkshire Geological Society,* Vol. 45, 85–89.

Fielding, C R. 1986. Fluvial channel and overbank deposits from the Westphalian of the Durham Coalfield, NE England. *Sedimentology,* Vol. 33, 119–140.

Fielding, C R, and Johnson, G A L. 1986. Sedimentary structures associated with extensional fault movement from the Westphalian of NE England. 511–516 *in* Continental extensional tectonics. Coward, M P, and Dewey, F J (editors). *Special Publication of the Geological Society of London,* No. 28.

Forster, T E. 1930. 25–56 *in* Collieries and the coal trade. Dodds, M H. *A history of Northumberland, Vol. 13. The parishes of Heddon on the Wall, Newburn, Long Benton and Wallsend; the Chapelries of Gosforth and Cramlington: the townships of Benwell, Elswick, Heaton, Byker, Fenham and Jesmond.* (Newcastle upon Tyne and London.)

Fortey, N J, Roberts, B, and Hirons, S R. 1993. Relationship between metamorphism and structure in the Skiddaw Group, English Lake District. *Geological Magazine,* Vol. 130, 631–638.

Fortey, R A, Owens, R M, and Rushton, A W A. 1989. The palaeogeographic position of the Lake District in the early Ordovician. *Geological Magazine,* Vol. 126, 9–17.

Fowler, A. 1936. The geology of the country around Rothbury, Amble and Ashington. *Memoir of the Geological Survey of Great Britain,* Sheets 9 and 10 (England and Wales).

Francis, E A. 1970. Quaternary. 134–152 *in* The geology of Durham county. Hickling, G (editor). *Transactions of the Natural History Society of Northumberland, Durham and Newcastle upon Tyne,* Vol. 41.

Francis, E H. 1982. Magma and sediment I. Emplacement mechanism of late Carboniferous tholeiite sills in northern Britain. *Journal of the Geological Society of London,* Vol. 139, 1–20.

Fraser, A J, Nash, D F, Steele, R P, and Ebdon, C C. 1990. A regional assessment of the intra-Carboniferous play of northern England. 417–440 *in* Classic petroleum provinces. Brooks, J (editor). *Special Publication of the Geological Society of London,* Vol. 50.

Frost, D V, and Holliday, D W. 1980. Geology of the country around Bellingham. *Memoir of the Geological Survey of Great Britain,* Sheet 13 (England and Wales).

Galloway, R L. 1882. *A history of coal mining in Great Britain.* (London.)

Gebski, J S, Wheildon, J, and Thomas-Betts, A. 1987. *Investigations of the UK heat flow field (1984–1987). Investigation of the geothermal potential of the UK.* (Keyworth, Nottingham: British Geological Survey.)

Giles, J R A. 1981. The sand and gravel resources of the country around Blaydon, Tyne and Wear. *Mineral Assessment Report, Institute of Geological Sciences,* No. 74.

Green, R. 1954. The Upper Limestone Group, 'Millstone Grit', and the Lower Coal Measures of the lower Tyne valley. Unpublished PhD thesis, University of Durham.

Gripp, K. 1975. 100 Jahre Untersuchungen über das Geschehen am Rande des nordeuropäischen Inlandeises. *Eiszeitalter und Gegenwart,* Vol. 26, 31–73.

Hair, T H. 1844. *Sketches of the coal mines in Northumberland and Durham.* (Camden Town: T H Hair.) (Facsimile Reprint, 1969, for Frank Graham by J and P Bealls Ltd. Newcastle upon Tyne.)

Haszeldine, R S. 1983a. Descending tabular cross-bed sets and bounding surfaces from a fluvial channel in the Upper Carboniferous coalfield of north-east England. *Special Publication, International Association of Sedimentologists,* Vol. 6, 449–456.

Haszeldine, R S. 1983b. Fluvial bars reconstructed from a deep, straight channel, Upper Carboniferous coalfield of north-east England. *Journal of Sedimentary Petrology,* Vol. 53, 1233–1247.

Haszeldine, R S. 1984a. Muddy deltas in freshwater lakes, and tectonism in the Upper Carboniferous coalfield of north-east England. *Sedimentology,* Vol. 31, 811–822.

Haszeldine, R S. 1984b. Carboniferous North Atlantic palaeogeography: stratigraphic evidence for rifting, not megashear or subduction. *Geological Magazine,* Vol. 121, 443–463.

Haszeldine, R S, and Anderton, R. 1980. A braidplain facies model for the Westphalian B Coal Measures of north-east England. *Nature, London,* Vol. 284, 51–53.

Hedley, W P. 1931. The stratigraphy of the Bernician and Millstone Grit of south Northumberland. *Transactions of the Natural History Society of Northumberland, Durham and Newcastle upon Tyne* (New Series), Vol. 7, 179–190.

Hedley, W P, and Waite, S T. 1929. The sequence of the Upper Limestone Group between Corbridge and Belsay. *Proceedings of the University of Durham Philosophical Society,* Vol. 8, 136–152.

Herdman, T. 1909. The glacial phenomena of the Vale of Derwent. *Proceedings of the University of Durham Philosophical Society,* Vol. 3, 109–119.

Hickling, G (editor). 1970. The geology of Durham County. *Transactions of the Natural History Society of Northumberland, Durham and Newcastle upon Tyne,* Vol. 41.

Hickling, H G A, and Robertson, T. 1949. Geology. 10–30 in *Scientific survey of north-eastern England.* (Newcastle upon Tyne: British Association for the Advancement of Science.)

Hodge, B L, and Dunham, K C. 1991. Clastics, coals, palaeodistributaries and mineralisation in the Namurian Great Cyclothem, Northern Pennines and Northumberland Trough. *Proceedings of the Yorkshire Geological Society,* Vol. 48, 323–337.

Holland, J G. 1967. Rapid analysis of the Weardale Granite. *Proceedings of the Yorkshire Geological Society,* Vol. 36, 91–113.

Holland, J G. 1980. The Weardale Granite. 61–62 in *The geology of north east England.* Robson, D A (editor). (Newcastle upon Tyne: Natural History Society of Northumbria.)

Holland, J G, and Lambert R St J. 1970. Weardale Granite. 108–118 *in* Geology of Durham County. Hickling, G (editor).

Transactions of the Natural History Society Northumberland, Durham and Newcastle upon Tyne, Vol. 41.

Holliday, D W. 1986. Devonian and Carboniferous basins. 84–110 in *Geothermal energy — the potential in the United Kingdom.* Downing, R A, and Gray, D A (editors). (London: HMSO for British Geological Survey.)

Holliday, D W. 1993. Mesozoic cover over northern England: interpretation of apatite fission track data. *Journal of the Geological Society of London,* Vol. 150, 657–660.

Holliday, D W, and Pattison, J. 1990. Carboniferous geology of Corbridge and Prudhoe: Geological notes and local details for sheets NZ06 and the eastern parts of NY96NE and SE (Northumberland). *British Geological Survey Technical Report,* No. WA/90/13.

Holmes, A, and Harwood, H F. 1929. The tholeiite dykes of the north of England. *Mineralogical Magazine,* Vol. 22, 1–52.

Hopkins, W. 1927. Further modifications of the correlation of the coal-seams of the Northumberland and Durham Coalfield. *Transactions of the Institution of Mining Engineers,* Vol. 74, 221–241.

Hopkins, W. 1928. The distribution of mussel bands of the Northumberland and Durham Coalfield. *Proceedings of the University of Durham Philosophical Society,* Vol. 8, 1–14.

Hopkins, W. 1929. The distribution and sequence of the non-marine lamellibranchs in the Coal Measures of Northumberland and Durham. *Transactions of the Institution of Mining Engineers,* Vol. 78, 126–144.

Hopkins, W. 1930. A revision of the Upper Carboniferous non-marine lamellibranchs of Northumberland and Durham and a record of their sequence. *Transactions of the Institution of Mining Engineers,* Vol. 80, 101–110.

Hopkins, W. 1933. Impoverished areas of the Maudlin and Busty seams in the Durham Coalfield. *Transactions of the Institution of Mining Engineers,* Vol. 85, 212–222.

Hopkins, W. 1934. *Lingula* horizons in the Coal Measures of Northumberland and Durham. *Geological Magazine,* Vol. 71, 183–189.

Howse, R. 1864. On the glaciation of the counties of Durham and Northumberland. *Transactions of the North of England Institute of Mining and Mechanical Engineers,* Vol. 13, 169–185.

Howse, R. 1890a. Contributions towards a catalogue of the flora of the Carboniferous System of Northumberland and Durham. Part I — Fossil plants of the Hutton collection. *Transactions of the Natural History Society of Northumberland,* Vol. 10, 19–151.

Howse, R. 1890b. Catalogue of the local fossils in the museum of the Natural History Society. *Transactions of the Natural History Society of Northumberland,* Vol. 10, 227–288.

Hughes, R A, Cooper, A H, and Stone, P. 1993. Structural evolution of the Skiddaw Group (English Lake District) on the northern margin of eastern Avalonia. *Geological Magazine,* Vol. 130, 621–629.

Hull, J H. 1968. The Namurian stages of north-eastern England. *Proceedings of the Yorkshire Geological Society,* Vol. 36, 297–308.

Hurst, T G. 1860. On some peculiarities of the Tyne Low Main Seam. *Transactions of the North of England Institution of Mining Engineers,* Vol. 8, 23–31.

Jackson, I, Lawrence, D J D, and Frost, D V. 1985. *Geological notes and local details for Sheet NZ 27. Cramlington, Killingworth and Wide Open, SE Northumberland.* (Newcastle upon Tyne: British Geological Survey.)

Johnson, G A L. 1959. The Carboniferous stratigraphy of the Roman Wall district in western Northumberland. *Proceedings of the Yorkshire Geological Society,* Vol. 32, 83–130.

Johnson, G A L. 1967. Basement control of Carboniferous sedimentation in northern England. *Proceedings of the Yorkshire Geological Society,* Vol. 36, 175–194.

Johnson, G A L. 1970. Carboniferous. 23–42 *in* The geology of Durham County. Hickling, G (editor). *Transactions of the Natural History Society of Northumberland, Durham and Newcastle upon Tyne,* Vol. 41.

Johnson, G A L. 1980. Carboniferous Dinantian and Namurian rocks. 11–23 in *The geology of north east England.* Robson, D A (editor). (Newcastle upon Tyne: Natural History Society of Northumbria.)

Johnson, G A L, Hodge, B L, and Fairbairn, R A. 1962. The base of the Namurian and of the Millstone Grit in north-eastern England. *Proceedings of the Yorkshire Geological Society,* Vol. 33, 341–362.

Jones, J M, and Creaney, S. 1977. Optical character of thermally metamorphosed coals of northern England. *Journal of Microscopy,* Vol. 109, 105–118.

Jones, J M, and Magraw, D. 1980. Carboniferous Westphalian (Coal Measures) rocks. 22–36 in *The geology of north east England.* Robson, D A (editor). (Newcastle upon Tyne: Natural History Society of Northumbria.)

Jones, J M, Magraw, D, Robson, D A, and Smith, FW. 1980. Movements at the end of the Carboniferous period. 79–85 in *The geology of north east England.* Robson, D A (editor). (Newcastle upon Tyne: Natural History Society of Northumbria.)

Kidston, R. 1922. List of fossil plants from the Upper Carboniferous rocks of the Northumberland and Durham Coalfield and their bearing on the age of the coalfield. *Summary of Progress of the Geological Survey of Great Britain for 1921,* 129–145.

Kimbell, G S, Chadwick, R A, Holliday, D W, and Werngren, O C. 1989. The structure and evolution of the Northumberland Trough from new seismic reflection data and its bearing on modes of continental extension. *Journal of the Geological Society of London,* Vol. 146, 775–787.

Kirkby, J W. 1860. On the occurrence of *Lingula credneri* Geinitz in the Coal Measures of Durham, and on the claim of the Permian rocks to be entitled a system. *Quarterly Journal of the Geological Society of London,* Vol. 16, 412–421.

Kirsopp, J. 1907. *Map and plotted vertical sections of strata of the Northumberland and Durham Coalfield.* (Newcastle upon Tyne.)

Land, D H. 1974. Geology of the Tynemouth district. *Memoir of the Geological Survey of Great Britain,* Sheet 15 (England and Wales).

Lawrence, D J D, and Jackson, I. 1986. Geology of the Ponteland–Morpeth district. *Research Report of the British Geological Survey.*

Lawrence, D J D, and Jackson, I. 1990. Geology and land-use planning: Morpeth–Bedlington–Ashington. Part 2: Geology. *British Geological Survey Technical Report,* WA/90/19.

Lebour, G A. 1875. On the Little Limestone and its accompanying coal in south Northumberland. *Transactions of the North of England Institute of Mining and Mechanical Engineers,* Vol. 24, 73–83.

Lebour, G A. 1876. On the larger divisions of the Carboniferous System in Northumberland. *Transactions of the North of England Institute of Mining and Mechanical Engineers,* Vol. 25, 225–237.

LEBOUR, G A. 1878. *Outlines of the geology of Northumberland and Durham.* (Newcastle upon Tyne: Lambert and Company Ltd.)

LEBOUR, G A. 1885. Brief notes on the geology of Corbridge, Northumberland. *Proceedings of the Berwickshire Naturalists' Club,* Vol. 10, 121–127.

LEBOUR, G A. 1886. *Outlines of the geology of Northumberland and Durham* (2nd edition). (Newcastle upon Tyne: Lambert and Company Ltd.)

LEBOUR, G A. 1906. Note on a small boulder found in the later glacial deposits in a 'wash out' near Low Spen in the Derwent valley. *Proceedings of the University of Durham Philosophical Society,* Vol. 2, 81–82.

LEEDER, M R. 1982. Upper Palaeozoic basins of the British Isles — Caledonide inheritance versus Hercynian plate margin processes. *Journal of the Geological Society of London,* Vol. 139, 479–491.

LEEDER, M R, FAIRHEAD, D, LEE, A, STUART, G, CLEMMEY, H, AL-HADDEH, B, and GREEN, C. 1989. Sedimentary and tectonic evolution of the Northumberland Basin. 207–223 in *The role of tectonics in Devonian and Carboniferous sedimentation in the British Isles.* ARTHURTON, R S, GUTTERIDGE, P, and NOLAN, S C (editors). (Leeds: Yorkshire Geological Society.)

LEEDER, M R, and STRUDWICK, A. 1987. Delta-marine interactions: a discussion of sedimentary models for Yoredale-type cyclicity in the Dinantian of northern England. 115–130 in *European Dinantian environments.* MILLER, J, ADAMS, A E, and WRIGHT, V P (editors). (Chichester: John Wiley and Sons.)

LOVELL, J H. 1981. The sand and gravel resources of the country around Hexham, Northumberland. *Mineral Assessment Report, Institute of Geological Sciences,* No. 65.

LUMSDEN, G I, TULLOCH, W, HOWELLS, M F, and DAVIES, A. 1967. The geology of the neighbourhood of Langholm. *Memoir of the Geological Survey of Great Britain,* Sheet 10 (Scotland).

LUNN, A G. 1980. Quaternary. 48–60 in *The geology of north east England.* ROBSON, D A (editor). (Newcastle upon Tyne: Natural History Society of Northumbria.)

MAGRAW, D, CLARKE, A M, and SMITH, D B. 1963. The stratigraphy and structure of part of the south-east Durham coalfield. *Proceedings of the Yorkshire Geological Society,* Vol. 34, 153–208.

MANDERS, F W D. 1973. *A history of Gateshead.* (Gateshead Corporation.)

MARSHALL, C E. 1936. The alteration of coal seams by the intrusion of some of the igneous dykes in the Northumberland and Durham Coal Field. *Transactions of the Institute of Mining Engineers,* Vol. 91, 235–260.

MAYNARD, J R, and LEEDER, M R. 1992. On the periodicity and magnitude of Late Carboniferous glacio-eustatic sea-level changes. *Journal of the Geological Society of London,* Vol. 149, 303–311.

MCCORD, N. 1979. *North-east England: the region's development 1760–1960.* (Batsford.)

MCCORD, N, and ROWE, D C. 1977. Industrialization and urban growth in north-east England. *International Review of Social History,* Vol. 22.

MCKERROW, W S, and SOPER, N J. 1989. The Iapetus Suture in the British Isles. *Geological Magazine,* Vol. 126, 1–8.

MERRICK, E. 1910. On the superficial deposits around Newcastle upon Tyne. *Proceedings of the University of Durham Philosophical Society,* Vol. 2, 141–152.

MERRICK, E. 1915. On the formation of the River Tyne drainage area. *Geological Magazine,* Vol. 52, 293–304 and 353–60.

MIDDLEBROOK, S. 1950. *Newcastle upon Tyne. Its growth and achievement.* (Newcastle upon Tyne: Newcastle Chronicle and Journal Ltd.)

MILLER, E. 1968. *Eye witness: The North East in the early C19.* (Sunderland College of Education: Harold Hill.)

MILLS, A B. 1974. Glacial rafting of coal in Durham. *Opencast Geologist,* Vol. 1, 16.

MILLS, D A C. 1982. *Geological notes and local details for NZ 15 NW, NE, SW and SE (Chopwell, Rowlands Gill, Consett and Stanley).* (Keyworth: Institute of Geological Sciences.)

MILLS, D A C, and HULL, J H. 1968. The Geological Survey Borehole at Woodland, Co. Durham. *Bulletin of the Geological Survey of Great Britain,* No. 28, 1–37.

MILLS, D A C, and HULL, J H. 1976. Geology of the country around Barnard Castle. *Memoir of the Geological Survey of Great Britain,* Sheet 32 (England and Wales).

MURTON, C J. 1892. Geology of the coal-field of Northumberland and Durham. *Transactions of the Institution of Mining Engineers,* Vol. 3, 620–631.

NATIONAL COAL BOARD. 1965. *Coal survey seam maps.* (National Coal Board Scientific Department.)

NEF, J U. 1932. *The rise of the British coal industry.* (London.)

NORTH OF ENGLAND INSTITUTE OF MINING AND MECHANICAL ENGINEERS, COUNCIL OF 1878–1910. *An account of the strata of Northumberland and Durham as proved by borings and sinkings.* 7 Vols. in 4 books. (Newcastle upon Tyne.)

OWENS, B. 1972. Palynological report on the IGS Throckley Borehole. *Report of the Palaeontological Department, Institute of Geological Sciences,* No. PDL/72/41 (unpublished).

OWENS, B. 1978a. Palynological report on samples from Aydon Castle. *Report of the Palaeontological Department, Institute of Geological Sciences,* No. PDL/78/80 (unpublished).

OWENS, B. 1978b. Palynological report on samples from Deadridge Quarry, Thornbrough Farm and Low Shilford. *Report of the Palaeontological Department, Institute of Geological Sciences,* No. PDL/ 78/81 (unpublished).

OWENS, B, and BURGESS, I C. 1965. The stratigraphy and palynology of the Upper Carboniferous outlier of Stainmore, Westmorland. *Bulletin of the Geological Survey of Great Britain,* No. 23, 17–44.

PATTISON, J. 1980. Namurian macrofossils from the 1:50 000 sheet 20 (Newcastle upon Tyne). *Report of the Palaeontological Department, Institute of Geological Sciences,* No. PDL/80/374 (unpublished).

PEEL, R F. 1949. A study of two Northumbrian spillways. *Transactions of the Institute of British Geographers,* Vol. 15, 73–89.

PEEL, R F. 1956. The profiles of glacial drainage channels. *Geographical Journal,* Vol. 122, 483–487.

PERCIVAL, C J. 1983. The Firestone Sill Ganister, Namurian, northern England — the A2 horizon of a podzolic palaeosol. *Sedimentary Geology,* Vol. 36, 41–49.

PERCIVAL, C J. 1986. Palaeosols containing an albic horizon: examples from the Upper Carboniferous of northern England. 87–111 in *Palaeosols, their recognition and interpretation.* WRIGHT, V P (editor). (Oxford: Blackwells.)

POLLARD, J E. 1966. A non-marine ostracod fauna from the Coal Measures of Durham and Northumberland. *Palaeontology,* Vol. 9, 667–697.

Pollard, J E. 1969. Three ostracod-mussel bands in the Coal Measures (Westphalian) of Northumberland and Durham. *Proceedings of the Yorkshire Geological Society*, Vol. 37, 239–276.

Raistrick, A. 1931. Glaciation. 281–291 *in* The geology of Northumberland and Durham. Carruthers, R G, et al. *Proceedings of the Geologists' Association*, Vol. 42.

Raistrick, A, and Marshall, C E. 1939. *The nature and origin of coal and coal seams.* (London: English Universities Press.)

Ramsbottom, W H C, Calver, M A, Eagar, R M C, Hodson, F, Holliday, D W, Stubblefield, C J, and Wilson, R B. 1978. A correlation of Silesian rocks in the British Isles. *Special Report of the Geological Society of London*, No. 10.

Randall, B A O. 1980a. The Great Whin Sill and its associated dyke suite. 67–75 in *The geology of north-east England.* Robson, D A (editor). (Newcastle upon Tyne: Natural History Society of Northumbria.)

Randall, B A O. 1980b. The Tertiary dykes. 75–77 in *The geology of north-east England.* Robson, D A (editor). (Newcastle upon Tyne: Natural History Society of Northumbria.)

Read, W A. 1981. Facies breaks in the Scottish Passage Group and their possible correlation with the Mississippian–Pennsylvanian hiatus. *Scottish Journal of Geology*, Vol. 17, 295–300.

Richardson, G. 1965. P.50 in *Summary of Progress of the Geological Survey of Great Britain for 1964.* (London: HMSO.)

Richardson, G. 1966. P.56 in *Summary of Progress of the Geological Survey of Great Britain for 1965.* (London: HMSO.)

Richardson, G. 1983. *Geological notes and local details for 1:10 000 sheets NZ 26 NW, NE, SW and SE (Newcastle upon Tyne and Gateshead).* (Keyworth: Institute of Geological Sciences.)

Richardson, G, and Francis, E H. 1971. Fragmental clayrock (FCR) in coal-bearing sequences in Scotland and north-east England. *Proceedings of the Yorkshire Geological Society*, Vol. 38, 229–260.

Ridd, M F, Walker, D B, and Jones, J M. 1970. A deep borehole at Harton on the margin of the Northumbrian Trough. *Proceedings of the Yorkshire Geological Society*, Vol. 38, 75–103.

Riley, N J. 1982. Micropalaeontology of a Kinderscoutian fauna from Acton Burn. *Report of the Palaeontological Department, Institute of Geological Sciences,* No. PDL 82/327. (unpublished).

Robson, D A (editor). 1980. *The geology of north east England.* (Newcastle upon Tyne: Natural History Society of Northumbria.)

Rogers, F. 1974. *Gateshead: An early Victorian boom town.* (Priory Press.)

Rollin, K E. 1987. *Catalogue of geothermal data for the land area of the United Kingdom. Third revision: April 1987. Investigation of the geothermal potential of the UK.* (Keyworth, Nottingham: British Geological Survey.)

Scott, J, and Colter, V S. 1987. Geological aspects of current onshore Great Britain exploration plays. 95–107 in *Petroleum geology of north west Europe.* Brooks, J, and Glennie, K W (editors). (London: Graham and Trotman.)

Sherwood, J M, and Younger, P L. In press. Modelling groundwater rebound after coalfield closure: an example from County Durham, UK. *Proceedings of the 5th International Minewater Congress, Nottingham, UK.*

Simpson, J B. 1904. The probability of finding workable seams of coal in the Carboniferous Limestone or Bernician Formation, beneath the regular Coal-Measures of Northumberland and Durham, with an account of a recent deep boring in Chopwell Woods, below the Brockwell seam. *Transactions of the Institute of Mining Engineers*, Vol. 24, 549–571.

Sissons, J B. 1958. Sub-glacial stream erosion in southern Northumberland. *Scottish Geographical Magazine*, Vol. 74, 163–174.

Sissons, J B. 1960. Erosion surfaces, cyclic slopes and drainage systems in southern Scotland and northern England. *Transactions of the Institute of British Geographers*, Vol. 28, 23–38.

Sladen, J A. 1979. Weathering and its effect on the geotechnical properties of tills in south-east Northumberland. Unpublished MSc thesis, University of Newcastle upon Tyne.

Smailes, A E. 1935. The development of the Northumberland and Durham Coalfield. *Scottish Geographical Magazine*, Vol. 51, 201–214.

Smith, D B. 1994. Geology of the country around Sunderland. *Memoir of the British Geological Survey*, Sheet 21 (England and Wales).

Smith, D B, and Francis, E A. 1967. Geology of the country between Durham and West Hartlepool. *Memoir of the Geological Survey of Great Britain*, Sheet 27 (England and Wales).

Smith, S. 1910. The faunal succession of the Upper Bernician. *Transactions of the Natural History Society of Northumberland*, New Series, Vol. 3, 591–645.

Smith, S. 1912. *Report of the committee appointed to report upon the Carboniferous Limestone Formation of the north of England with special reference to its coal resources.* (Newcastle upon Tyne: North of England Institution of Mining and Mechanical Engineers.)

Smith, S. 1917. *Aulina rotiformis* gen. et sp. nov., *Phillipsastrea hennahi* (Lonsdale) and *Orionastraea* gen. nov. *Quarterly Journal of the Geological Society of London*, Vol. 72, 280–307.

Smith, S. 1923. Lead and zinc ores of Northumberland and Alston Moor. *Economic Memoir of the Geological Survey of Great Britain*, Vol. 25.

Smith, S, and Yü, C C. 1943. A revision of the coral genus *Aulina* Smith and descriptions of new species from Britain and China. *Quarterly Journal of the Geological Society of London*, Vol. 99, 37–61.

Smith, S A, and Holliday, D W. 1991. The sedimentology of the Middle and Upper Border groups (Viséan) in the Stonehaugh Borehole, Northumberland. *Proceedings of the Yorkshire Geological Society*, Vol. 48, 435–446.

Smythe, J A. 1908. Glacial phenomena of the country between the Tyne and the Wansbeck. *Transactions of the Natural History Society of Northumberland*, Vol. 3, 141–153.

Smythe, J A. 1912. Glacial geology of Northumberland. *Transactions of the Natural History Society of Northumberland*, Vol. 4, 86–116.

Soper, N J, England, R W, Snyder, D B, and Ryan, P D. 1992. The Iapetus suture zone in England, Scotland and eastern Ireland: a reconciliation of geological and deep seismic data. *Journal of the Geological Society, London*, Vol. 149, 697–700.

Soper, N J, Webb, B C, and Woodcock, N H. 1987. Late Caledonian (Acadian) transpression in north-west England: timing, geometry and tectonic significance. *Proceedings of the Yorkshire Geological Society*, Vol. 46, 175–192.

Stone, P, Floyd, J D, Barnes, R P, and Lintern, B C. 1987. A sequential back-arc and foreland basin thrust duplex model for the Southern Uplands of Scotland. *Journal of the Geological Society of London*, Vol. 144, 753–764.

STUBBLEFIELD, C J, and TROTTER, F M. 1957. Divisions of the Coal Measures on Geological Survey maps of England and Wales. *Bulletin of the Geological Survey of Great Britain,* No. 13, 1–5.

TAYLOR, B J, BURGESS, I C, LAND, D H, MILLS, D A C, SMITH, D B, and WARREN, P T. 1971. *British regional geology; northern England* (4th edition). (London: HMSO for Institute of Geological Sciences.)

TEALL, J J H. 1884. Petrological notes on some north of England dykes. *Quarterly Journal of the Geological Society of London,* Vol. 40, 209–247.

THABET, K M A. 1973. Geotechnical properties and sedimentation characteristics of tills in south-east Northumberland. Unpublished PhD thesis (2 vols), University of Newcastle upon Tyne.

TOMKEIEFF, S I. 1953. The tholeiite dyke at Cowgate, Newcastle upon Tyne. *Proceedings of the University of Durham Philosophical Society,* Vol. 11, 91–94.

TROTTER, F M. 1929. The Tertiary uplift and resultant drainage of the Alston Block and adjacent areas. *Proceedings of the Yorkshire Geological Society,* Vol. 21, 161–180.

TROTTER, F M. 1954. Genesis of the high rank coals. *Proceedings of the Yorkshire Geological Society,* Vol. 29, 267–303.

TROTTER, F M, and HOLLINGWORTH, S E. 1927. On the Upper Limestone Group and 'Millstone Grit' of north-east Cumberland. *Summary of Progress of the Geological Survey of Great Britain for 1926,* 98–107.

TROTTER, F M, and HOLLINGWORTH, S E. 1932. The geology of the Brampton district. *Memoir of the Geological Survey of Great Britain,* Sheet 18 (England and Wales).

WHITTAKER, A. 1963. The geology of the Namurian and Lower Coal Measures of the Tyne valley and adjoining areas. Unpublished PhD thesis, University of Newcastle upon Tyne.

WILSON, K S. 1980. Geophysical surveys for buried valleys in north-east England. III. Consett area. *Report of the Applied Geophysics Unit, Institute of Geological Sciences,* 37/3.

WINCH, N J. 1817. Observations on the geology of Northumberland and Durham. *Transactions of the Geological Society of London,* Vol. 4, 1–101.

WINGFIELD, R T R. 1990. The origin of major incisions within the Pleistocene deposits of the North Sea. *Marine Geology,* Vol. 91, 538–548.

WOOD, N, and BOYD, E F. 1864. On a Wash or Drift through a portion of the Coalfield at Durham. *Transactions of the North of England Institution of Mining Engineers,* Vol. 13, 285–287.

WOOLACOTT, D. 1905. The superficial deposits and pre-glacial valleys of the Northumberland and Durham coalfields. *Quarterly Journal of the Geological Society of London,* Vol. 61, 64–95.

WOOLACOTT, D. 1913. The geology of north-east Durham and south-east Northumberland. *Proceedings of the Geologists' Association,* Vol. 24, 87–107.

WOOLACOTT, D. 1921. The interglacial problem and the glacial and post glacial sequence in Northumberland and Durham. *Geological Magazine,* Vol. 58, 21–32 and 60–69.

WOOLACOTT, D. 1923. A boring at Roddymoor Colliery, near Crook, Co. Durham. *Geological Magazine,* Vol. 60, 50–62.

YOUNGER, P L. 1993. Possible environmental impact of the closure of two collieries in County Durham. *Journal of the Institution of Water and Environmental Management,*Vol. 7, 521–531.

YOUNGER, P L. 1994. Minewater pollution: the revenge of Old King Coal. *Geoscientist,* Vol. 4, No. 5, 6–8.

YOUNGER, P L, and SHERWOOD, J M. 1993. The cost of decommissioning a coalfield: potential environmental problems in County Durham. *Mineral Planning,* Vol. 57, 26–29.

YOUNGER, P L, and BRADLEY, K M. In press. Application of geological mineral exploration techniques to the cataloguing of problematic discharges from abandoned mines in north-east England. *Proceedings of the 5th International Minewater Conference, Nottingham, UK.*

APPENDIX 1

List of boreholes and shafts

The following is a selection, by National Grid 1:10 000 scale sheets, of some of the more important and representative borehole and shaft records in this district, with five from just beyond its limits. Their locations are indicated in Figure 30. The list includes only a small proportion of the records for any given National Grid sheet, which in some instances number many hundreds of entries. The complete records of most of the above, and many other records, may be inspected at the Edinburgh Office of the British Geological Survey; copies of the records can also be obtained at a standard tariff. Listings of records held for any given sheet can also be obtained on enquiry.

For each record quoted below, the British Geological Survey permanent record number, location and broad stratigraphical range is given together with starting level (where known), the drift thickness, and total depth. Records marked with an asterisk* are quoted in more detail in Appendix 2.

The column headed 'B & S number.' refers to the appropriate numbered entry in: BORINGS and SINKINGS 1878–1910. *An account of the strata of Northumberland and Durham as proved by borings and sinkings.* 7 volumes in 4 books. (Newcastle upon Tyne: Council of the North of England Institute of Mining and Mechanical Engineers.)

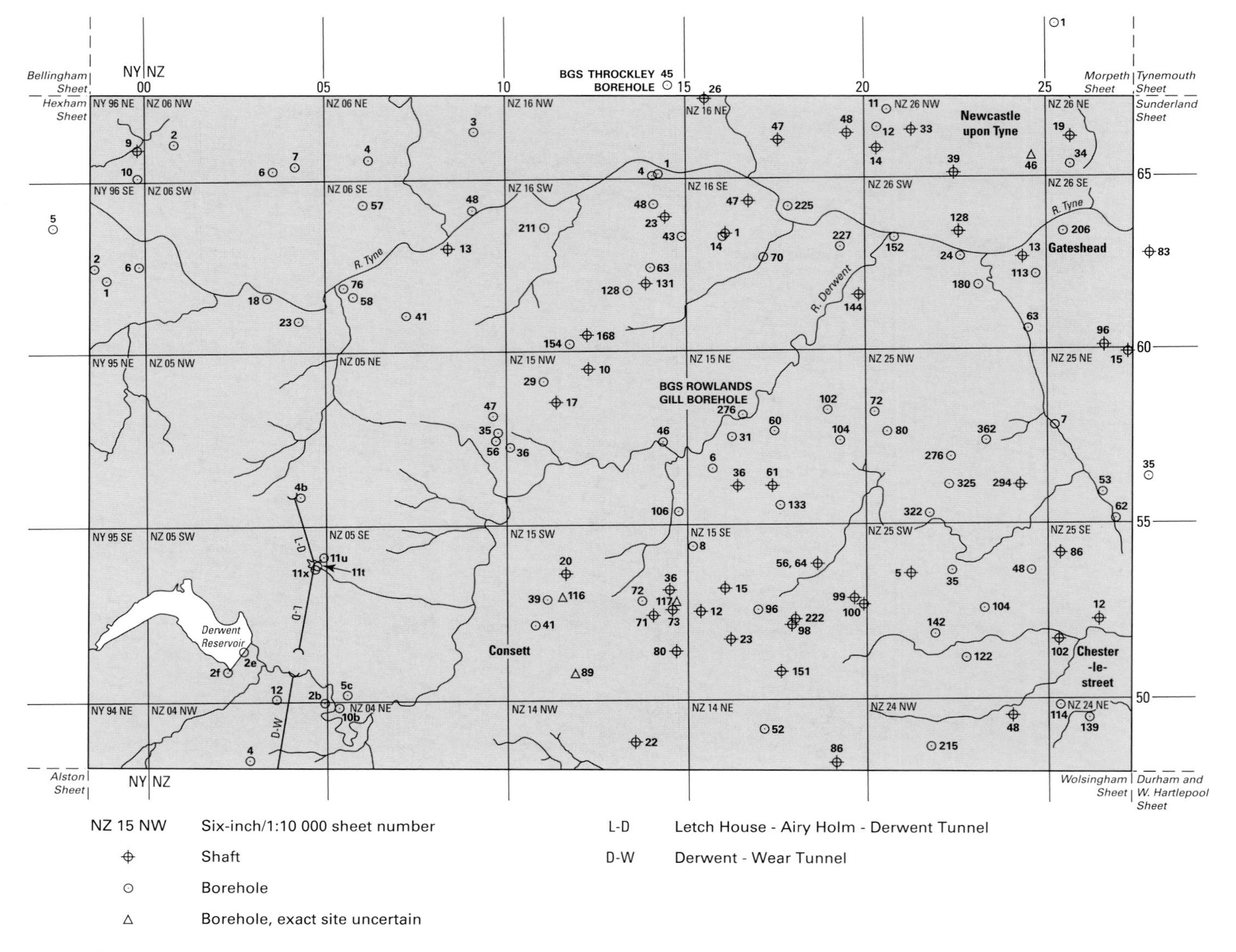

Figure 30 Location map of selected boreholes and shafts listed in Appendix 1. The BGS registered number takes the form NZ25NE/52: the number of the 1:10 000 National Grid sheet on which it is situated, followed by the serial number shown against the site on the map.

Name	BGS reference number	National Grid reference	B & S number	Starting level (m) above or below (-) OD	Drift thickness (m)	Total depth (m)	Stratigraphical range; notes
Cor Burn Pit	NY96NE/9	NY 9962 6583	544	c.94.49	4.88	42.26	Stainmore Group; Belsay Dene Limestone
Ramshawe Borehole	NY96NE/10	NY 9989 6507	—	c.88.39	7.24	40.23	Stainmore Group; Belsay Dene Limestone
Temperley Grange No. 1 Borehole	NY96SE/1	NY 9898 6211	—	c.185.93	0.83	60.96	Stainmore Group; Whitehouse and Styford limestones
Temperley Grange No. 2 Borehole	NY96SE/2	NY 9859 6250	—	c.146.30	0.76	60.96	Stainmore Group; Whitehouse and Styford limestones
Dilston No. 1 Borehole*	NY96SE/5	NY 9750 6364	—	c.31.09	10.49	272.87	Stainmore Group; Coalcleugh Shell Bed to Little Limestone
Farnley Hill Borehole	NY96SE/6	NY 9985 6257	—	c.118.87	8.64	87.17	Stainmore Group; below Whitehouse Limestone to below Grindstone Limestone
Derwent Valley Reservoir No. 1 Dam Site, No. 9 Borehole	NZ04NW/2b	0493 4990	—	c.205.74	42.06	67.67	Drift and Stainmore Group; Upper Felltop Limestone
DW/C Borehole*, Lambshield Moss	NZ04NW/4	0282 4834	—	326.26	2.82	202.50	Stainmore Group; 'Second Grit' to Crag Limestone
Derwent Valley Reservoir No. 1 Dam Site, No. 5 Borehole	NZ04NE/10b	0539 4983	—	c.205.13	56.54	75.59	Drift and Stainmore Group
Letch House No. 2 Borehole*	NZ05NW/4b	0421 5580	—	240.24	1.25	36.54	Lower Coal Measures; below Brockwell Coal
Whittonstall New Drift Borehole 213	NZ05NE/35	0973 5767	—	155.45	33.52	82.90	Lower Coal Measures; Busty Coal to Brockwell Coal
Chopwell No. 188 Borehole	NZ05NE/47	0968 5813	—	186.23	11.43	37.48	Lower Coal Measures; Harvey Marine Band to Harvey Coal
Chopwell No. 197 Borehole	NZ05NE/56	0976 5742	—	160.32	52.12	91.44	Lower Coal Measures; Three-Quarter Coal to Victoria Coal
Derwent Dam Edmundbyers TW/3 Borehole	NZ05SW/2e	0267 5142	—	194.89	52.42	83.06	Drift and Stainmore Group
Derwent Dam Edmundbyers TW/4 Borehole	NZ05SW/2f	0229 5089	—	c.230.00	7.77	42.67	Drift and Stainmore Group, including Upper Felltop Limestone
Airy Holm No. 18 Borehole*	NZ05SW/11t	0477 5386	—	211.93	1.60	33.50	Stainmore Group; 'First' and 'Second' grits
Airy Holm No. 19 Borehole	NZ05SW/11u	0497 5412	—	227.30	3.35	19.85	Stainmore Group; 'Second Grit'
Airy Holm No. 22 Borehole *	NZ05SW/11x	0470 5376	—	221.22	2.45	31.34	Lower Coal Measures and Stainmore Group
Muggleswick, DW/B Borehole	NZ05SW/12	0352 5008	—	248.13	0.80	100.00	Stainmore Group to Grindstone Sill
Derwent Valley Reservoir No. 1 Dam Site, No. 3 Borehole	NZ05SE/5c	0558 5015	—	205.74	90.83	97.23	Drift and Stainmore Group

Name	BGS reference number	National Grid reference	B & S number	Starting level (m) above or below (-) OD	Drift thickness (m)	Total depth (m)	Stratigraphical range; notes
Aydon South Farm Borehole	NZ06NW/2	0083 6607	—	c.140.20	8.03	42.37	Stainmore Group; Coalcleugh Shell Bed
Round Hill Borehole	NZ06NW/6	0359 6521	—	c.116.12	8.69	38.71	Stainmore Group; Whitehouse and Styford limestones
Newton Hall Borehole	NZ06NW/7	0417 6534	—	c.158.50	5.33	46.63	Stainmore Group; 'First Grit' and Whitehouse Limestone
Horsley Borehole	NZ06NE/3	0922 6631	—	c.131.06	3.20	68.58	Stainmore Group; below Whitehouse Limestone to below Grindstone Limestone
Nafferton Farm Borehole	NZ06NE/4	0625 6563	—	c.106.68	27.74	91.44	Lower Coal Measures to Stainmore Group
Low Shilford Borehole	NZ06SW/18	0330 6159	—	25.6	16.6	17.0	Drift
Shilford East Wood Borehole	NZ06SW/23	0426 6090	—	68.4	23.8	23.8	Drift
Eltringham Shaft and Borehole	NZ06SE/13	0845 6298	2679	c.51.82	6.96	111.61	Lower Coal Measures; Tilley Coal to below Ganister Clay Coal
Tilley No. 3 Borehole, Mickley	NZ06SE/41	0729 6110	—	170.53	0.69	74.06	Lower Coal Measures; Harvey Marine Band to Top Busty Coal
Ovingham Borehole	NZ06SE/48	0907 6411	2888	c:21.34	10.67	82.02	Lower Coal Measures; Brockwell Coal to below Ganister Clay Coal
Ovington Nurseries Borehole	NZ06SE/57	0605 6428	—	c.114.30	9.91	96.55	Lower Coal Measures and Stainmore Group
Stocksfield Dairy Borehole	NZ06SE/58	0575 6168	—	27.43	38.10	77.01	Stainmore Group; 'First' and 'Second' grits
Stocksfield Hall Borehole	NZ06SE/76	0552 6189	—	21.0	20.3+	20.3	River terrace deposits on boulder clay
Stockerley Pit, Consett	NZ14NW/22	c.1353 4873	—	c.164.89	4.87	27.3	Lower Coal Measures; Three-Quarter Coal to Brockwell Coal
Lanchester 'C' Borehole	NZ14NE/52	1712 4924	—	208.53	c.9.75	179.83	Middle and Lower Coal Measures; Maudlin Coal to Victoria Coal
'Annie' Pit, Burnhope	NZ14NE/86	1918 4811	2467	c.243.84	c.6.17	222.90	Middle and Lower Coal Measures; High Main Coal to Brockwell Coal
Conclusion Shaft, Chopwell	NZ15NW/10	1225 5938	488	251.15	c.1.52	103.32	Middle and Lower Coal Measures; Hutton Coal Coal to Three-Quarter Coal
Chopwell No. 2 Pit	NZ15NW/17	1143 5860	—	192.33	3.04	119.60	Middle and Lower Coal Measures; Brass Thill Coal to Brockwell Coal
Chopwell No. 3 Borehole	NZ15NW/29	1102 5913	—	236.83	3.04	115.82	Middle and Lower Coal Measures; Durham Low Main Coal to Busty Coal
Broad Oak Farm Borehole	NZ15NW/36	1001 5733	—	137.67	46.33	78.33	Lower Coal Measures; Three-Quarter Coal to Victoria Coal
Chopwell Borehole*	NZ15NW/46	1438 5743	2532	c.57.91	5.03	401.72	Lower Coal Measures to Alston Group; below Brockwell Coal to Four Fathom Limestone
Low Ewehurst Farm Borehole	NZ15NW/106	1482 5535	—	96.62	32.89	85.95	Lower Coal Measures; Busty Coal to Victoria Coal

Name	BGS reference number	National Grid reference	B & S number	Starting level (m) above or below (-) OD	Drift thickness (m)	Total depth (m)	Stratigraphical range; notes
Priestfield Farm Borehole	NZ15NE/6	1575 5663	—	124.05	3.35	79.40	Lower Coal Measures; Harvey Marine Band to Victoria Coal
No. 1 South Garesfield Borehole	NZ15NE/31	1633 5760	—	66.45	1.12	33.76	Lower Coal Measures; Victoria Coal to Marshall Green Coal
Anna Pit, Lintz	NZ15NE/36	1646 5614	—	192.94	5.79	143.63	Middle and Lower Coal Measures; Durham Low Main Coal to Brockwell Coal
B_2 Borehole, Busty Bank	NZ15NE/60	1739 5784	—	68.58	5.79	35.96	Lower Coal Measures; Brockwell Coal to Victoria Coal
Hobson Pit, Burnopfield	NZ15NE/61	1738 5616	—	219.05	c.6.09	199.74	Middle and Lower Coal Measures; Five-Quarter Coal to Brockwell Coal
Fellside Borehole	NZ15NE/102	1899 5830	—	174.12	6.19	171.67	Middle and Lower Coal Measures; Durham Low Main Coal to Victoria Coal
High Marley Hill Borehole	NZ15NE/104	1928 5743	—	195.19	3.04	59.13	Middle Coal Measures; Main Coal to Maudlin Coal
Town Head Farm Borehole*	NZ15NE/133	1740 5527	—	216.20	7.16	198.68	Middle and Lower Coal Measures; Main Coal to Victoria Coal
Rowlands Gill Heat Flow Borehole*	NZ15NE/276	1664 5815	—	43.11	26.70	242.89	Lower Coal Measures; Stainmore Group to below Lower Felltop Limestone
Isabella Pit, Medomsley	NZ15SW/20	1160 5355	2839	262.74	8.40	145.36	Middle and Lower Coal Measures; Durham Low Main Coal to Brockwell Coal
Annie Pit, S. Medomsley	NZ15SW/36	1436 5303	1831	246.88	6.63	191.72	Middle and Lower Coal Measures; Five-Quarter Coal to Brockwell Coal
Medomsley No. 2 Borehole	NZ15SW/39	1101 5281	—	248.10	0.30	63.09	Middle and Lower Coal Measures; above Harvey Marine Band to Busty Coal
Medomsley Underground Borehole	NZ15SW/41	1086 5235	—	177.60	—	17.70	Lower Coal Measures; Brockwell Coal to Victoria Coal
Eden Mine Shaft and Borehole	NZ15SW/71	1342 5202	—	254.38	2.44	150.44	Middle and Lower Coal Measures; Durham Low Main Coal to Brockwell Coal
Billingside No. 1 Borehole	NZ15SW/72	1376 5284	—	227.38	2.13	116.43	Middle and Lower Coal Measures; Maudlin Coal to Tilley Coal
North (Windsor) Pit, Pontop	NZ15SW/73	1447 5248	1557	c.293.52	1.22	213.74	Middle and Lower Coal Measures; High Main Coal to Three-Quarter Coal
Stony Heap Sinking	NZ15SW/80	1469 5150	—	199.64	c.1.83	146.60	Middle and Lower Coal Measures; Durham Low Main Coal to Brockwell Coal
No. 9 Borehole, Stockerley House Pit, Crook Hall	NZ15SW/89	c.1186 5068	608	c.237.74	1.22	156.13	Lower Coal Measures and Stainmore Group; Top Busty Coal to below Quarterburn Marine Band
German Bands Borehole, Consett	NZ15SW/116	c.115 528	540	c.262.13	2.08	136.09	Middle and Lower Coal Measures; Durham Low Main Coal to Victoria Coal
Pontop Colliery Sinking	NZ15SW/117	c.147 527	1552	c.300	1.83	267.72	Middle and Lower Coal Measures; Ryhope Little Coal to Brockwell Coal

Name	BGS reference number	National Grid reference	B & S number	Starting level (m) above or below (-) OD	Drift thickness (m)	Total depth (m)	Stratigraphical range; notes
Collierly Farm Borehole	NZ15SE/8	1516 5438	—	165.81	1.22	103.78	Middle and Lower Coal Measures; Durham Low Main Coal to Tilley Coal
New Pit, Dipton	NZ15SE/12	1551 5249	723	251.15	4.06	210.70	Middle and Lower Coal Measures; High Main Coal to Busty Coal
Harelaw Pit, Pontop	NZ15SE/15	1609 5310	1559	254.20	5.03	131.47	Middle Coal Measures; Ryhope Little Coal to Durham Low Main Coal
South Pit, Pontop	NZ15SE/23	1626 5177	1551	247.80	4.88	211.43	Middle and Lower Coal Measures; High Main Coal to Three-Quarter Coal
Old Engine & Wind Pits, Tanfield	NZ15SE/56, 64	1856 5390	2995/ 2998	157.88	19.76	239.29	Middle and Lower Coal Measures; Five-Quarter Coal to Victoria Coal
Tanfield Lea Borehole, West Kyo	NZ15SE/96	1700 5257	—	196.52	10.67	235.86	Middle and Lower Coal Measures; Five-Quarter Coal to Victoria Coal
'A' Pit, S. Tanfield Colliery	NZ15SE/98	1793 5218	1847/ 2952	c.234.09	15.44	309.63	Middle and Lower Coal Measures; Ryhope Little Coal to below Marshall Green Coal
Kettledrum Pit, W. Stanley	NZ15SE/99	1977 5290	2192/ 2193	228.60	3.35	259.74	Middle and Lower Coal Measures; Ryhope Five-Quarter Coal to Busty Coal
New Pit*, W. Stanley	NZ15SE/100	1996 5272	3048	c.223.72	5.48	316.38	Middle and Lower Coal Measures; Ryhope Five-Quarter Coal to Victoria Coal
East Busty Pit, Annfield Plain	NZ15SE/151	1758 5091	—	227.99	8.20	245.36	Middle and Lower Coal Measures; High Main Coal to Brockwell Coal
Sinking, W. Stanley Colliery	NZ15SE/222	c.180 520	—	223.72	4.27	326.99	Middle and Lower Coal Measures; Ryhope Little Coal to Marshall Green Coal
Stella & Towneley Borehole	NZ16NW/1	1423 6513	1883	9.75	20.34	101.14	Middle and Lower Coal Measures; Ruler Coal to Three-Quarter Coal
Clara Vale Under-ground Borehole No. 2	NZ16NW/4	1407 6501	—	-85.34	—	77.03	Lower Coal Measures; Busty Coal to Marshall Green Coal
Throckley Borehole*	NZ16NW/45	1456 6762	—	102.18	1.52	604.85	Lower Coal Measures to Upper Liddesdale Group including the Great Whin Sill
Maria Pit, Throckley	NZ16NE/26	1569 6726	—	84.13	1.83	124.85	Middle and Lower Coal Measures; Hutton Coal to Brockwell Coal
Wellington (Blucher) Pit, Throckley	NZ16NE/47	1769 6606	2101	88.39	1.06	160.33	Middle and Lower Coal Measures; Maudlin Coal to Brockwell Coal
Francis Pit, Montagu Main, East Denton	NZ16NE/48	1969 6631	1220/ 1374	c.103.63	—	c.208.48	Middle and Lower Coal Measures; High Main Coal to Tilley Coal
Emma Pit, Stella and Towneley Colliery, Ryton	NZ16SW/23	1440 6388	1875	85.34	5.00	161.44	Middle and Lower Coal Measures; Durham Low Main Coal to Brockwell Coal
Stargate Colliery, Underground Borehole	NZ16SW/43	1497 6326	—	41.45	—	39.62	Lower Coal Measures; Brockwell Coal to Marshall Green Coal
Barmoor Borehole, Ryton	NZ16SW/48	1407 6441	—	78.84	0.20	117.35	Middle and Lower Coal Measures; Durham Low Main Coal to Hodge Coal
Emma Colliery, Surface Borehole	NZ16SW/63	1396 6230	—	112.30	5.94	46.84	Middle Coal Measures; Main Coal to Hutton Coal

Name	BGS reference number	National Grid reference	B & S number	Starting level (m) above or below (-) OD	Drift thickness (m)	Total depth (m)	Stratigraphical range; notes
Underground Borehole, Success Pit, Greenside	NZ16SW/128	1346 6176	—	—	—	146.30	Middle and Lower Coal Measures; Main Coal to Three-Quarter Coal
'A' Pit, Greenside	NZ16SW/131	1395 6195	2969	151.48	5.79	171.93	Middle and Lower Coal Measures; Durham Low Main Coal to below Victoria Coal
Bucks Nook Farm Borehole	NZ16SW/154	1181 6010	—	217.22	4.72	32.61	Lower Coal Measures; Harvey Marine Band to Tilley Coal
North Pit, Chopwell	NZ16SW/168	1228 6054	489	c.224	c.1.30	86.08	Lower Coal Measures; Harvey Marine Band to Brockwell Coal
Eastwoods Farm Borehole	NZ16SW/211	1109 6357	—	55.4	25.0+	25.0	Glacial sand and gravel
New Winning Pit, Stargate	NZ16SE/1	1610 6337	1868	85.34	12.01	151.48	Middle and Lower Coal Measures; Durham Low Main Coal to Brockwell Coal
Stargate Colliery No. 1 Underground Borehole*	NZ16SE/14	1604 6334	—	- 65.10	—	53.31	Lower Coal Measures; Brockwell Coal to Ganister Clay Coal
Addison Shaft, Blaydon	NZ16SE/47	1685 6426	1888	c.18.29	c.7.31	86.87	Lower Coal Measures; Harvey Coal to Brockwell Coal
Blaydon Burn No. 2 Borehole	NZ16SE/70	1724 6275	—	89.00	1.83	113.99	Middle and Lower Coal Measures; Ruler Coal to Victoria Coal
Engine or Axwell Pit, Swalwell	NZ16SE/144	1988 6167	2381	c.24.38	16.46	197.66	Lower Coal Measures and Stainmore Group; Three-Quarter Coal to 'Second Grit'
Newburn Haughs Borehole	NZ16SE/225	1787 6421	—	4.9	25.0+	25.0	Alluvium on buried valley deposits
Blaydon Borehole	NZ16SE/227	1932 6301	—	3.8	22.0+	22.0	Alluvium on buried valley deposits
Byron Pit, Waldridge	NZ24NW/48	2409 4950	3871	24.97	2.44	188.80	Middle and Lower Coal Measures; High Main Coal to Busty Coal
Holmside Borehole	NZ24NW/215	2180 4858	—	158.19	7.92	225.80	Middle and Lower Coal Measures; Five- Quarter Coal to Victoria Coal
Waldridge 'D' Pit Under-ground Borehole No. 60	NZ24NE/114	2538 4989	—	73.60	—	60.96	Lower Coal Measures; Busty Coal to Victoria Coal
Waldridge Lane Borehole	NZ24NE/139	2618 4941	—	78.64	c.45.72	173.43	Middle and Lower Coal Measures; Durham Low Main Coal to Tilley Coal
Marley Hill Upover Borehole	NZ25NW/72	2029 5817	—	20.92	—	57.94	Lower Coal Measures; Harvey Coal to Brockwell Coal
Marley Hill Downover Borehole	NZ25NW/80	2078 5760	—	129.44	—	76.20	Middle and Lower Coal Measures; Hutton Coal to Tilley Coal
Birkland Lane Borehole	NZ25NW/276	2242 5687	—	162.28	12.06	214.80	Middle and Lower Coal Measures; Five-Quarter Coal to Victoria Coal
Glamis Shaft, Kibblesworth	NZ25NW/294	2432 5619	—	c.71.63	8.28	228.90	Middle and Lower Coal Measures; Five-Quarter Coal to Brockwell Coal

Name	BGS reference number	National Grid reference	B & S number	Starting level (m) above or below (-) OD	Drift thickness (m)	Total depth (m)	Stratigraphical range; notes
Beamish Mary Pit Underground Borehole	NZ25NW/322	2183 5523	—	- 117.01	—	18.59	Lower Coal Measures; Brockwell Coal to Victoria Coal
Kibblesworth No. 1 Downover Borehole	NZ25NW/325	2238 5611	—	- 104.10	—	45.72	Lower Coal Measures; Busty Coal to Victoria Coal
Kibblesworth Borehole*	NZ25NW/362	2340 5731	—	94.72	4.57	188.21	Middle and Lower Coal Measures; Five-Quarter Coal to Brockwell Coal
Lamesley Church Borehole	NZ25NE/7	2523 5789	—	12.87	48.08	50.77	Drift
Isabella Pit, Sheriff Hill	NZ25NE/15	2730 5966	1743	c.155.45	c.7.92	224.58	Middle Coal Measures; Kirkby's Marine Band to Hutton Coal
Ravensworth Anne Pit Underground Borehole No. 6	NZ25NE/35	2792 5640	—	c.- 139.90	—	111.78	Lower Coal Measures; Harvey Coal to Marshall Green Coal
Birtley No. 4 Borehole	NZ25NE/53	2657 5583	2430	c.24.38	39.01+	39.01	Drift
Birtley Slag Heaps Borehole	NZ25NE/62	2689 5526	—	26.38	56.39	57.83	Drift
Mary Pit, Beamish	NZ25SW/5	2102 5356	2395	179.83	5.71	256.72	Middle and Lower Coal Measures; High Main Coal to Brockwell Coal
Mary Pit, Beamish Downover Borehole No. 145	NZ25SW/35	2175 5399	—	- 52.17	—	51.81	Lower Coal Measures; Top Busty Coal to Victoria Coal
Kings Lane Borehole, Pelton	NZ25SW/48	2469 5360	—	81.46	6.10	248.72	Middle and Lower Coal Measures; High Main Coal to Victoria Coal
Handen Hold Downover Borehole*	NZ25SW/104	2331 5260	—	- 84.65	—	43.89	Lower Coal Measures; Busty Coal to Victoria Coal
Grange Villa/Craghead Road Borehole	NZ25SW/122	2262 5117	—	140.64	2.74	236.52	Middle and Lower Coal Measures; High Main Coal to Victoria Coal
Craghead Borehole	NZ25SW/142	2192 5179	—	139.19	27.43	218.59	Lower Coal Measures; Harvey Marine Band to Victoria Coal
South Pelaw Pit	NZ25SE/12	2652 5224	2950	56.39	18.29	189.99	Middle and Lower Coal Measures; Five-Quarter Coal to Busty Coal
Urpeth 'D' Pit	NZ25SE/86	2549 5414	2084 3026 3027	74.07	c.5.49	244.88	Middle and Lower Coal Measures; Five-Quarter Coal to Brockwell Coal
Low Main Shaft, Pelton Fell	NZ25SE/102	2534 5178	1524	62.48	4.27	280.67	Middle and Lower Coal Measures; Five-Quarter Coal to 'Third Grit'
Black Swine Borehole, Westerhope	NZ26NW/11	2066 6691	—	98.93	1.07	73.15	Middle Coal Measures; High Main Coal to Main Coal
Westerhope No. 4 Borehole	NZ26NW/12	2036 6646	—	c.105.16	4.87	103.63	Middle Coal Measures; Moorland Coal to Five-Quarter Coal
Caroline Pit, Montagu Main, E. Denton	NZ26NW/14	2035 6589	1375	c.79.25	—	174.62	Middle and Lower Coal Measures; Durham Low Main Coal to Brockwell Coal

Name	BGS reference number	National Grid reference	B & S number	Starting level (m) above or below (-) OD	Drift thickness (m)	Total depth (m)	Stratigraphical range; notes
Lady Pit, Montagu Main, Blakelaw	NZ26NW/33	2138 6631	1373	c.91.44	4.57	225.62	Middle and Lower Coal Measures; Main Coal to Victoria Coal
Fenham Pit, Newcastle	NZ26NW/39	2253 6510	—	c.100.58	12.19	127.91	Middle and Lower Coal Measures; High Main Coal to Harvey Coal
Town Moor Borehole	NZ26NW/46	c.247 655	—	c.66.45	20.57	113.46	Middle Coal Measures; High Main Coal to Low Main Coal
Heathery Lane No. 11 Borehole*, North Gosforth	NZ26NE/1	2524 6923	—	c.48.77	36.42	498.39	Middle and Lower Coal Measures; Hebburn Fell Coal to Brockwell Coal
Middle Pit, Jesmond	NZ26NE/19	2572 6615	1211	c.48.77	—	191.87	Middle Coal Measures; Five-Quarter Coal to Durham Low Main Coal
Stone Brewery Borehole	NZ26NE/34	2570 6530	—	c.44.20	21.79	55.93	Middle Coal Measures; High Main Marine Band
Redheugh Pit, Gateshead	NZ26SW/13	2443 6263	—	c.29.87	15.85	222.68	Lower Coal Measures; Harvey Coal to Marshall Green Coal
Dunstanhaugh No. 1 Borehole	NZ26SW/24	2267 6264	—	c.3.96	43.89	43.89	Drift
Team Valley Trading Estate Borehole	NZ26SW/63	2452 6053	—	7.77	65.07	68.27	Drift
Bensham Road Borehole, Gateshead	NZ26SW/113	2482 6223	—	56.49	2.44	92.96	Middle Coal Measures; Five-Quarter Coal to Hutton Coal
Beaumont Pit, South Elswick	NZ26SW/128	2266 6332	788	35.97	4.27	178.38	Middle and Lower Coal Measures; Maudlin Coal to Brockwell Coal
Delta Ironworks Borehole*, Gateshead	NZ26SW/152	2075 6327	—	3.05	44.60	233.78	Lower Coal Measures and Stainmore Group
Dunston No. 3 Borehole	NZ26SW/180	2314 6188	—	6.90	25.60	27.12	Drift
Venture Pit, Felling	NZ26SE/83	2797 6277	866	22.86	40.23	229.62	Middle Coal Measures; Ryhope Five-Quarter Coal to Hutton Coal
Fanny Pit, Sheriff Hill	NZ26SE/96	2673 6018	1745	163.07	—	245.05	Middle and Lower Coal Measures; Kirkby's Marine Band to Harvey Coal
West Street Borehole, Gateshead	NZ26SE/206	2545 6345	—	31.29	4.57	19.10	Middle Coal Measures; High Main Coal and Five-Quarter Coal

APPENDIX 2

Summary logs of selected boreholes and shafts

Abbreviated logs or synopses of a selection of boreholes and shafts (arranged alphabetically) are given below. These have been chosen to illustrate the geological sequence encountered within the district, and also to amplify stratigraphical details not recorded in the text. A few of the boreholes lie just outside the boundaries of the district (Figure 30). Prior to 1971, all depths were recorded in Imperial Units, and these have been converted to the nearest metric equivalent. Levels quoted are relative to Ordnance Datum. Numbers in the form NZ05SW/11 identify the boreholes in the BGS 1:10 000 sheet registration system. National Grid references are in square brackets; all lie in square NZ unless otherwise stated.

Airy Holm No. 18 Borehole, Shotleyfield (NZ 05 SW/11t)

Surface level + 211.93 m [0477 5386].
Drilled 1970 for Northumbrian River Authority. Logged by D A C Mills and P G Carter (Babtie Group).

	Thickness m	*Depth* m
Drift		
Alluvium — boulders and clay	1.60	1.60
Stainmore Group		
'Second Grit'		
Sandstone, thin- to medium-bedded, medium- to coarse-grained, often pebbly, hard to moderately hard; sporadic micaceous carbonaceous partings; very pebbly and conglomeratic between 6.20 and 6.80 m, locally jointed (open holed to 3.00 m)	9.40	11.00
Mudstone, laminated to very thin-bedded, micaceous and carbonaceous; sporadic sandy layers	1.82	12.82
Coal	0.10	12.92
Seatearth, irregularly thin-bedded, rooty, micaceous and carbonaceous; grades into rooty sandstone towards base; passes into	1.44	14.36
Sandstone, thin- to medium-bedded, fine-grained, micaceous and carbonaceous; rooty in part	0.77	15.13
Mudstone, sandy and silty, thin-bedded, micaceous and carbonaceous; sandy interlaminae	0.47	15.60
'First Grit'		
Sandstone, predominately medium- bedded, medium-grained with sub-horizontal micaceous carbonaceous partings; below 17.10 m becomes more coarse grained; 0.24 m pebbly conglomeratic beds at 21.60 m, with coal traces, and 0.24 m bed at 22.68 m; rather finer-grained towards base	7.51	23.11
Mudstone, often silty, laminated to very thin-bedded, finely micaceous and carbonaceous; between 29.70 and 31.62 m, sandy with thin beds and interlaminae of fine-grained sandstone	10.39+	33.50

Airy Holm No. 22 Borehole, Shotleyfield (NZ 05 SW/11x)

Surface level + 221.22 m [0470 5376].
Drilled 1970–71 for Northumbrian River Authority. Logged by D A C Mills and P G Carter (Babtie Group).

	Thickness m	*Depth* m
Drift		
Boulder clay	2.45	2.45
? Lower Coal Measures		
Sandstone, thin-bedded, coarse-grained, shattered; open-holed to 3.50 m	2.58	5.03
Siltstone, laminated, micaceous and carbonaceous with iron-coated laminae, rooty and mushy at base: assumed position of Subcrenatum (Quarterburn) Marine Band	2.46	7.49
Stainmore Group		
Coal	0.40	7.89
Shale and mudstone, rooty, micaceous; disintegrated, poor recovery	2.87	10.76
'Second Grit'		
Sandstone, thin- to medium-bedded, medium- and medium- to coarse-grained with micaceous carbonaceous patches and blebs; below 16.26 m more regularly shaped micaceous carbonaceous partings; vertical to subvertical joints	9.42	20.18
Mudstone, laminated, micaceous and carbonaceous, finely jointed; silty towards base	1.68	21.86
Coal	0.19	22.05
Seatearth, silty, muddy, micaceous and carbonaceous; sandy towards base	0.60	22.65
'First Grit'		
Sandstone, thin- to medium-bedded, predominantly medium- to coarse-grained, micaceous carbonaceous blebs and sporadic partings	3.88	26.53
Mudstone, sandy, silty, laminated, micaceous, carbonaceous; locally rooty	2.71	29.24
Sandstone, medium-bedded, medium- to coarse-grained and coarse-grained; micaceous carbonaceous blebs	2.10+	31.34

Chopwell Borehole (NZ 15 NW/46)

Surface level approx. + 57.91 m [1438 5743].
Drilled 1897 by Wm Coulson, Durham.
Shown graphically in Figures 8–11. Further details published in Borings and Sinkings (1878–1910) No. 2532 and in Simpson (1904). Several local, old miners' terms were used in the original record and have been replaced here by their modern equivalents.

	Thickness m	*Depth* m
Drift		
Clay, sand and soil	1.37+	1.37
Boulder clay	3.66	5.03
Lower Coal Measures		
Sandstone, broken	0.20	5.23
Shale, very dark grey	0.46	5.69
Coal	0.10	5.79
Shale, dark grey	0.61	6.40
Sandstone, grey; shale partings towards base	8.08	14.48
Shale, grey; sandstone ribs near top	2.74	17.22
Sandstone, grey	0.64	17.86
Shale, dark grey, soft	4.69	22.55
GANISTER CLAY COAL		
Coal	0.03	22.58
Seatearth	0.76	23.34
Shale, dark grey	1.45	24.79
Sandstone, grey	3.35	28.14
Shale, dark grey	1.17	29.31
Sandstone, grey	1.88	31.19
Shale: approximate position of RODDYMOOR MARINE BAND	3.35	34.54
'THIRD GRIT'		
Sandstone, grey, laminated	18.77	53.31
Shale, dark grey: approximate position of SUBCRENATUM (QUARTERBURN) MARINE BAND	6.56	59.87
Stainmore Group		
'SECOND GRIT'		
Sandstone, grey	4.87	64.74
Seatearth and shale, grey, soft	0.69	65.43
Sandstone, grey	3.05	68.48
Shale, dark grey	2.13	70.61
Sandstone, dark grey	1.58	72.19
Shale, dark grey	5.23	77.42
Sandstone, grey; coal streaks	0.91	78.33
Approximate position of WOODLAND SHELL BEDS		
Seatearth	0.71	79.04
Shale	1.25	80.29
Sandstone, grey	0.30	80.59
Shale, grey; sandstone ribs towards base	3.03	83.62
'FIRST GRIT'		
Sandstone, grey; mudstone partings	7.46	91.08
Shale, dark grey	4.42	95.50
WHITEHOUSE LIMESTONE		
Limestone, grey	0.58	96.08
Seatearth	0.67	96.75
Sandstone, grey; shale partings	2.46	99.21
Shale, grey	2.44	101.65
Seatearth	0.61	102.26
Sandstone, grey; shale laminae	3.53	105.79
Shale, grey	1.55	107.34
Sandstone, grey	1.88	109.22
Seatearth	0.63	109.85
Sandstone, grey; shale partings	10.06	119.91
Shale, grey; sandstone partings	3.10	123.01
Sandstone, grey, hard	2.16	125.17
GRINDSTONE (NEWTON) LIMESTONE		
Shale, grey, and limestone, shelly	1.85	127.02
GRINDSTONE SILL		
Sandstone, grey, very coarse-grained	10.04	137.06
Shale, grey; sandstone ribs	6.83	143.89
UPPER FELLTOP (THORNBROUGH) LIMESTONE		
Limestone; shale partings	5.99	149.88
Sandstone, grey; median shale bed; shaly towards base	10.60	160.48
Shale, dark grey; ironstone ribs	6.40	166.88
Sandstone, grey	2.33	169.21
?COALCLEUGH SHELL BED		
Limestone, grey	0.82	170.03
?HIGH GRIT SILL		
Sandstone, grey, coarse-grained	15.52	185.55
Sandstone, grey; coal streaks	1.65	187.20
Shale, dark grey	1.55	188.75
LOWER FELLTOP (CORBRIDGE) LIMESTONE		
Limestone, grey	4.77	193.52
Sandstone, grey; shale partings	5.87	199.39
Shale, grey; sandstone ribs	3.33	202.72
Sandstone, grey	4.80	207.52
Coal, coarse	0.02	207.54
Sandstone, grey; shale partings	7.16	214.70
Shale, grey	4.07	218.77
Sandstone, grey, and limestone (?AYDON SHELL BED)	0.41	219.18
?LOW GRIT SILL		
Sandstone, grey; occasional shale partings	30.22	249.40
Shale grey; ironstone nodules	5.87	255.27
CRAG (OAKWOOD) LIMESTONE		
Limestone, blue; shelly	2.18	257.45
Sandstone, grey; shaly at base	6.79	264.24
Shale, dark grey	1.03	265.27
Sandstone, grey; shaly partings	1.58	266.85
Shale, dark grey, soft	0.48	267.33
Sandstone, grey	11.97	279.30
Shale, grey	2.69	281.99
Sandstone, grey; shale partings	7.77	289.76
Shale, grey and dark grey; sandstone rib 0.6 m near top	5.79	295.55
Sandstone, grey; shale partings	2.98	298.53
Shale, grey; iron pyrites	5.76	304.29
LITTLE LIMESTONE		
Limestone, blue	3.71	308.00
Sandstone, grey; shale partings	1.65	309.65
Shale, grey; sandstone layers	5.39	315.04
Sandstone, grey; shale partings	6.14	321.18
Coal	0.10	321.28
Sandstone, grey	0.38	321.66
Shale, grey; pyrites	6.78	328.44
GREAT LIMESTONE		
Limestone, blue	2.09	330.53
Shale, dark grey (?Tumbler Beds)	4.11	334.64
Limestone, blue	15.32	349.96
Alston Group		
Sandstone, dark grey	0.41	350.37
Shale, grey; some pyrites	0.33	350.70
Ganister	0.30	351.00
Sandstone, grey	10.52	361.52
Coal, coarse	0.05	361.57
Sandstone, predominantly very coarse-grained; coal streaks	17.32	378.89
Shale, dark grey	12.24	391.13

	Thickness m	Depth m
FOUR FATHOM LIMESTONE		
Limestone, blue	6.15	397.28
Sandstone, grey	4.44+	401.72

Delta Iron Works Water Borehole, Derwenthaugh, Gateshead (NZ 26 SW/152)

Surface level c. + 3.05 m [2075 6327].
Drilled 1942 for Messrs Raine & Co. Ltd. Examined in part by R G Carruthers and G Burnett.
Part synopsis to presumed Marshall Green Coal at 155.55 m.

	Thickness m	Depth m
Drift		
Made ground, ashes, timbers, etc.	2.13	2.13
Silt, grey	4.73	6.86
Sand; a few stones	1.52	8.38
Silt, grey (includes timber between 8.86 and 8.94 m)	1.83	10.21
Silt, black; stones	0.92	11.13
Sand and gravel	4.06	15.19
Clay, leafy; silty sand partings	4.16	19.35
Clay, silty	1.38	20.73
Clay, leafy; thin sand partings	6.24	26.97
Clay, stony, brown	17.63	44.60
Lower Coal Measures		
Strata to ? Marshall Green Coal, including Harvey, Tilley group, Top Busty horizon, Bottom Busty, Three-Quarter, Brockwell and Victoria coals; recovery of coals poor; dark grey shale containing *Anthraconauta* sp., ostracods and *Spirorbis* at 106.01 m possibly constitutes Brockwell Ostracod Band	110.95	155.55
?MARSHALL GREEN COAL		
Coal	0.51	156.06
Fireclay	0.91	156.97
Sandstone; shale partings	4.14	161.11
Shale, sandy	1.65	162.76
Shale; sandstone ribs	2.19	164.95
?GANISTER CLAY COAL		
Coal	0.05	165.00
Fireclay	0.46	165.46
Sandstone	1.75	167.21
Fireclay	0.81	168.02
Coal; thin stone bands	0.20	168.22
Fireclay	0.49	168.71
'THIRD GRIT'		
Sandstone, gritty	14.12	182.83
Shale, grey; fish debris	0.38	183.21
Ganister; upper 0.30 m good quality, remainder inferior, passing down into	1.21	184.42
Sandstone, bedded, micaceous	5.78	190.20
Shale, sandy; sandstone beds	1.06	191.26
Fireclay; shale partings	0.87	192.13
Coal	0.22	192.35
Seatearth-mudstone; shale partings containing *Lingula* sp., fish teeth, plant debris	1.22	193.57
Coal	0.10	193.67
Fireclay, soapy, marly	1.61	195.28
Shale, muddy, nodular	3.35	198.63
SUBCRENATUM (QUARTERBURN) MARINE BAND (inferred position)		
Shale, dark grey; fish scale at 198.72 m; sporadic *Lingula* sp.at 199.64 m and 200.20 m; ?gastropod	1.62	200.25
Stainmore Group		
Fireclay	1.38	201.63
Coal	0.17	201.80
Fireclay, ganisteroid	0.89	202.69
'SECOND GRIT'		
Sandstone, grey, micaceous	8.38	211.07
Shale, grey	5.03	216.10
Ganister, inferior	0.69	216.79
Marl, porcellanous; spherulitic nodules	0.22	217.01
Sandstone, greenish grey, silty	0.16	217.17
Marl, porcellanous	0.15	217.32
Marl, sandy, carbonaceous; slickensided in lower 0.76 m	1.06	218.38
Sandstone, silty, fine-grained	0.16	218.54
Sandstone, cross-bedded, fine-grained; passes down into light grey fine-grained sandstone with micaceous patches and 'worm' burrows	2.51	221.05
Sandstone, very fine-grained, micaceous partings; silty at top	0.92	221.97
Sandstone, white, sugary, calcareous	0.22	222.19
?'FIRST GRIT'		
Sandstone, coarse-grained	1.83	224.02
Grit, soft; faulted at base	1.98	226.00
Shale, grey, sandy, micaceous; shaly micaceous sandstone beds; *Lingula* sp. at 230.73 m, *Productus* sp. at 231.31 m	5.34	231.34
Sandstone, yellow, strong	0.30	231.64
Shale, dark grey, micaceous; *Lingula* sp. and *Orbiculoidea* sp.	1.70	233.34
Sandstone, strong, grey	0.44+	233.78

Dilston No. 1 Borehole, near Corbridge (NY 96 SE/5). 1:50 000 Sheet 19 (Hexham).

Surface level c. + 31.09 m [NY 9750 6364].
Drilled 1922. Logged, in part, by R G Carruthers.
Synopsis. Shown graphically in Figures 8 and 9.

	Thickness m	Depth m
Drift		
Sand, gravel and boulders	10.49	10.49
Stainmore Group		
Strata; shale at base with fragments of *Productus* sp.; assumed position of COALCLEUGH SHELL BED (PIKE HILL LIMESTONE)	21.53	32.02
Coal	0.15	32.17
Strata, including coal 0.07 m at 40.56 m	20.51	52.68
LOWER FELLTOP (CORBRIDGE) LIMESTONE		
Limestone, blue and dark grey, with shale beds; largely unfossiliferous but with some crinoidal debris; small productoids in shale beds near top	5.61	58.29
Strata; largely sandstone, including coarse-grained gritty sandstone 3.13 m at 77.85 m	20.86	79.15
BELSAY DENE LIMESTONE		
Limestone, sandy; some coral debris	0.12	79.27

	Thickness m	Depth m
Shale, sandy	0.79	80.06
Limestone, sandy, hard; shelly with brachiopods and crinoid ossicles	1.83	81.89
Shale, dark blue	0.99	82.88
Shale; brachiopods including productoids and rhynchonellids	0.99	83.87
Shale, black; *Lingula* sp. and fish scales	0.61	84.48
Coal	0.10	84.58
Strata, with thin limy sandstone containing productoids at base	18.41	102.99
AYDON SHELL BED		
Limestone, dark grey, shelly	0.68	103.67
Sandstone, grey	0.30	103.97
Coal	0.15	104.12
Sandstone, predominantly massive, cross-bedded, quartzo-feldspathic	9.62	113.74
Strata, including pale shelly limestone 1.82 m at 120.20 m	21.28	135.02
Shale, blue; *Lingula* sp., some fish scales; assumed position of Plankey Shell Bed	3.00	138.02
Strata, including shelly calcareous sandstone with productoids and lamellibranchs at base	8.99	147.01
Shale, rooty; abundant *Lingula* sp. at base	1.04	148.05
Strata, including coal 0.15 m at 148.20 m, coal 0.05 m at 150.38 m and mudstones, shales and thin sandstones towards base containing rhynchonellids, small productoids and crinoid debris	18.04	166.09
CRAG (OAKWOOD) LIMESTONE		
Limestone, hard, dark grey; crinoidal	1.04	167.13
Strata; largely arenaceous at top with three thin coals; argillaceous towards base; basal 24 m contain shelly layers with productoid debris; basal 16.45 m contain limestone ribs	83.14	250.27
LITTLE LIMESTONE		
Limestone, blue and dark grey, compact, hard	4.83	255.10
Strata, including coal 0.76 m at 270.86 m	17.77+	272.87

Handen Hold Downover Borehole, near West Pelton (NZ 25 SW/104)

Underground level – 84.65 m [2331 5260].
Drilled 1951 for NCB. Logged by R H Price. Shown graphically in Figure 15.

	Thickness m	Depth m
Lower Coal Measures		
Floor of BUSTY COAL		
Open hole	1.07	1.07
Fireclay, sandy	0.15	1.22
Mudstone, sandy; sandstone ribs and partings	0.99	2.21
Sandstone, grey, fine-grained, shaly; slump structures	5.41	7.62
Mudstone, predominantly black, carbonaceous; thin ironstone bed	1.07	8.69
Sandstone, fine-grained, irony	0.30	8.99
THREE-QUARTER COAL		
Coal	0.38	9.37
Fireclay, sandy, micaceous	0.84	10.21
Sandstone, fine-grained, micaceous; ganisteroid at top	1.12	11.33
Mudstone, sandy, with ironstone beds; *Calamites* sp. and plant debris	0.69	12.02
Mudstone with ironstone; small mussels and abundant plants	0.78	12.80
Mudstone, sandy; partings of pyritous carbonaceous shale	0.87	13.67
Shale, black; ironstone nodules; small mussel	0.33	14.00
Coal	0.02	14.02
Fireclay and ganisteroid sandstone	0.51	14.53
Coal	0.02	14.55
Fireclay, sandy, nodular; carbonaceous and slickensided at top	0.87	15.42
Sandstone, ganisteroid; argillaceous at base	0.73	16.15
Mudstone, sandy, and shale, carbonaceous	0.21	16.36
TOP BROCKWELL COAL		
Coal, three thin seams with two dirt partings	0.56	16.92
Fireclay, black, sandy, carbonaceous	0.30	17.22
Sandstone, fine- to medium-grained, flaggy; thin beds of ankeritic cement; massive and coarser to base	5.34	22.56
Mudstone, sandy; thin sandstone partings	0.76	23.32
Mudstone, sandy; sandstone partings at top	0.45	23.77
BOTTOM BROCKWELL COAL		
Coal	0.59	24.36
Fireclay, sandy, nodular; slickensided; shale rib at top	0.94	25.30
Coal, four thin seams and three dirt partings	0.39	25.69
Fireclay, slickensided	0.52	26.21
Mudstone, sandy; beds of sandstone	1.22	27.43
Sandstone, predominantly fine-grained, flaggy, ankeritic; shale fragments at top	1.07	28.50
Mudstone, sandy; partings of flaggy sandstone	0.91	29.41
Mudstone, sandy; partings of sandstone; sporadic, commonly fragmentary *Curvirimula* sp.	2.29	31.70
Mudstone with ironstone rib; poorly preserved *Carbonicola* sp.	1.57	33.27
Shale, splintery; abundant fish spines and scales	0.26	33.53
Mudstone, very sandy, micaceous; poorly preserved *Carbonicola* sp.	0.61	34.14
Sandstone, fine-grained, flaggy	0.15	34.29
Mudstone; ironstone beds; *Naiadites* sp. at base (base of VICTORIA SHELL BED)	0.76	35.05
Sandstone, argillaceous, flinty; 0.25 m shale bed at base	0.71	35.76
TOP VICTORIA COAL		
Coal	0.03	35.79
Shale and fireclay	0.76	36.55
Shale with coal parting	0.01	36.56
Fireclay, sandy and nodular to base	0.83	37.39
Sandstone, rubbly, earthy, flaggy, fine-grained; ganisteroid at top	0.86	38.25
Mudstone, ankeritic; slickensided at base	0.84	39.09
Mudstone, sandy; alternating with fine-grained sandstone	0.91	40.00
Mudstone, slickensided	0.49	40.49
BOTTOM VICTORIA COAL		
Coal	0.23	40.72
Sandstone, rubbly, ganisteroid; black and carbonaceous at top, shaly at base	0.68	41.40
Mudstone, sandy, rooty, nodular; beds and partings of sandstone	0.36	41.76
Sandstone, flaggy, fine- to medium-grained; micaceous; ankerite veins	2.13+	43.89

Heathery Lane No. 11 Borehole, North Gosforth (NZ 26NE/1). 1:50 000 Sheet 14 (Morpeth).

Surface Level + c.48.77 m [2524 6923].
Drilled 1953 for NCB. Shown graphically in Figures 16, 18, 20 and 22. Logged by R H Price and J B W Day. This borehole is included as it constitutes the only modern continuous record of the higher Westphalian strata occurring within the Newcastle district. It is 498 m deep and proves strata from above the Hebburn Fell Coal to below the Brockwell Coal. Synopsis below Bottom Main Coal at 291.28 m. The borehole is located some 2 km to the north of the district.

	Thickness m	*Depth* m
Drift		
Boulders, sandy clay and boulder clay	36.42	36.42
Middle Coal Measures		
Strata not present in the Newcastle district, including HEBBURN FELL and USWORTH coals; GRINDSTONE POST absent	43.43	79.85
Mudstone, grey, sandy	4.42	84.27
Mudstone; occasional irony bands; ?ostracods	10.11	94.38
Shale, rooty, micaceous, pyritic; worm tracks. *Lingula* spp. at 94.48 m, ribbed marine shell at base (RYHOPE MARINE BAND)	0.26	94.64
Sandstone, predominantly massive, coarse-grained, kaolinitic; pyritous pellets, dark partings; fireclay rib and rooty micaceous sandstone 0.45 m at top (part of SEVENTY FATHOM POST)	18.39	113.03
?BURRADON COAL		
Coal	0.17	113.20
Fireclay	0.08	113.28
Coal	0.14	113.42
Coal, shaly	0.14	113.56
Fireclay, grading to mudstone at base	2.11	115.67
Sandstone, predominantly medium-grained, wispy bedded; shaly at top and base, slumped with shale pellets at base; uneven base (part of SEVENTY FATHOM POST)	8.38	124.05
Mudstone and shale, sandy (base at horizon of HYLTON MARINE BAND)	0.68	124.73
HYLTON MARINE BAND COAL		
Coal, shaly	0.26	124.99
Fireclay	0.28	125.27
Mudstone, sandy	0.89	126.16
Sandstone, flaggy	0.96	127.12
Mudstone	0.71	127.83
Sandstone, flaggy, medium-grained; occasional shale beds; coarse-grained at base	7.42	135.25
Coal	0.10	135.35
Fireclay, sandy to base	0.38	135.73
Sandstone, predominantly flaggy, fine-grained; some massive beds	7.57	143.30
Shale, sandy, micaceous; sandy partings and patches	2.08	145.38
Mudstone; sandy ribs and partings	1.43	146.81
Sandstone, fine-grained, flaggy; mudstone beds	0.49	147.30
Mudstone; occasional sandy beds; fragmentary mussels at base	0.47	147.77
Mudstone, dark grey, mottled; occasional sandy micaceous films and flaggy sandstone partings	0.41	148.18
Mudstone, sandy; micaceous sandstone partings in basal 0.81 m	1.12	149.30
Mudstone, smooth; *'Estheria'* at 149.50 m and 150.11 m, ?crustacean at 149.80 m	0.96	150.26
KIRKBY'S MARINE BAND		
Shale, black; *Lingula* sp.	0.31	150.57
Shale, sandy, micaceous; pyritous tubules	0.16	150.73
Mudstone, sandy; mussels	0.60	151.33
Shale, black; *Lingula* sp	0.30	151.63
Coal; shale partings (?CROW COAL)	0.23	151.86
Fireclay-mudstone	0.31	152.17
Sandstone, rooty, flaggy; shaly partings	0.45	152.62
Mudstone, sandy, micaceous	0.33	152.95
Sandstone, argillaceous; wisps of flaggy sandstone	0.44	153.39
Mudstone, rooty; occasional ironstone nodules; grades into sandy mudstone towards base	1.37	154.76
Sandstone, medium-grained, ankeritic; flaggy at base	4.06	158.82
Mudstone, sandy; grades down into smooth irony mudstone	0.86	159.68
Shale, black; *Naiadites* sp. and ostracods at base	1.07	160.75
Fireclay and fireclay-mudstone; sandy at base; grades to argillaceous sandstone	2.62	163.37
Mudstone, sandy, with *Neuropteris* sp. and *Calamites* sp.; alternates with argillaceous sandstone	6.40	169.77
Mudstone, sandy; *Neuropteris* sp.	2.03	171.80
Shale, carbonaceous	0.10	171.90
RYHOPE FIVE-QUARTER COAL		
Coal	0.21	172.11
Shale and fireclay-mudstone	0.61	172.72
Coal	0.73	173.45
Fireclay and fireclay-mudstone	2.57	176.02
Mudstone, sandy; ironstone ribs	4.01	180.03
Mudstone; ostracods and mussels	0.10	180.13
Sandstone, argillaceous	0.21	180.34
Mudstone, sandy; mussels	0.10	180.44
Mudstone, predominantly sandy; thin sandstone partings; sporadic mussels	1.83	182.27
Shale, black, with ironstone nodules; mussels (assumed position of LITTLE MARINE BAND)	0.61	182.88
Coal traces and shale	0.22	183.10
Shale, black; pyritous at base	0.59	183.69
RYHOPE LITTLE COAL		
Coal	0.56	184.25
Fireclay, sandy, and sandstone, rooty	1.06	185.31
Sandstone, rooty	0.89	186.20
Sandstone, flaggy, medium-grained; locally massive with ankeritic cement; shaly partings	5.06	191.26
Mudstone, sandy; sandstone rib at top; *?'Estheria'* at 192.78 m	3.02	194.28
Mudstone; ironstone nodules	0.69	194.97
Shale, black and dark grey; fish scales and remains at 195.32 m and 195.85 m	0.88	195.85
MOORLAND COAL		
Coal	0.23	196.08
Fireclay; sandy beds	1.98	198.06
Mudstone, sandy; sandstone partings	2.14	200.20
Sandstone, medium-grained, kaolinitic; flaggy at base	2.69	202.89
Sandstone, flaggy, wispy; alternating with sandy mudstone	1.02	203.91
Mudstone, sandy, wispy; slickensided at base	1.52	205.43

	Thickness m	Depth m
Sandstone, predominantly flaggy, wispy bedded, locally shaly	4.39	209.82
Shale; coal traces	0.18	210.00
Mudstone, sandy, with thin sandstone partings; mussels at base	0.61	210.61
Shale and mudstone; poorly preserved mussels	1.02	211.63
Shale; coal traces	0.02	211.65
Sandstone, rooty, micaceous	0.64	212.29
Sandstone, flaggy, wispy bedded; black micaceous partings	2.41	214.70
Mudstone, sandy; mussels including *Naiadites* sp. towards base	1.53	216.23
Mudstone; scattered mussels	0.91	217.14
HIGH MAIN SHELL BED		
Mudstone, dark grey to black; ironstone nodules; mussels	4.52	221.66
Shale, black, micaceous; coaly debris (horizon of HIGH MAIN MARINE BAND)	0.03	221.69
HIGH MAIN MARINE BAND COAL		
Coal	0.33	222.02
Fireclay, rubbly, sandy; grades to ganisteroid sandstone	0.38	222.40
HIGH MAIN POST		
Sandstone, medium- to coarse-grained becoming finer downwards with micaceous partings	18.82	241.22
Broken strata	0.13	241.35
HIGH MAIN COAL		
Coal	0.91	242.26
Fireclay, soft, shaly	1.55	243.81
Mudstone, blocky; small ironstone nodules	0.53	244.34
Sandstone, fine-grained, shaly and wispy; argillaceous to base	1.12	245.46
Mudstone, sandy; irony ribs; sandstone partings; *Neuropteris* sp.	2.52	247.98
Shale, black, sandy; irony ribs; fish debris at 248.64 m	0.91	248.89
Shale, sandy; thin sandstone wisps	3.12	252.01
Shale, black, canneloid; ankerite veins; fish scales	0.05	252.06
Mudstone, greenish grey, blocky; mussels at base	1.73	253.79
Shale, black; fish scales	0.03	253.82
TOP FIVE-QUARTER COAL		
Coal	0.56	254.38
Fireclay, shaly; ironstone nodules	0.73	255.11
Mudstone, sandy; ironstone nodules	1.27	256.38
Shale, greenish grey, and mudstone, sandy	0.59	256.97
Mudstone, sandy; alternates with shaly and wispy bedded sandstone with slump structures and irony beds	6.98	263.95
Shale, black, sandy, micaceous; ?mussel fragment	1.50	265.45
BOTTOM FIVE-QUARTER (STONE) COAL		
Coal	0.25	265.70
Shale; coal traces and films, rooty	0.16	265.86
Coal	0.02	265.88
Seatearth, sandy, shaly; ironstone nodules	1.55	267.43
Shale, black, sandy, micaceous, pyritous; mussels	0.31	267.74
Sandstone, argillaceous, rooty; coarser partings	0.61	268.35
Sandstone, shaly, alternating with mudstone, sandy	1.14	269.49
Mudstone, sandy, wispy bedded; micaceous sandstone partings	0.71	270.20
Sandstone, medium-grained, ankeritic cement; massive at top, flaggy below	2.72	272.92
Mudstone, sandy; irony beds; mussels at 273.40 m and 274.87 m	2.92	275.84
Mudstone with ribs of ironstone; sporadic mussels	1.83	277.67
Mudstone, smooth; ghost mussels	0.12	277.79
TOP MAIN (BENTINCK) COAL		
Coal	0.26	278.05
Shale, black, carbonaceous, rooty; coal streaks	0.35	278.40
Coal	0.16	278.56
Fireclay, sandy	0.61	279.17
Mudstone, sandy; sandstone ribs	3.63	282.80
Mudstone	0.07	282.87
Sandstone, argillaceous at top; coarser and flaggy to base	2.65	285.52
Sandstone, medium-grained, shaly at top; more massive below with flaggy beds	3.65	289.17
Mudstone, sandy, alternating with sandstone, argillaceous; irony ribs	1.10	290.27
BOTTOM MAIN (YARD) COAL		
Coal	0.33	290.60
Shale, black, pyritous	0.07	290.67
Coal	0.38	291.05
Shale, black, carbonaceous	0.05	291.10
Coal	0.18	291.28
Strata including MAUDLIN (BENSHAM), DURHAM LOW MAIN, BRASS THILL, and PLESSEY COALS, with base of **Middle Coal Measures** at c. 404.16 m, HARVEY, TILLEY group, TOP BUSTY, BOTTOM BUSTY, THREE-QUARTER and BROCKWELL coals in **Lower Coal Measures**	207.11	498.39
Total depth		498.39

Kibblesworth Borehole (NZ 25 NW/362)

Surface level + 94.72 m [2340 5731].
Drilled 1965 for NCB. Shown graphically in Figures 17, 19 and 20.

	Thickness m	Depth m
Drift		
Sand, pebbles and boulders	1.82	1.82
Clay	1.22	3.04
Gravel	1.53	4.57
Middle Coal Measures		
Shale, including ?old workings of TOP FIVE-QUARTER COAL (open holed to 6.40 m)	1.83	6.40
Sandstone, medium-grained, laminated and wispy bedded; occasional siltstone laminae; cavity at 12.72 m may represent old workings of BOTTOM FIVE-QUARTER COAL	7.56	13.96
Mudstone, silty; irony ribs and ironstone nodules; shattered at base	3.10	17.06
Old workings of MAIN COAL	0.56	17.62
Seatearth, silty, and sandy at base	2.49	20.11
Sandstone, medium-grained, rooty at top	1.22	21.33
Seatearth	0.20	21.53
Siltstone, abundant roots, irregularly laminated; grades to fine-grained laminated sandstone below 22.55 m; erosive base	3.99	25.52

	Thickness m	*Depth* m
Mudstone, slightly silty and shaly; local poorly preserved mussels and some fish debris (?Blackhall 'Estheria' Band)	0.69	26.21
Mudstone, flinty and rooty at top; silty beds and some ironstone beds; sporadic mussels	1.67	27.88
Siltstone; sandstone beds	0.77	28.65
Mudstone with leached appearance; abundant mussels at base	1.22	29.87
Sandstone, irregularly bedded at top, massive at base	0.61	30.48
Mudstone; irony patches and beds; mussels at base	1.39	31.87
Coal	0.13	32.00
Seatearth and coaly shale; 0.15 m ganister at base	1.37	33.37
Sandstone, ripple-laminated, medium-grained; sporadic beds of siltstone and mudstone; fine-grained and rooty at top	5.72	39.09
Mudstone; thin ironstone ribs; sporadic small mussels; *Planolites* sp.	0.73	39.82
Maudlin Coal		
Coal	0.41	40.23
Seatearth	0.30	40.53
Coal	0.21	40.74
Seatearth, silty at base	0.71	41.45
Siltstone, rooty	0.40	41.85
Durham Low Main Post		
Sandstone, predominantly pale grey, medium-grained, irregularly laminated, micaceous below 42.36 m; massive below 46.63 m; beds of mudstone breccia below 60.35 m; erosive base	23.63	65.48
Durham Low Main Coal		
Coal	1.11	66.59
Seatearth	0.11	66.70
Coal	0.05	66.75
Seatearth, shaly, carbonaceous; coal beds up to 0.02 m	1.37	68.12
Top Brass Thill Coal		
Coal, inferior and shaly	0.35	68.47
Seatearth	0.11	68.58
Bottom Brass Thill Coal		
Goaf	1.52	70.10
Seatearth	0.05	70.15
Coal	0.05	70.20
Seatearth; coal veins	0.36	70.56
Siltstone, micaceous and carbonaceous	0.53	71.09
Sandstone, predominantly medium-grained, micaceous; sporadic thin siltstone beds; 0.38 m siltstone at 72.84 m	5.36	76.45
Mudstone, silty; pyritic ironstone ribs at base; *Planolites* sp.	0.91	77.36
Hutton Coal		
Coal	1.58	78.94
Seatearth, grading to rooty siltstone	2.13	81.07
Siltstone	0.46	81.53
Sandstone, very fine-grained with small-scale slump structures; becomes massive and medium-grained to base	2.89	84.42
Seatearth, silty with pale sandstone laminae; grades to siltstone towards base	3.66	88.08
Mudstone, grey, thin ironstone ribs; mussels, abundant at certain levels (?Plessey Shell Bed)	2.90	90.98
Sandstone, predominantly fine-grained, irregularly bedded, locally flaser-bedded	7.54	98.52
Mudstone; *Planolites* sp.	0.10	98.62
Ruler Coal		
Coal	0.03	98.65
Seatearth	0.05	98.70
Coal	0.46	99.16
Seatearth; coal veins at top	0.30	99.46
Sandstone, fine-grained grading to medium-grained, irregularly laminated; siltstone beds near base	7.22	106.68
Mudstone, silty at top; mussels towards base	1.21	107.89
Sandstone, fine-grained, laminated	0.61	108.50
Mudstone, black, shaly, often silty and micaceous with ironstone ribs; mussels abundant in lower part (Harvey Shell Bed)	3.20	111.70
Harvey (Vanderbeckei) Marine Band		
Mudstone, black; abundant *Lingula* sp.; ?foraminifera	0.16	111.86
Lower Coal Measures		
Coal	0.10	111.96
Seatearth grading to fine-grained rooty sandstone; mudstone pellets at base	0.66	112.62
Mudstone, rooty	0.15	112.77
Sandstone, fine-grained	0.92	113.69
Mudstone, silty, with medium-grained sandstone beds at top	2.03	115.72
Sandstone, fine-grained; ironstone nodules at top	0.81	116.53
Mudstone; irregular irony layers and nodules; listric surfaces at base	1.96	118.49
Sandstone, fine-grained, rooty	0.38	118.87
Siltstone and silty mudstone; irony beds at top; sandstone beds and laminae below	1.83	120.70
Mudstone; ironstone ribs and nodules; abundant mussels at several horizons; abundant ostracods in 0.12 m bed at 122.14 m; fragmental clay rock in 0.05 m bed at 122.19 m (Hopkins Band)	1.52	122.22
Harvey Coal		
Coal	0.38	122.60
Seatearth; irony nodules; silty below 123.44 m	1.60	124.20
Sandstone, fine-grained, wispy bedded	4.17	128.37
Coal (?Hodge Coal)	0.05	128.42
Seatearth; abundant small pellets at top	0.71	129.13
Sandstone, rooty at top, otherwise medium-grained, wispy bedded	5.54	134.67
Coal	0.05	134.72
Seatearth, silty, with fine-grained sandstone beds	2.13	136.85
Mudstone, silty	0.13	136.98
Tilley Coal		
Coal	0.30	137.28
Fragmental clay rock	0.03	137.31
Coal	0.20	137.51
Seatearth, silty at top; some ironstone nodules	1.30	138.81
Sandstone, fine-grained, rooty at top; below 141.57 m becomes medium-grained, wispy bedded; erosive base	5.05	143.86
Coal	0.03	143.89
Seatearth, silty; grades to seatearth-siltstone with fine-grained sandstone layers and ironstone nodules	1.19	145.08

	Thickness m	Depth m
(Core largely missing 145.08 m to 149.96 m; fragments of siltstone, sandstone and mudstone)	4.88	149.96
Mudstone, silty; siltstone beds	1.09	151.05
Sandstone, fine-grained, micaceous, grading to siltstone at base	0.41	151.46
Mudstone, silty, grading to mudstone	0.07	151.53
TOP BUSTY COAL		
Coal	0.38	151.91
Mudstone	0.05	151.96
Coal	0.16	152.12
Seatearth, rapidly becoming silty; grades to medium-grained, irregularly bedded sandstone; some ironstone nodules towards base	2.48	154.60
Sandstone, medium-grained	1.45	156.05
Siltstone	1.40	157.45
Mudstone	0.18	157.63
BOTTOM BUSTY COAL		
Coal	0.02	157.65
Seatearth	0.05	157.70
Coal	0.61	158.31
Seatearth, coal veins at top; ironstone nodules at base	1.40	159.71
Siltstone, rooty at top, finely micaceous; sandstone beds	2.13	161.84
Mudstone, silty, finely micaceous; fine-grained sandstone beds	1.45	163.29
Sandstone, coarse-grained, massive at top, becoming medium-grained, laminated and wispy bedded below 163.98 m; conglomeratic with mudstone pellets at base	6.53	169.82
THREE-QUARTER COAL		
Coal	0.18	170.00
Seatearth, silty; some ironstone nodules	1.14	171.14
Siltstone, rooty, with medium- to fine-grained rooty sandstone beds	1.37	172.51
Mudstone, rooty; siltstone beds at base	1.98	174.49
Coal and coaly shale	0.05	174.54
Seatearth; grades to medium-grained sandstone	1.02	175.56
Siltstone; sandstone beds	1.52	177.08
Mudstone; mussels preserved in pyrites	0.08	177.16
Mudstone, silty	0.18	177.34
Mudstone with thin ironstone beds; mussels preserved in pyrites	0.68	178.02
Mudstone; some ironstone laminae; mussels, ostracods, *Spirorbis* sp. and ?fish debris	0.11	178.13
Seatearth	0.30	178.43
TOP BROCKWELL COAL		
Coal, four thin seams up to 0.35 m, and three partings of mudstone	0.84	179.27
Seatearth, silty; sandstone beds at base	0.43	179.70
Sandstone, medium-grained, wispy bedded with siltstone beds; ironstone nodules at 180.44 m; grades to siltstone with fine sandstone beds	2.95	182.65
BOTTOM BROCKWELL COAL		
Coal	0.40	183.05
Seatearth; sphaerosiderite at base	2.11	185.16
Siltstone; sandstone beds; mudstone rib at base	1.22	186.38
Coal	0.15	186.53
Seatearth, silty; ironstone nodules (core missing at top)	1.07	187.60
Siltstone; fine-grained sandstone beds	0.61+	188.21

DW/C Borehole, Lambshield Moss (NZ 04 NW/4)

Surface level + 326.26 m [0282 4834].
Drilled 1970. Cored below 60.00 m. Logged by D A C Mills and P G Carter (Babtie Group). Shown graphically in Figures 10 and 11.

	Thickness m	Depth m
Drift		
Open hole (boulder clay)	2.82	2.82
Stainmore Group		
Open hole (interbedded sandstones and shales; sandstones mainly medium- to coarse-grained)	57.18	60.00
Mudstone, calcareous, laminated to thin-bedded, finely micaceous and carbonaceous; fossil debris including brachiopods at base	1.00	61.00
Mudstone, calcareous, splintery	0.20	61.20
Seatearth-siltstone, rooty, micaceous and carbonaceous	0.20	61.40
Sandstone, thin-bedded, fine-grained, micaceous and carbonaceous; silty and muddy at top	2.40	63.80
Siltstone, laminated to thin-bedded; locally grades to seatearth	0.60	64.40
Sandstone, thin- to medium-bedded, fine-grained; muddy, micaceous and carbonaceous at base	2.15	66.55
Mudstone, sandy, micaceous and carbonaceous	1.30	67.85
Sandstone, predominantly muddy, thin- to medium-bedded, fine- to medium-grained; very muddy at base with increasing mudstone fraction	5.27	73.12
Mudstone, sandy, laminated to thin-bedded; fossiliferous	3.08	76.20
Limestone, sandy, thin-bedded, fine calcarenite; fossiliferous	0.40	76.60
Mudstone rib	0.34	76.94
Mudstone, calcareous, and limestone, impure; calcareous sandstone interlaminae	0.83	77.77
Mudstone, calcareous, laminated to very thin-bedded; fossiliferous	0.79	78.56
GRINDSTONE LIMESTONE		
Limestone, muddy, impure, medium-bedded, fine-grained; fossiliferous	2.34	80.90
Mudstone, sandy, laminated; fine-grained sandstone interlaminae; locally calcareous and fossiliferous	0.31	81.21
GRINDSTONE SILL		
Sandstone, grey, locally greenish, brown and purple mottled, predominantly muddy, impure, thin- to medium-bedded, fine-grained; calcareous patches and blebs; below 82.50 m rock becomes generally more homogeneous, compact, siliceous, medium-grained	7.28	88.49

	Thickness m	*Depth* m
Mudstone, often sandy, laminated to thin-bedded, highly micaceous and carbonaceous; interlaminae of muddy, micaceous, carbonaceous sandstone at top; becomes highly fissile to base	13.91	102.40
UPPER FELLTOP LIMESTONE		
Limestone, muddy, thin- to medium-bedded, fine- and very fine-grained; comminuted fossil debris; muddy and shaly towards base	2.04	104.44
HIPPLE SILL		
Sandstone, thin- to medium-bedded, predominantly fine-grained; calcareous and bioturbated at top; 1.41 m sandy mudstone at 107.95 m; 0.30 m siliceous rooty ganisteroid bed at 107.95 m; becomes thin-bedded, micaceous and carbonaceous towards base	5.89	110.33
Mudstone and sandstone, interbedded and interlaminated, very fine- and fine-grained, micaceous and carbonaceous	2.67	113.00
Mudstone, laminated to very thin-bedded, very finely micaceous and carbonaceous; scattered fossil debris throughout, but particularly at 129.32 to 130.38 m (?COALCLEUGH SHELL BED)	19.31	132.31
?HIGH GRIT SILL		
Sandstone, thin- to thick-bedded, predominantly fine- to medium-grained, micaceous and carbonaceous; medium-grained towards base	3.31	135.62
LOWER FELLTOP LIMESTONE		
Limestone, grey, sandy, thin-bedded; fine- to medium-grained; shell debris; inclined base	0.42	136.04
Sandstone, largely as at 135.62 m; sporadic shaly beds	6.26	142.30
Mudstone, sandy, laminated to thin-bedded, micaceous and carbonaceous; sandy interlaminae and beds; 0.03 m calcareous bed with fossil debris at 142.91 m; sporadic small ironstone nodules and laminae; shaly and highly fissile at base	6.28	148.58
HIGH SLATE SILL		
Sandstone, medium- to thick-bedded, predominantly medium- to coarse-grained, irregular micaceous blebs and partings; sporadic shaly partings and beds	12.51	161.09
Mudstone, silty, laminated to thin-bedded, highly micaceous and carbonaceous; rare irony blebs	5.02	166.11
LOW GRIT SILL		
Sandstone, medium- to thick-bedded, fine- to medium-grained, locally bioturbated, micaceous and carbonaceous; upper 0.25 m calcareous (? position of Aydon Shell Bed); below 167.11 m more coarse-grained; basal 4.10 m interbedded and interlaminated with micaceous carbonaceous mudstone	11.81	177.92
Mudstone, as at 166.11 m	8.05	185.97
KNUCTON SHELL BEDS (?PLANKEY SHELL BED)		
Sandstone, locally calcareous, medium-bedded, fine-grained	0.76	186.73
Sandstone, medium-bedded, fine-grained, siliceous; upper 0.25 m bioturbated; becomes muddy towards base with mudstone interlaminae	2.62	189.35
Mudstone, silty, laminated to very thin-bedded; shelly at 190.53 m, locally calcareous; sporadic ironstone nodules towards base	8.68	198.03
CRAG LIMESTONE		
Limestone, medium- to thick-bedded, finely crystalline, muddy and argillaceous at top and base; 0.1 m calcareous mudstone at 199.53 m; abundant comminuted fossil debris	2.00	200.03
Mudstone, laminated, fossiliferous; coaly at base	0.54	200.57
FIRESTONE SILL		
Sandstone, predominantly medium-bedded, fine-grained, siliceous, micaceous and carbonaceous	1.93+	202.50

Letch House No. 2 Borehole, Greymare Hill (NZ 05 NW/4b)

Surface level + 240.24 m [0421 5580].
Drilled 1971 for Northumbrian River Authority. Logged by D A C Mills and P G Carter (Babtie Group).

	Thickness m	*Depth* m
Drift		
Boulder clay, sandy	1.25	1.25
Lower Coal Measures		
Sandstone, thin-bedded, predominantly medium-grained, micaceous and carbonaceous, vertically jointed; below 5.80 m grades down into very thin-bedded siltstone (open holed to 2.0 m)	5.35	6.60
Mudstone, laminated, finely micaceous and carbonaceous	1.89	8.49
VICTORIA GROUP (to 25.14 m)		
TOP VICTORIA COAL		
Coal	0.15	8.64
Seatearth-mudstone, irregularly bedded, slickensided; grades to mudstone, micaceous and carbonaceous, rooty	5.47	14.11
TOP VICTORIA COAL		
Coal	0.34	14.45
Shale	0.28	14.73
Coal	0.24	14.97
Seatearth-mudstone and siltstone, irregularly thin-bedded, rooty, micaceous and carbonaceous; grades to thin-bedded, grey, muddy siltstone at base	1.46	16.43
Sandstone, thin-bedded to shaly, fine-grained, micaceous and carbonaceous; becomes medium-bedded below 17.31 m	2.28	18.71
Mudstone and siltstone, sandy, laminated to very thin-bedded, micaceous and carbonaceous; coaly to base	1.62	20.33
BOTTOM VICTORIA COAL		
Coal	0.27	20.60
Seatearth-mudstone and -siltstone, irregularly thin-bedded, carbonaceous, micaceous; dark grey to black at base	1.40	22.00

	Thickness m	Depth m
BOTTOM VICTORIA COAL		
Coal	0.09	22.09
Seatearth and shale	0.10	22.19
Sandstone, thin- to medium-bedded, fine- to medium-grained, micaceous and carbonaceous	2.15	24.34
Mudstone	0.56	24.90
BOTTOM VICTORIA COAL		
Coal	0.24	25.14
Seatearth	0.10	25.24
Sandstone, medium-bedded, medium- to coarse-grained, locally coarse-grained; subhorizontal to subvertical joints	11.30+	36.54

New Pit, West Stanley (NZ 15 SE/100)

Surface level + 223.72 m [1996 5272].
Sunk 1876. Synopsis below 114.75 m (Main Coal).
Shown graphically in Figures 19–21.

	Thickness m	Depth m
Drift		
Outset, soil and gravelly clay	3.96	3.96
Sand, loamy	0.61	4.57
Clay, blue	0.91	5.48
Middle Coal Measures		
Sandstone, rubbly	0.92	6.40
Sandstone, yellow	6.40	12.80
RYHOPE FIVE-QUARTER COAL		
Coal	0.38	13.18
Seatearth	1.45	14.63
Mudstone and shale, grey and blue	5.48	20.11
Sandstone, brown	2.19	22.30
Shale, black	0.40	22.70
Seatearth	1.37	24.07
Mudstone, blue	0.61	24.68
Sandstone, brown	6.71	31.39
Mudstone, blue	0.69	32.08
RYHOPE LITTLE COAL		
Coal	0.69	32.77
Seatearth	0.75	33.52
Mudstone, blue	1.22	34.74
Sandstone, hard, white	1.83	36.57
Mudstone, grey	0.76	37.33
Shale, blue	10.21	47.54
?CROW NO. 2 COAL		
Coal	0.77	48.31
Seatearth	1.37	49.68
Sandstone, grey, laminated	2.28	51.96
Mudstone and shale, blue: base at assumed position of HIGH MAIN MARINE BAND	5.39	57.35
HIGH MAIN MARINE BAND COAL		
Coal	0.10	57.45
Seatearth and seatearth-sandstone	0.76	58.21
HIGH MAIN POST		
Sandstone, brown	12.19	70.40
HIGH MAIN COAL		
Coal	0.56	70.96
Seatearth	0.16	71.12
Coal	1.39	72.51
Seatearth	2.16	74.67
Sandstone, white, hard	2.74	77.41
Mudstone, grey	4.12	81.53
?METAL COAL		
Coal	0.15	81.68
Seatearth	0.92	82.60
Mudstone, grey	2.74	85.34
Sandstone, grey, jointed; some ironstone	2.59	87.93
Mudstone, grey	8.23	96.16
FIVE-QUARTER COAL		
Coal	1.22	97.38
Seatearth	1.22	98.60
Mudstone and shale	11.12	109.72
Shale, blue	3.51	113.23
MAIN COAL		
Coal	1.52	114.75
Strata (mostly sandstone towards base)	39.32	154.07
MAUDLIN COAL		
Coal	0.46	154.53
Strata (including DURHAM LOW MAIN POST)	15.54	170.07
DURHAM LOW MAIN and BRASS THILL COAL		
Coal	1.38	171.45
Strata	7.01	178.46
HUTTON COAL		
Coal	1.14	179.60
Strata, including position of HARVEY (VANDERBECKEI) MARINE BAND at base	40.69	220.29
Lower Coal Measures		
Seatearth	1.83	222.12
Coal	0.20	222.32
Strata	3.96	226.28
HARVEY COAL		
Coal	0.61	226.89
Strata (including 0.15 m coal at 236.70 m)	16.84	243.73
TILLEY COAL		
Coal	0.99	244.72
Seatearth	0.11	244.83
Coal, cannel	0.25	245.08
Strata (mostly sandstone)	8.33	253.41
BUSTY COAL (combined)		
Coal	0.15	253.56
Shale, black	0.23	253.79
Coal	1.02	254.81
Seatearth and clay	0.76	255.57
Coal	0.92	256.49
Strata (mostly sandstone in lower part)	23.82	280.31
THREE-QUARTER COAL		
Coal (NCB depth recorded as 280.11 m)	0.36	280.67
Strata	9.49	290.16
TOP BROCKWELL COAL		
Coal (NCB depth recorded as 285.60 m)	0.39	290.55
Strata	5.76	296.31
BOTTOM BROCKWELL COAL		
Coal; 0.05 m cannel at base (NCB depth recorded as 294.13 m)	0.61	296.92
Strata, including 0.12 m coal at 298.01 m	13.34	310.26
?VICTORIA COAL		
Coal, with 0.10 m dirt parting near base	0.63	310.89
Strata	5.49+	316.38

Rowlands Gill Heat Flow Borehole (NZ 15 NE/276)

Surface level + 43.11 m [1664 5815].
Drilled 1986 on behalf of British Geological Survey.
Cored below 28.36 m. Logged by E W Johnson, I Jackson and D J D Lawrence. Shown graphically in Figure 12.

	Thickness m	*Depth* m
Drift		
Sand, gravel and boulder clay (open hole)	26.70	26.70
Lower Coal Measures		
Open hole (sandstone)	1.66	28.36
Sandstone, medium- and thin-bedded, fine- to medium-grained	3.61	31.97
Coal	0.08	32.05
Sandstone, medium- and thin-bedded, fine- to medium-grained; 0.7 m seatearth at top	2.59	34.64
Siltstone, grey, laminated, in part sandy, micaceous; scattered plant debris	3.86	38.50
Mudstone, silty; plant fragments	2.32	40.82
Sandstone, thin- and medium-bedded; silty micaceous partings; coal 0.1 m at 40.90 m; 0.57 m seatearth at 41.85 m	2.68	43.50
Mudstone and siltstone, interbedded; coarsens upwards into overlying sandstone; sparse fauna below 44.50 m including *Lingula* sp.; turritellid gastropods and shell fragments; sharp base	1.85	45.35
Coal (?SALTWICK)	0.08	45.43
Sandstone-seatearth; thin siltstone interbeds towards base	1.96	47.39
Mudstone and siltstone, interbedded, dark grey; plant fragments abundant above 49.50 m; ironstone laminae and lenses below 49.00 m	4.28	51.67
Coal, bright	0.13	51.80
Siltstone, grey, coarse-grained; seatearth at top	2.91	54.71
Conglomerate (intraformational); clasts of sandstone in siltstone matrix; thin coal layer; abundant plant fragments at top	0.60	55.31
Siltstone, grey, laminated; abundant plant fragments; *Planolites* sp. and bivalves at 55.50 to 55.60 m – assumed position of QUARTERBURN (SUBCRENATUM) MARINE BAND	1.77	57.08
Stainmore Group		
Sandstone, predominantly medium-bedded and fine- to medium-grained but sporadic coarser beds; 0.27 m seatearth at 57.38 m; productoid at 57.09 m	2.72	59.80
Siltstone, laminated, sandy at top; fauna includes brachiopods, and bivalves below 60.80 m	2.02	61.82
Sandstone, mainly thin-bedded; silty and micaceous	3.48	65.30
Siltstone, sandstone laminae at top; *Lingula* sp., fish fragments and shell debris 69.97 to 70.26 m; septarian limestone with shell fragments 70.31 to 70.49 m; *Lingula* sp., ?goniatite and fish fragments 70.49 to 70.74 m; 0.12 m cannel coal at 70.86 m; seatearth-siltstone 71.95 to 72.40 m; sharp irregular base	10.64	75.94
Sandstone, thin- to medium-bedded, fine-grained; siltstone laminae increase to base and gradational to	2.44	78.38
Siltstone, laminated, in part sandy; *Planolites* sp. at 81.90 m	3.71	82.09
Sandstone, thin- to medium-bedded, fine-grained; 0.25 m sandy limestone bed with shell fragments at 82.60 m	1.11	83.20
Sandstone, siltstone and mudstone, interlaminated and interbedded; sparse fauna including *Planolites* sp. and bivalves from 89.10 to 89.30 m	6.41	89.61
Sandstone, thin- and medium-bedded, fine- to medium-grained; silty micaceous partings	2.14	91.75
Siltstone, laminated, in part sandy; abundant plant fragments	1.50	93.25
Coal and mudstone, carbonaceous	0.12	93.37
Sandstone, siltstone and mudstone, interlaminated and interbedded; plant debris locally abundant; fauna of brachiopods between 99.06 and 99.20 m; sandstone interbeds range between fine- and very coarse-grained; soft sediment deformation and loading common in sandstone; seatearth between 101.71 and 102.70 m	11.25	104.62
Sandstone, thin- and medium-bedded, fine- to very coarse-grained; silty micaceous partings; quartz pebbles at base (?'First Grit')	6.53	111.15
Sandstone and siltstone, interbedded; plant debris; brachiopods at 112.45 m	5.05	116.20
Siltstone and mudstone, sparse ironstone nodules below 118.60 m; fauna including *Planolites* sp., brachiopods and bivalves below 118.40 m	2.81	119.01
WHITEHOUSE LIMESTONE		
Siltstone, calcareous, laminated; abundant fauna of brachiopods and crinoids; 0.05 m limestone rib at base	1.64	120.65
Sandstone, siltstone and mudstone, interlaminated and interbedded with mudstone, micaceous with abundant plant debris; *Lingula* sp. and bivalves in 0.70 m bed at 128.60 m	13.55	134.20
GRINDSTONE SILL		
Sandstone, medium- and thick-bedded, mainly coarse-grained, feldspathic; siltstone clasts and quartz pebbles in basal 2.0 m; sharp base	8.26	142.46
Sandstone, medium- and thin-bedded, medium-grained; silty micaceous laminae increasing towards base	3.74	146.20
Sandstone and siltstone, interbedded; two seatearths, 1.00 m and 1.21 m at 148.60 m and 151.00 m respectively; soft sediment deformation common in more argillaceous beds	5.20	151.40
Sandstone, predominantly thin- and medium-bedded, fine- to medium grained; mainly silty and micaceous	6.20	157.60
Sandstone and siltstone, interbedded, fine-grained, silty, micaceous	4.85	162.45
Siltstone, laminated; sporadic platy ironstone nodules; calcareous at base with brachiopods	4.33	166.78

	Thickness m	*Depth* m
UPPER FELLTOP LIMESTONE		
Limestone, thin- and medium-bedded; argillaceous laminae at top; abundant shell debris and crinoid fragments	7.34	174.12
Sandstone, thin- and medium-bedded, fine- to medium-grained, silty	0.58	174.70
Siltstone, laminated, in part sandy; scattered plant debris; ironstone nodules below 176.40 m	2.66	177.36
Coal, bright, pyritic	0.07	177.43
Sandstone, laminated and thin-bedded, medium-grained; seatearth in top 1.02 m; becomes interbedded with siltstone towards base	3.37	180.80
Siltstone, laminated; sandy at top; sparse fauna of fish scales and bivalves below 183.50 m	3.55	184.35
Sandstone, thin- and medium-bedded, fine- to medium-grained; bioturbated below 187.44 m	4.72	189.07
Siltstone, laminated; sparse plant debris; ironstone nodules from 190.10 to 192.20 m	3.58	192.65
COALCLEUGH SHELL BED		
Limestone, mainly thin-bedded laminated calcarenite; fauna includes productoids and corals	2.55	195.20
Sandstone and siltstone, interbedded and interlaminated, fine- to medium-grained; thin seatearth beds; seatearth-siltstone bed 0.85 m at 197.20 m	5.05	200.25
Coal, bright, pyritic	0.05	200.30
Siltstone and seatearth	0.15	200.45
Coal, bright, pyritic	0.13	200.58
Seatearth	0.92	201.50
Sandstone and siltstone, interbedded and interlaminated; sandstone predominantly fine- to medium-grained	4.80	206.30
Siltstone, grey, partly sandy; occasional ironstone nodules at top	1.95	208.25
LOWER FELLTOP LIMESTONE		
Limestone, grey, thin- and medium-bedded, medium and coarse calcarenite; argillaceous at top; abundant crinoid fragments, sparse brachiopods	4.63	212.88
Sandstone, thin-bedded, fine- to medium-grained, micaceous with silty partings; thin seatearth at top	1.87	214.75
Sandstone and siltstone, interbedded; platy ironstone nodules below 217.50 m	5.51	220.26
Siltstone, coarse, micaceous, sandy in part; sparse plant fragments and scattered ironstone nodules	2.97	223.23
Limestone, thin- to thick-bedded, medium and coarse calcarenite; sparse crinoid and shell fragments (?BELSAY DENE LIMESTONE)	2.77	226.00
Sandstone, laminated to medium-bedded, fine- to medium-grained; silty with seatearth at top; 0.83 m coarse-grained feldspathic bed at 226.22 m; 0.15 m siltstone bed with brachiopods at 226.37 m; 0.02 m coal at 226.39 m; siltstone beds at base	8.88	234.88
Siltstone, laminated, sandy at top and base; bivalve at 238.45 m	4.41	239.29
Siltstone, medium- and thin-bedded, fine- to medium-grained, quartzose; silty micaceous partings; seatearth at top	3.60+	242.89

Stargate Colliery No. 1 Underground Borehole (NZ 16 SE/14)

Underground level -65.10 m [1604 6334].
Drilled 1954 for NCB. Logged by G Armstrong and D E White.
Shown graphically in Figure 15.

	Thickness m	*Depth* m
Lower Coal Measures		
Floor of BROCKWELL COAL	0.00	0.00
Sandstone	0.91	0.91
Coal	0.15	1.06
Mudstone, grey	2.39	3.45
Sandstone, predominantly fine- to medium-grained, cross-bedded, commonly massive; locally argillaceous and micaceous; scattered shale pellets present in lower part; 0.15 m shale bed at 8.07 m	12.42	15.87
Shale, grey at top, black at base, micaceous; sporadic hard irony beds; scattered fish debris at base	1.68	17.55
Shale, black, canneloid, and coal, shaly	0.05	17.60
Fireclay, sandy, grading to mudstone, rooty, sandy	0.94	18.54
Shale, sandy, ankeritic, micaceous, grading to mudstone, shaly	2.33	20.87
Fireclay, nodular, slickensided; ankeritic beds	1.22	22.09
VICTORIA COAL		
Coal	0.36	22.45
Seatearth-mudstone	0.86	23.31
Sandstone, massive, fine-grained, argillaceous	1.35	24.66
Shale, sandy, micaceous; grades down to shale, slickensided; finely micaceous at base	2.56	27.22
Coal	0.16	27.38
Fireclay, irony at base	1.27	28.65
Shale, sandy, with plant debris including *Calamites* sp. and *Stigmaria* at top; 0.22 m argillaceous sandstone bed at 30.25 m	2.31	30.96
Coal	0.18	31.14
Fireclay, sandy	0.40	31.54
Shale, sandy	0.41	31.95
Sandstone, massive, medium-grained, cross-bedded; fine-grained at base	3.61	35.56
Shale, sandy, finely micaceous	2.48	38.04
MARSHALL GREEN COAL		
Coal	0.41	38.45
Fireclay, sandy to base	0.58	39.03
Shale, very sandy	0.46	39.49
Sandstone, massive, medium- to coarse-grained	5.03	44.52
Shale, slightly sandy, finely micaceous; fish debris and ?*Lingula* sp. at 45.41 m (?WELL HILL MARINE BAND)	1.27	45.79
Shale, silty and sandy, micaceous; argillaceous fine-grained sandstone beds and partings	2.16	47.95
Sandstone, medium- to coarse-grained, kaolinitic; coalified plant remains	2.11	50.06
Shale, silty, micaceous	0.07	50.13
GANISTER CLAY COAL		
Coal	0.18	50.31
Fireclay, rooty; slickensided at base	0.28	50.59
Sandstone, micaceous; rooty at top	2.11	52.70
Shale, sandy, micaceous	0.61+	53.31

Throckley Borehole (NZ 16 NW/45).
1:50 000 Sheet 14 (Morpeth).

Surface level + 102.18 m [1456 6762]
Drilled 1964-65 for Geological Survey of Great Britain.
Logged by G Richardson. Open-holed to 5.49 m. Abbreviated log. Shown graphically in Figures 8–12 and 14.

	Thickness m	*Depth* m
Drift		
Soil and sandy clay	1.52	1.52
Lower Coal Measures		
Strata	1.83	3.35
TOP VICTORIA COAL		
Coal, thin	0.31	3.66
Strata	7.50	11.16
BOTTOM VICTORIA COAL		
Coal, thin	c.0.40	11.56
Strata	3.84	15.40
MARSHALL GREEN COAL		
Coal, thin	c.0.30	15.70
Strata	6.93	22.63
GANISTER CLAY COAL		
Coal, thin	c.0.30	22.93
Strata	7.85	30.78
Coals, thin, and seatearth-mudstone beds	0.97	31.75
Strata	0.86	32.61
'THIRD GRIT' (upper leaf)		
Sandstone, light grey and buff, massive, fine-grained at top; otherwise medium- to coarse-grained, feldspathic; small quartz pebbles	24.31	56.92
Strata	12.12	69.04
Shale, canneloid; *Lingula* sp. and fish scales	0.04	69.08
Strata	4.27	73.35
'THIRD GRIT' (lower leaf)		
Sandstone, massive, coarse- and very coarse-grained, pebbly	4.25	77.60
QUARTERBURN (SUBCRENATUM) MARINE BAND (inferred position)		
Mudstone, silty with iron-cemented layers and ironstone nodules; *Lingula* sp., fish scales and worm tubes	2.03	79.63
Stainmore Group		
Strata	14.88	94.51
'SECOND GRIT'		
Sandstone, massive and cross-bedded, medium- to coarse-grained, locally pebbly	11.33	105.84
Strata	4.27	110.11
Mudstone, pyritous, with irony patches; *Lingula* sp., ostracods and brachiopods	0.38	110.49
Strata	1.32	111.81
'FIRST GRIT' (upper leaf)		
Sandstone, predominantly massive, coarse-grained; pyritous at top and base	5.08	116.89
Strata	12.65	129.54
'FIRST GRIT' (lower leaf)		
Sandstone, fine- to medium-grained; siltstone and mudstone ribs and beds	11.76	141.30
Strata	7.87	149.17
WHITEHOUSE LIMESTONE		
Limestone, argillaceous, with silty beds; brachiopods	0.48	149.65
Strata, including limestone 0.15 m at 156.89 m, and 0.33 m at 162.91 m	25.10	174.75
STYFORD LIMESTONE		
Limestone, argillaceous; shell fragments	0.81	175.56
Strata	28.09	203.65
GRINDSTONE (NEWTON) LIMESTONE		
Limestone, finely crystalline, locally recrystallised; nodular in upper part within a matrix of calcareous mudstone; silty towards base with bituminous patches; crinoids and ribbed brachiopods	2.62	206.27
Strata	38.31	244.58
UPPER FELLTOP (THORNBROUGH) LIMESTONE		
Limestone, argillaceous; calcareous mudstone ribs; crinoids and brachiopods	1.62	246.20
Mudstone, calcareous, with limestone ribs: crinoids and ribbed brachiopods	0.68	246.88
Limestone, argillaceous, silicified, with irony beds; 'cauda-galli' in top 0.30 m; crinoids, coral, brachiopods and lamellibranchs	3.15	250.03
Strata	32.27	282.30
COALCLEUGH SHELL BED (PIKE HILL LIMESTONE)		
Limestone, argillaceous; pyritous with ironstone nodules at top; irregular mudstone laminae to base; rare corals and brachiopods	1.37	283.67
Mudstone, calcareous; irregular layers of argillaceous limestone; polyzoa, corals and brachiopods	1.16	284.83
Limestone, argillaceous, finely crystalline, nodular; partings of calcareous mudstone; crinoids and brachiopods	2.36	287.19
Strata	22.16	309.35
LOWER FELLTOP (CORBRIDGE) LIMESTONE		
Limestone, argillaceous, finely crystalline, nodular; calcareous mudstone ribs and laminae; polyozoa, crinoids and brachiopods	7.34	316.69
Strata (assumed position of BELSAY DENE LIMESTONE at 336.65 m; AYDON SHELL BED at 364.49 m; PLANKEY SHELL BED at 387.12 m; and ?CROWHALL COALS, 0.23 m at 396.80 m, 0.03 m at 399.26 m and 0.05 m at 402.49 m)	95.07	411.76
CRAG (OAKWOOD) LIMESTONE		
Limestone, argillaceous, silty; silty calcareous mudstone rib near top; crinoids and shell debris	0.64	412.40
Strata	5.25	417.65
CRAG COAL		
Coal	0.33	417.98
Strata	72.90	490.88
LITTLE LIMESTONE		
Limestone, argillaceous, finely crystalline, slightly metamorphosed, stylolitic, bituminous, locally pyritous; very nodular below 492.56 m; sporadic calcareous siltstone ribs; crinoids, corals (at 494.69 m), brachiopods, gastropods, and other small shell debris.	5.03	495.91
Strata	9.35	505.26
WHIN SILL		
Quartz-dolerite, light to dark grey and grey-green, predominantly fine- to medium-crystalline; pyroxene phenocrysts at various levels; calcite vesicles and calcite- and chlorite-filled joints	38.53	543.79
Strata (including 0.57 m limestone at 545.08 m)	32.94	576.73

	Thickness m	Depth m
GREAT LIMESTONE		
Limestone, predominantly thick-bedded to massive, argillaceous, often silty, partly recrystallised; nodular and stylolitic; sporadic carbonaceous mudstone and siltstone beds but especially in upper 2.50 m; crinoids, sporadic corals, brachiopods and bivalves; much comminuted shell debris	13.82	590.55
Upper Liddesdale Group		
Strata (including coals 0.29 m at 590.84 m, 0.04 m at 596.79 m and 0.10 m at 598.62 m)	14.30	604.85
Total depth		604.85

Note: the palynology of the Throckley Borehole was described by Owens (1972).

Town Head Farm Borehole, near Burnopfield (NZ 15 NE/133)

Surface level + 216.20 m [1740 5527]
Drilled 1958–59 for NCB. Abbreviated log. Shown graphically in Figures 17 and 19.

	Thickness m	Depth m
Drift		
Predominantly sand and boulders (Driller's log)	7.16	7.16
Middle Coal Measures		
Sandstone, thin- and wispy bedded, with micaceous partings; shaly at top	2.13	9.29
Mudstone, silty; siltstone partings	7.01	16.30
Probable old workings of MAIN COAL	0.87	17.17
Mudstone, silty, rooty; grades to	1.42	18.59
Sandstone, siltstone and shale, predominantly alternating in thin beds; 2.28m thin- to thick-bedded fine-grained sandstone at 24.68 m	10.13	28.72
Mudstone; abundant mussels	1.30	30.02
Mudstone, silty; scattered mussels and some *Spirorbis*; becomes grey, sandy and rooty at base	1.78	31.80
Coal (?MAUDLIN COAL)	0.10	31.90
Shale, black, on mudstone, silty and rooty	0.30	32.20
Sandstone, thin- to thick-bedded, cross-bedded, predominantly coarse-grained, but becoming finer-grained with increasing depth; often micaceous and carbonaceous with local shaly partings; sporadic mudstone inclusions; 0.30 m grey silty mudstone bed at 58.14 m; seatearth beds near top	31.65	63.85
Mudstone, broken	1.68	65.53
Suspected old workings of DURHAM LOW MAIN and TOP BRASS THILL COAL	2.44	67.97
Sandstone, thin and wispy bedded, fine-grained, micaceous, rooty at top; sandy shaly mudstone beds to base	6.60	74.57
Mudstone, shaly, sandy, silty	2.23	76.80
HUTTON COAL		
Coal, sandstone and rooty mudstone (old workings)	0.92	77.72
Mudstone, grey, predominantly sandy and silty, often rooty; micaceous and carbonaceous; 0.53 m thick-bedded fine-grained sandstone band at 82.82 m	6.78	84.50
Sandstone, white with grey layers, predominantly thick-bedded to massive, fine-grained with medium-grained beds at top, micaceous; occasional inclusions of ironstone, mudstone and coal	18.92	103.42
RULER COAL		
Coal	0.31	103.73
Shale, grey and black, silty and sandy, interbedded with sandstone, fine-grained	4.72	108.45
Shale, grey; 0.07 m mussel band at 108.73 m	0.56	109.01
Mudstone, shaly, with silty and sandy beds; *Naiadites* sp. at 109.80 m; ?*Anthracosia* sp. at 110.10 m; mussels at several horizons between 110.71 and 114.30 m; nodular structures in shales at base	6.63	115.64
HARVEY (VANDERBECKEI) MARINE BAND		
Shale, dark grey, sandy; abundant *Lingula* sp. and fish scales	0.15	115.79
Lower Coal Measures		
Coal	0.13	115.92
Seatearth-mudstone	0.89	116.81
Mudstone and shale, alternating with sporadic beds of fine-grained, thin-bedded sandstone	7.09	123.90
Shale, grey, silty	0.61	124.51
HARVEY COAL		
Coal	0.20	124.71
Shale	0.05	124.76
Coal	0.38	125.14
Seatearth-mudstone, grading to mudstone	0.25	125.39
Sandstone, fine-grained, rooty	2.54	127.93
Mudstone, sandy, shaly	0.23	128.16
Seatearth-mudstone	0.38	128.54
Shale and siltstone	3.66	132.20
Mudstone, with ironstone nodules; poorly preserved mussels	0.71	132.91
HODGE COAL		
Coal	0.13	133.04
Sandstone, fine-grained; shale rib at top	2.82	135.86
Mudstone, sandy; sandstone beds and partings	0.84	136.70
Sandstone, fine-grained, rooty at top	1.98	138.68
Shale, sandy; sandstone beds	0.71	139.39
TILLEY COAL		
Coal	0.30	139.69
Seatearth-mudstone	0.15	139.84
Coal	0.67	140.51
Mudstone, silty, predominantly rooty	1.45	141.96
Sandstone, massive at top, thin-bedded below; fine-grained	4.90	146.86
BUSTY COAL		
Old workings; shattered strata	3.35	150.21
Mudstone, silty; rooty at top and base	2.95	153.16
Siltstone, grey; 0.15 m bed of silty mudstone containing *Neuropteris* sp. at 155.29 m	2.28	155.44
Sandstone, fine-grained, thick-bedded; occasional silty and shaly beds; inclusion of mudstone at base	6.78	162.22
Mudstone, sandy	0.26	162.48
THREE-QUARTER COAL		
Coal with dirt parting	0.89	163.37

	Thickness m	*Depth* m
Sandstone, thick-bedded to massive, fine-grained at top, medium- to coarse-grained at base; sporadic beds of sandy mudstone; 0.30 m rooty mudstone bed at 170.07 m; abrupt base with mudstone inclusions	11.12	174.49
Mudstone, sandy; sandstone beds; silty shale bed near base	4.30	178.79
BROCKWELL COAL		
Coal	0.61	179.40
Seatearth-mudstone; sandstone beds	0.83	180.23
Coal	0.26	180.49
Seatearth-mudstone and mudstone, rooty	0.25	180.74
Sandstone, predominantly thick- and cross-bedded, medium-grained; thinner bedded towards base	7.77	188.51
Shale, dark grey to black; mussels and fish debris; mudstone rib at base	0.54	189.05
Sandstone, white and grey, fine-grained; passes down into siltstone	2.82	191.87
Mudstone, shaly, silty; ironstone ribs; very shaly and dark at base	2.64	194.51
VICTORIA (?TOP) COAL		
Coal	0.46	194.97
Seatearth-mudstone, and mudstone, rooty	1.37	196.34
Sandstone, predominantly thin- to thick-bedded, cross-bedded; micaceous partings	1.95	198.29
Mudstone, shaly, silty	0.21	198.50
Sandstone, coarse-grained, feldspathic	0.18+	198.68

APPENDIX 3

Geological section of Letch House–Airy Holm–Derwent Tunnel and part of Derwent–Wear Tunnel

The Kielder Water Scheme is designed to satisfy the water requirements of the north-east of England into the next century. The Kielder transfer works enable water from the River Tyne to be abstracted and transferred southwards to allow regulation of the rivers Wear and Tees. They include the three Kielder Tunnels, the northern part of which lie within the Newcastle district (Figure 30).

1. The Letch House–Airy Holm–Derwent Tunnel, 4.5 km long, connects Letch House on the northern flanks of Greymare Hill with the northern end of the buried pipeline section crossing the Derwent Valley. It incorporates a median shaft to the Airy Holm Headpond, which acts as a balancing storage to the tunnel system.
2. The Derwent to Wear Tunnel, 13.4 km long, connects the southern end of the buried pipeline crossing the Derwent Valley with the Wear Valley crossing (Wolsingham district) between Stanhope and Frosterley. It includes an intermediate air shaft near Waskerley.
3. The Wear to Tees Tunnel, 14.4 km long, connects the Wear Valley crossing and the Tees outlet below Eggleston (Wolsingham and Barnard Castle districts) and also includes an intermediate air shaft at Sharnberry.

The Kielder Water Scheme has been described by Coats and Ruffle (1982), and Coats et al. (1982), while aspects of the geology of the scheme have been described by Carter and Mills (1976), Coats et al. (1977), and Davies et al. (1981).

The following descriptions cover those parts of the tunnels that fall within the Newcastle district (see Figure 30); these are the whole of the Letch House–Airy Holm–Derwent Tunnel (Figure 31) and the northern end of the Derwent to Wear Tunnel (Figure 32). The information is summarised from tunnel record sheets prepared during construction by the Consulting Engineers to the Kielder Water Scheme, the Babtie Group.

LETCH HOUSE–AIRY HOLM–DERWENT TUNNEL

The Letch House to Derwent Tunnel (Figure 31) is 4511 m in length. It consists of two sections, Letch House to Airy Holm, 2020 m long, and Airy Holm to Derwent, 2491 m long; an intermediate shaft links the two parts of the tunnel to the Airy Holm Headpond Reservoir. Cover to surface along most of the tunnel is relatively low, attaining a maximum of rather over 80 m in the Greymare Hill area in the north and rather less than 80 m north of Carterway Heads in the south. The northern and central parts of the Letch House to Airy Holm section penetrate very gently dipping or horizontal Lower Coal Measures. Coals in the sequence, more particularly on the north side of Greymare Hill, were worked opencast prior to tunnel construction. The southern part of the Letch House to Airy Holm section and the whole of the Airy Holm to Derwent section (with the exception of a short stretch near the North Derwent portal) were driven in gently dipping strata high in the Stainmore Group ('First' and 'Second' grits). The sequence is not continuous, due to faulting. Near the North Derwent portal, nearly 150 m of tunnel were driven in superficial deposits on the northern flank of the buried valley of the River Derwent. Abridged geological data for this tunnel (Figure 31) are given below.

Chainages (Ch) refer to distances along the tunnel in metres from the north end at the Letch House Shaft.
Coordinates of tunnel at Letch House Shaft at Ch 1000 m: [0413 5584].
Invert level at Letch House Shaft: 200.13 m OD.
Direction of tunnel: S. 14°E.
Coordinates of tunnel at Airy Holm Headpond Shaft at Ch 3020 m: [0470 5392].
Invert level at Airy Holm Shaft: 202.29 m OD.
Direction of tunnel: S. 10°W.
Invert level at north Derwent portal: 199.15 m OD.
Coordinates of tunnel at north Derwent portal at Ch 5511 m: [0424 5148].
Tunnel Driven October 1977 to June 1979.
Reference: Tunnel Record Drawings Letch House–Airy Holm–Derwent Works prepared by Babtie Group. Also Coats et al. (1977), Davies et al. (1981), Coats et al. (1982) and references therein.
Chainage measurements normally refer to the position at which strata intersect the tunnel roof.

Tunnel chainage (metres) (Datum + 1000 m)	*Classification and description*	*Notes and comments*
	Lower Coal Measures	(Letch House to Airy Holm section) Ch 1000–3020
1000–1663	Sandstone, faintly to slightly weathered, moderately strong to strong, cross-bedded, fine- to medium-grained, siliceous, light grey, ferruginous stained; occasional discontinuous shaly beds. Major joint set at 070°–080°; several minor sets. Joints generally tight, some iron- stained and clay- smeared; occasionally completely weathered on open joints. A few faults, all tight, with throws of less than 2.5 m.	Strata virtually horizontal or very gentle dip to S. Sandstone (informally termed the Letch House Sandstone) occupies a position between Victoria group of coals and the underlying Marshall Green Coal.
1663	Fault; 1.0 m gouge, breccia and associated broken ground. Strike 065°, downthrow to N. Throw not known but not more than 8–10 m.	Greymare Hill Fault.
1663–1686	Sandstone, as at Ch 1000–1663; occupies most of tunnel profile at Ch 1663.	Moderate dip to N (base of Letch House Sandstone).
1686–1708	Mudstone, grey, predominantly fresh to faintly weathered, weak to very weak, 1.0 m, *on* Marshall Green Coal, c. 0.25 m between Ch 1678 (floor of tunnel) and Ch 1700 (roof), *on* seatearth, rooty, very weak, and mudstone, 0.60 m, as above. These beds underlain by sandstone (as at Ch 1708 to 2289) rising from Ch 1690 (floor) to Ch 1708 (roof). Dominant joint set 050°–060°, but some scatter; spacing generally moderate to close.	
1708–2289	Sandstone, fresh to faintly weathered, moderately strong to strong, fine- to medium-grained, light grey, micaceous and carbonaceous, cross-bedded, with occasional shaly laminae and coal streaks. Bed of seatearth and mudstone (below the Marshall Green Coal, see above) up to 0.80 m, reappearing in roof of tunnel between Ch 1714 and 1776. Dominant joint strike 050°–080° with broad scatter, supplemented by set at 145° approaching fault at Ch 2289. Joints generally tight, clean, calcite or pyrite-filled, only occasional iron staining; joints generally moderately closely spaced, but locally very closely spaced. Minor faults, locally with clay smearing, at Ch 1795, strike 147°; Ch 1825, strike 015°; Ch 1843, strike 074°; Ch 1986, strike 027°; Ch 2143, strike 159°; Ch 2197, strike 178°; Ch 2256–2261, broken faulted zone strikes between 140° and 152°.	N of Ch 1710, moderate dip to N into Greymare Hill Fault. Very gentle dip to N between Ch 1710 and Ch 1920; gentle or very gentle dip to S between Ch 1920 and Ch 2289. Sandstone (informally termed Marshall Green Sandstone) occupies a position between Marshall Green Coal and Ganister Clay Coal (or its assumed horizon).
2289	Fault; up to 0.3 m gouge and associated 3 m zone of broken ground. Strike 145°, downthrow to S, throw not known but estimated to be c. 60 m. Fault hading S. present between Ch 2289 and 2299.	Unthank Fault; reversed fault throwing Lower Coal Measures against Stainmore Group to S.
	Stainmore Group	
2289–2325	Mudstone, up to 2.0 m, *on* coal, up to 0.2 m, *on* seatearth and mudstone, up to 1.0 m, *on* sandstone 1.2 m (seen); mudstone is fresh, weak, dark grey, and is closely jointed near fault at Ch 2289.	Dip horizontal to gently S. Coal disappears from tunnel floor at Ch 2402.
2325–3226	'SECOND GRIT' (to Ch 3226) Sandstone, fresh to slightly weathered, becoming highly to completely weathered at major discontinuities. Moderately strong to very strong in competent zones, medium- and fine-grained, light grey, micaceous and locally carbon-aceous, thick-bedded to laminated; base of sandstone falls irregularly from Ch 2325 to 2470 beyond which it forms full face; lens of mudstone up to 1 m thick in upper part of section between Ch 2383 and 2398; siltstone and mudstone up to 0.60 m thick overlying sandstone in roof of tunnel between Ch 2762 and 2891 m; occasional argillaceous beds between Ch 3023 and 3226. Joints between Ch 2325 and 3000 moderately close to	Dip very gentle to S in N, otherwise horizontal to subhorizontal over greater part of section. Dip increases with N and E components towards S end of section at Ch 3226. Sandstone informally termed Airy Holm Sandstone. Airy Holm Shaft at Ch 3020. (Airy Holm–Derwent section) Ch 3020–5511.

Tunnel chainage (metres) (Datum + 1000 m)	*Classification and description*	*Notes and comments*
	widely spaced, with set at 075° prominent. Joints generally iron-stained, locally with traces of sand and clay; some decomposed pockets. Between Ch 3000 and 3226, dominant joint sets 065° and 155°, moderately close, frequently open, decomposed and stained.	
3226–3240	Argillaceous strata including mudstone, broken, up to 1.8 m, *on* coal, c.0.10 m, *on* seatearth, up to 0.40 m, *on* slightly to moderately weathered sandstone up to 3.4 m (?'FIRST GRIT') abutting against fault at Ch 3240.	Strata dip to N and pass below tunnel floor at Ch 3185.
3240	Fault; strike 161°, downthrow N at least 2.5m.	Fault zone between Ch 3240–3281 informally termed Fine House Fault. Overall throw N, but not large.
3240–3268	Sandstone, (?'FIRST GRIT') slightly weathered with moderately weathered base up to 1.0 m, in tunnel roof, overlying mudstone with some coaly laminae, up to 3.4 m.	Dip gentle to N to fault at Ch 3240.
3268	Fault; strike 164°, downthrow probably S, less than 1 m.	
3268-3281	Argillaceous measures including broken mudstone and slickensided mudstone/seatearth, 0.20–0.40m in roof of tunnel overlying sandstone ('FIRST GRIT') (as at Ch 3281–3416), up to 3.4 m.	Slight upward flexure of strata.
3281	Fault strike 148°, downthrow probably N.	
3281–3416	'FIRST GRIT' (to Ch 3429) Sandstone, fresh to slightly weathered, moderately weak to strong, locally very strong, medium- to fine-grained, grey and brown; occasional completely weathered bedded zones and soft argillaceous bands. Dominant joint set at 095°, with minor sets at 070° and 155°. Joints tight, or slightly open with some pyrite and calcite. Joint spacing close or moderately close. Argillaceous strata dipping N, underlying sandstone, present in lower part of tunnel between Ch 3400–3416.	Dip subhorizontal to level. Dip increases to N between Ch 3380 and 3416.
3416–3429	Fault; strike 160° and Fault; strike 150° between which disturbed zone of strata with dips to N at up to 45°. Sequence of argillaceous strata includes thin coal underlain by sandstone with argillaceous layers. Upper surface of sandstone dips from Ch 3428 (roof) to Ch 3416 (floor).	Fault zone between Ch 3416 and 3436 probably a continuation of the Healeyfield Fault to the NW.
3429–3436	Sandstone probably as at Ch 3416–3428, and S of fault at Ch 3436.	
3436	Fault; strike 155°, downthrow probably N.	
3436–3835	Mixed beds overlying sandstone ('FIRST GRIT'). Mixed beds consist of sandstone, slightly to moderately weathered, with ribs (up to 0.1 m) of soft, clayey, highly weathered shale and soft, very weak, brown-grey rooty seatearth. Mixed beds range up to 1.4 m thick forming roof and upper part of tunnel finally disappearing above roof at Ch 3835. Beyond Ch 3835 tunnel in full face of fresh sandstone ('FIRST GRIT'). Joint sets at 050°, 095° and 150°. Joints tight or slightly open, locally decomposed and stained, spacing close to moderately close.	Strata horizontal or subhorizontal dipping up to 2°N.
3835–4609	'FIRST GRIT' (full face to Ch 4609) Sandstone, fresh, moderately strong to strong, medium- to coarse-grained, light and dark grey, moderate- to thick-bedded, cross-bedded, locally slightly weathered at joints and with some poorly cemented very weak zones; occasional bands of mixed beds. Beyond c.Ch 4000 and to 4609, sandstone, fresh to slightly weathered, moderately	Dip very gentle with both N and E components.

Tunnel chainage (metres) (Datum + 1000 m)	*Classification and description*	*Notes and comments*
	weak to strong, with highly to completely weathered bedded zones. Scattered joint sets with 150° predominating. Some joints open with local iron staining and decomposition. Joint spacing moderately close.	
4609–4847	Mixed beds, overlying sandstone ('First Grit') — sequence broadly similar to that at Ch 3436–3835. Fresh to moderately weathered, very weak to strong, light and medium grey mudstone, siltstone and sandstone — these beds generally range between 1.5 and 2.0 m thick and form upper and middle parts of tunnel, but between Ch 4687 and 4800 strata fall locally to within less than 0.6 m of the floor. Base of mixed beds on sandstone irregular and showing much lateral variation. Broad scatter of joints but dominant between 100–170°.	Dip very gentle with both N and E components and some downward flexuring.
4847–5002	'First Grit' (full face to Ch 5002) Sandstone, fresh to faintly weathered, moderately weak to moderately strong, medium-grained, grey, and iron-stained with thin, soft argillaceous beds and highly to completely weathered zones Fault at Ch 4876 with zone of fracturing and close stained joints, some brecciation, strike 050°, downthrow estimated 0.7 m N; major discontinuities, possibly with some movement surfaces, between Ch 4955–4980.	Dip gently undulating but northerly component into fault at Ch 4876.
5002	Fault, strike 030°, 0.6m brecciation.	
5002–5295	'First Grit' (partial face over bulk of length) Sandstone, as at Ch 4847–5002, overlying largely argillaceous mixed beds (up to 1.0 m), at base of section between Ch 5002 and5053. Full section in sandstone between Ch 5053 and 5110. Base of 'First Grit' rises gently south from floor at Ch 5110 becoming almost horizontal to Ch 5295 giving a tunnel section of sandstone on mudstone. Up to 0.40 m lens of mixed beds forming tunnel roof between Ch 5242 and 5283. Variable jointing, 050°, 110°, and 140° sets present. Joints commonly open and decomposed, some with black carbonaceous material. Joint spacing moderately close.	Dip gently undulating.
5295	Fault; 0.6m wide broken zone, strike 040°, downthrow S c.0.7 m.	
5295–5361	'First Grit' Sandstone, as at Ch 5002–5295; 0.4 m of argillaceous strata underlying 'First Grit' and dipping S in floor of tunnel immediately S of fault at Ch 5295; minor fault near S end of section. Irregular interface of bedrock and superficial deposits Ch 5361 (roof) to Ch 5376 (floor).	Dip slightly inclined to S, steepening towards S end of section.
5361–5511	Superficial Deposits (on northern margin of buried valley of River Derwent) Boulder clay, stiff, dark grey/brown with layers and lenses of sand/clay/silt with gravel and cobbles; irregular interface with rockhead formed by sand and gravel with mudstone and sandstone cobbles.	North Derwent Portal (Ch 5510.83)

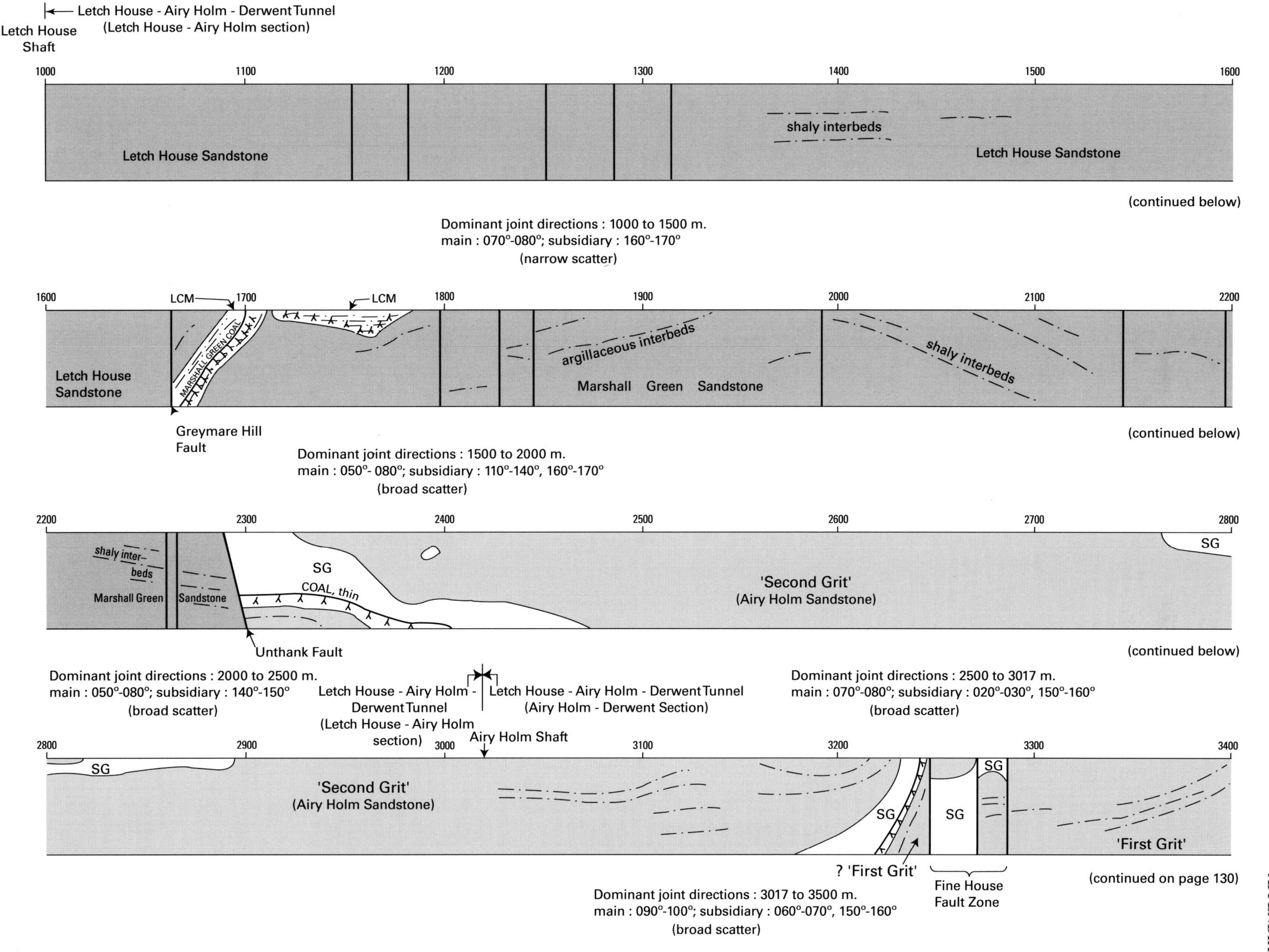

Figure 31 Geological section of the Letch House–Airy Holm–Derwent Tunnel.

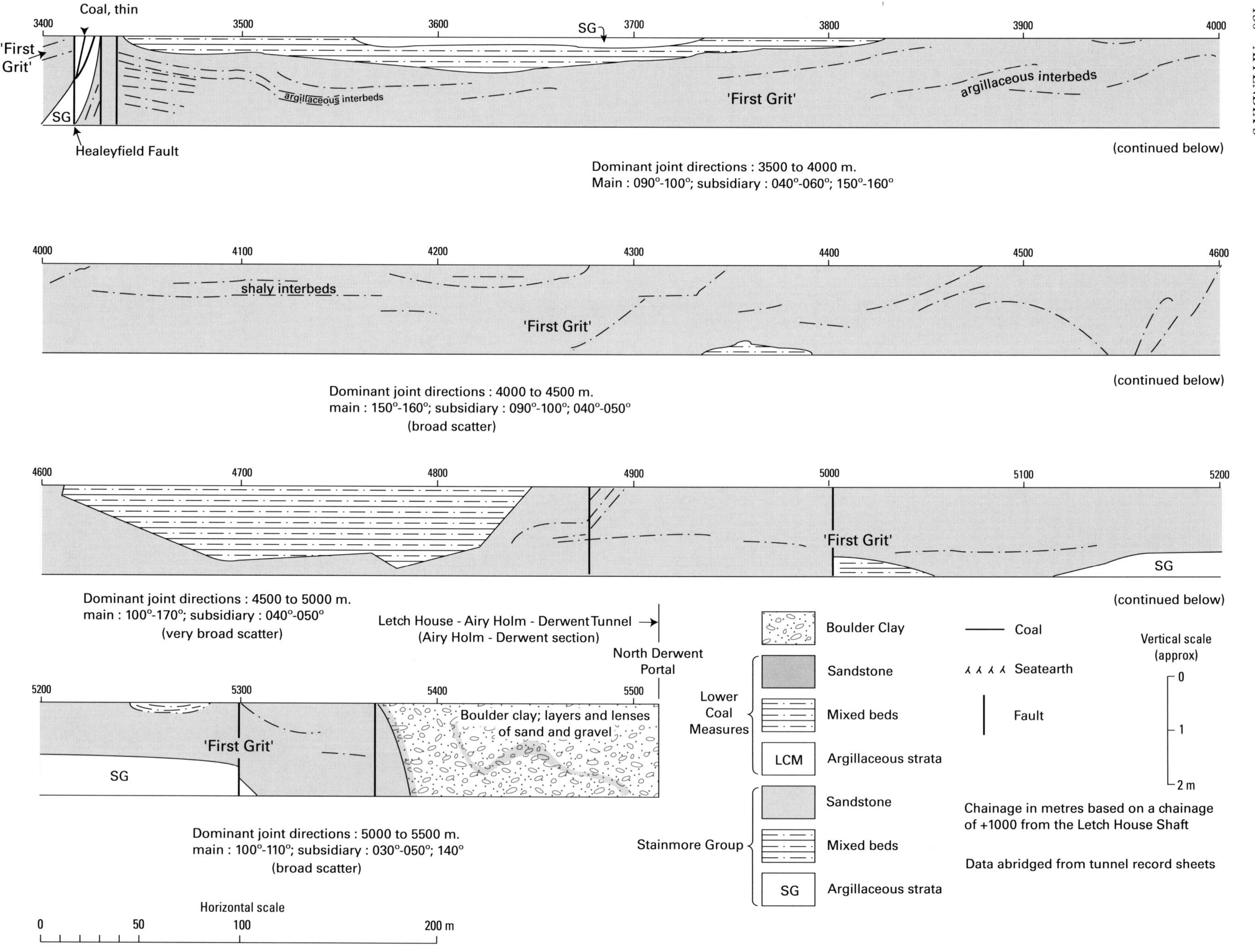

Figure 31 Geological section of the Letch House–Airy Holm–Derwent Tunnel.

DERWENT–WEAR TUNNEL

The Derwent–Wear Tunnel is 13 374 m in length [0405 5073 to 0121 3768]. An intermediate air shaft, the Waskerley Shaft [0296 4439], 138 m in depth, is located east of the Waskerley Reservoir. Cover to surface along most of the tunnel is over 100 m, rising in the southern section to between 200 and 300 m. The tunnel penetrates most of the Stainmore Group and a small part of the underlying Alston Group.

Only the northernmost part (2650 m) of the tunnel falls within the Newcastle district; the abridged geological data for this section (referred to as the Derwent to Waskerley section, Figure 32) are given below.

Chainages (Ch) refer to distances along the tunnel in metres from 0 datum.
Coordinates of tunnel at Ch 0: [0405 5073].
Finished tunnel invert at Ch 0: 185.17 m OD.
Direction of tunnel: 09° 44'.
Tunnel driven (this section) August 1976–September 1979.
Reference: Tunnel Record Drawings Derwent–Wear–Tees Works prepared by Babtie Group. See also Davies et al. (1981), Coats et al. (1982) and references therein.
Chainage measurements normally refer to the position at which strata intersect the tunnel roof.

Tunnel chainage (metres)	*Classification and description*	*Notes and comments*
	Stainmore Group	
0–36	'First Grit' (lower part) – (to Ch 200) Mixed beds; interbedded silty mudstone, and sandstone, slightly to moderately weathered, very weak to moderately weak, dark to medium rusty grey, very thin to moderately bedded, fine- to medium-grained; includes thin bed of moderately strong sandstone, up to 1.0 m, near top of tunnel and between Ch 0 and 21.	Dip of strata between Ch 0 and 325 ranges between 2 and 6°N with tendency to increase northwards.
36–200	Sandstone, slightly to moderately weathered, moderately strong to strong, rusty grey, thin to moderately bedded, medium- to coarse-grained; occasional shaly beds and partings (5–30 mm thick).	Two major joint sets predominantly ranged between 050° and 060° and 140° and 150°. Joint frequency often high, locally intense, in argillaceous strata. Joints in sandstone tight, to 10 mm open, ferruginous-stained, clay-smeared or occasionally clay-filled.
200–230	Mudstone, predominantly moderately weathered, very weak to moderately strong, medium to dark rusty grey.	
230–325	Mixed beds, moderately weathered, very weak to strong, medium to dark grey, laminated to thin-bedded; bed of sandstone in roof of tunnel Ch 319–325 with ? erosive base.	Dip 2–3° to N.
325	Fault, 0.3–0.6m gouge, strike 175°, downthrow c.30 m to N.	Fault appears to cut out much of sequence between base of 'First Grit' and Grindstone Sill.
325–440	Grindstone Sill (to Ch 440) Sandstone, highly to moderately weathered to Ch 332, weathering thereafter decreasing to slight to faint, moderately strong to strong,grey, thin- to moderately bedded, medium- to coarse-grained. Between Ch 386 and 419 sharp downward buckling of sandstone brings mixed beds overlying Grindstone Sill into upper and middle part of tunnel section. Top of Grindstone Sill disappears below floor at Ch 460. Minor fault at Ch 386, strike 100°, downthrow 1–2 m to S.	Two major joint sets largely as above, with predominantly medium to high frequency.
440–465	Mixed beds, faintly to slightly weathered, very weak to moderately strong, dark grey, laminated to very thin-bedded, medium to dark grey; thin limestone bed overlain by mudstone at Ch 458–465 in upper part of tunnel section.	Moderate dip to S into fault at Ch 465.
465	Fault, minor; 1.0 m gouge, shattered strata. Strike 110°, downthrow S, 0.8 m.	
465–643	Mixed beds with thin limestone, (as at Ch 458–465 m) between Ch 465 and c.510 (tunnel floor); thereafter mixed beds and mudstone, fresh to faintly weathered, moderately weak to strong, dark grey, laminated to moderately bedded; argillaceous limestone (?Grindstone Limestone) up to 1.25 m towards base of sequence between Ch 470 and 490 (roof) and Ch c.567 and 635 (floor).	Dip 2–5° S with minor upward flexure at c.Ch 610–630. Two to three major joint sets at 050°–060°, 140°–150°, and 020°–30°. Joints predominantly tight, fresh, partly clay-smeared.

Tunnel chainage (metres)	*Classification and description*	*Notes and comments*
643–714	Sandstone, faintly weathered, strong to very strong, light grey, fine- to medium-grained; 0.4 m siliceous layer at top. Sandstone becomes progressively more weathered and weaker into fault at Ch 714; dark grey mudstone overlying sandstone in roof between Ch 700 and 714.	
714	Fault; 0.1m gouge with shattered strata. Strike 230°.	Possibly a component of the Muggleswick Fault at depth.
714–1103	Mudstone and mixed beds; long sequence of strata, fresh, moderately weak to moderately strong, dark grey, fine-grained, thinly laminated to thin-bedded; sequence predominantly argillaceous between Ch 714 and c.825, thereafter with sandy and silty ribs. Sequence contains two thin argillaceous limestone beds; lower up to 0.5 m thick, between Ch 714 and c.725 (floor) passes laterally into thin sandstone; upper, up to c.1 m thick, (?GRINDSTONE LIMESTONE), between Ch 714 and 975 (tunnel floor), forms roof of tunnel between Ch 749 and 785.	Two main joint sets, 040°–070° and 140–150° normally developed, predominantly medium to low intensity, tight and clean. Gentle upward roll of strata between Ch 725 and 800; thereafter strata dip very gently S. Sequence between Ch. 440 and 714 closely resembles that between Ch 714 and 1174.
c.1103–1174	Sandstone, fresh, strong, light grey, moderate to thick-bedded, cross-bedded, micaceous. Erosive base cuts down between Ch 1110 and 1120.	Joints 055°, 150° and 120° mainly tight and clean.
1174–1177	Fault; 2.0–2.5 m wide zone containing large sandstone blocks in matrix of sheared mudstone and clay. Strike c.145°, inclined NE; throw N at least 50 m, possibly much greater.	Fault has the effect of cutting out the Upper Felltop Limestone and adjacent sequence from tunnel section. A possible extension to a component of the Muggleswick Fault at depth.
1177–2100	Mudstone, fresh, moderately weak to moderately strong, dark grey, laminated to thin-bedded, locally silty with occasional ironstone nodules; siltstone bands; minor shears and displacements at Ch 1197 and 1237 with traces of quartz on joints; minor shear at Ch 1617. Limestone appears in floor of tunnel at Ch 1805.	Dominant joint sets, variably developed at 060° and 160° but showing some scatter, and generally widely spaced. Dip gently undulating but stabilizes in the S at about 2° to N.
2100–2165	COALCLEUGH SHELL BED (PIKE HILL LIMESTONE) Limestone, fresh, moderately strong to strong, muddy, dark grey, laminated to thin-bedded, up to 1.5 m thick; occasional shell fragments more especially at base.	Few joints; tight and clean fractures only. Dip 2° N.
2165–2320	Mixed beds, fresh, moderately strong, dark grey, laminated to thin-bedded; coal up to 0.05 m at base (Ch 2126 to 2320).	Dip to N low and variable, few joints. Small monoclinal flexure at discontinuity at Ch 2272.
2320–2540	GRIT SILLS (?HIGH GRIT SILL) Sandstone, fresh, strong, thin- to thick-bedded, locally cross-bedded, fine- to medium-grained; dark grey siltstone layers. Minor faults at Ch 2443 to 2445 including zone of fractures and some brecciation, strike c.120°, downthrow c.0.7 m N; two minor faults at Ch 2474–2476, strike c.130°, downthrow c.0.5 m S. Slightly irregular base to sandstone at Ch 2533–2540.	Scattered jointing with variable direction 070–090°, 120°; 140–150°; and 160–170°; quartz and calcite on joints. Strata horizontal or very gentle dip to N.
2540–2650 m+	Mixed beds, fresh, moderately strong to strong, laminated to thin-bedded, locally sandy.	Jointing variable, but dominant sets striking 055° and 150°. Strata horizontal or inclined very gently N. (edge of Newcastle district) [at c.0360 4815]. Cover from surface to tunnel c.118 m.

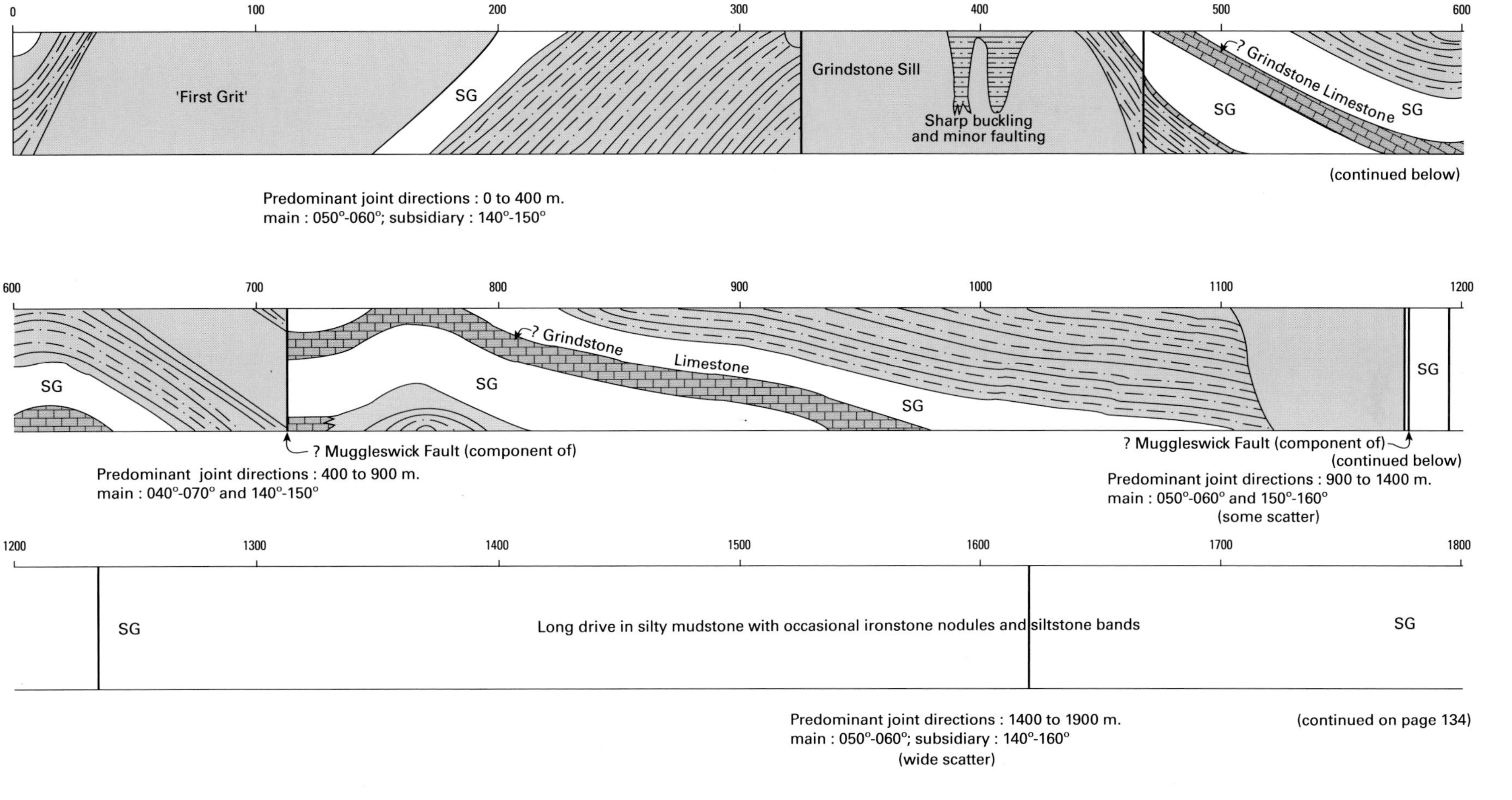

Figure 32 Geological section of the Derwent–Wear Tunnel (part of Derwent–Waskerley section).

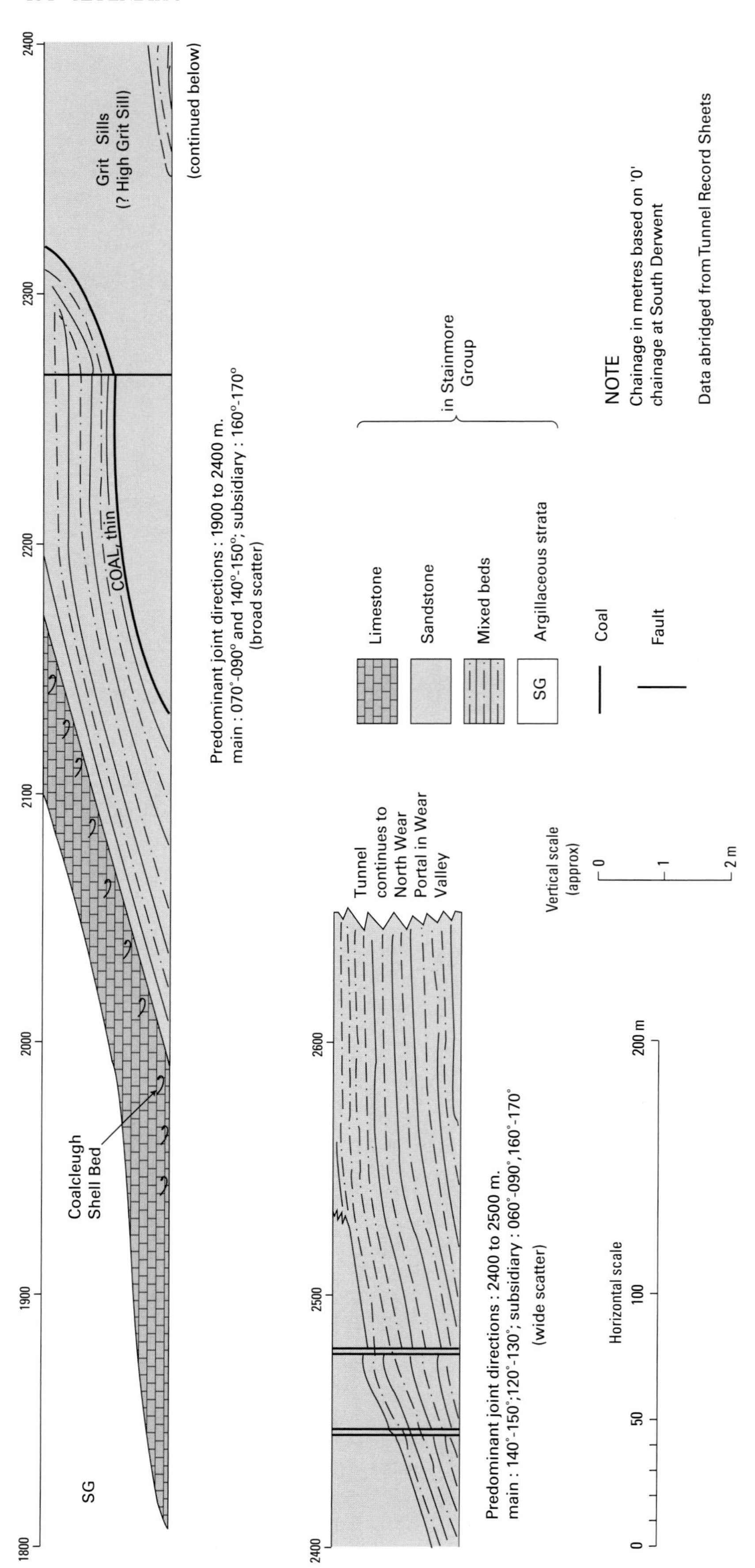

Figure 32 Geological section of the Derwent–Wear Tunnel (part of Derwent–Waskerley section).

APPENDIX 4

List of Geological Survey photographs

Copies of these photographs are deposited in the library of the British Geological Survey, Keyworth, Nottingham, NG12 5GG. They belong to Series L and may be supplied as prints and slides at a standard tariff.

STAINMORE GROUP

1721, 1722 Deadridge Quarry [NY 9959 6554] near Corbridge. Lower Felltop (Corbridge) Limestone, together with adit into underlying fireclay.

1723 Corbridge bypass [0000 6540]. Cross-bedded sandstones above the Lower Felltop (Corbridge) Limestone.

1724 Corbridge bypass [0112 6438]. Cross-bedded sandstone and overlying shale above Upper Felltop (Thornbrough) Limestone.

1725 Crossedge Quarry [0369 6569], near Newton. Grindstone (Newton) Limestone; type locality of the Newton Limestone.

1726 Temperley Grange Quarry [NY 9855 6239]. Quarry face in cross-bedded coarse-grained sandstone; 'First Grit'.

1733 Derwent Valley near Eddy's Bridge [0407 5085]. Strata near base of 'First Grit' at north portal of the Derwent-Wear Tunnel (part of Transfer Works of the Kielder Water Scheme).

1741 River Derwent [0881 5221] near Shotley Bridge. Coarse-grained sandstone ('Second Grit').

1742 River Derwent [0876 5220] near Shotley Bridge. Coarse-grained sandstone ('Second Grit').

1755 Quarry [0087 5143] near Pow Hill, Edmundbyers. Thick-bedded medium- and coarse-grained sandstone ('Second Grit').

1756 Edmundbyers to Blanchland road [at 0086 5140]. Escarpment of 'Second Grit' with morainic material in foreground.

1768–71 River Derwent [0881 5223; 0882 5223; 0884 5224] near Shotley Bridge. Various photos of coarse-grained sandstone ('Second Grit') in bed and banks of river.

2536 View [0416 5224] from near Manor House Inn, A68 road, looking towards Edmundbyers Common, showing plateau and sandstone escarpment on south side of Derwent Valley.

STAINMORE GROUP/COAL MEASURES

2538 Shotleyfield Burn [0631 5304] upstream from Hammermill Cottage. Seatearth with capping bed of sandstone containing the trace fossil *Zoophycos.*

2539 Burnhouse Gill [0790 5274] 260 m downstream from road bridge for B6278. Loose block of sandstone containing the trace fossil *Zoophycos.*

2540 Mere Burn [0886 5484] 270 m downstream from Mereburn Bridge. Highly fossiliferous sandstone post capping seatearth at position of Quarterburn (Subcrenatum) Marine Band.

COAL MEASURES

1738 St Andrew Opencast Site (1977) [0422 5566]. Victoria, Brockwell, Three-Quarter and Busty coals in process of being worked. Note old workings in the Brockwell Coal.

1739 Greymare Hill [0395 5605]. Backfilled opencast area of Victoria to Busty coals at St Andrew Opencast Site.

1747–50 Hown's Quarry [0974 4893; 0976 4890; 0978 4890] near Consett. Sandstone quarries on east side of Hown's Gill, with galleries into main rock face.

1754 Quarry [0770 4950] north-west of Castleside exposing predominantly thick-bedded medium- to coarse-grained sandstone, locally rotted.

1759 Mountsett Quarry [166 551] near Tantobie. Flaggy, sandstone at top of Durham Low Main Post and near top of former quarry face. Joints gaping due to old coal workings beneath.

1760 Mountsett Quarry [166 551] near Tantobie. Disturbed sandstone and fault line near top of former quarry face.

1761 Busty Bank Quarry [1748 5720] near Burnopfield. Quarry on west side of road between Rowlands Gill and Burnopfield, exposing medium- and coarse-grained sandstone, locally pebbly.

1762 Causey Burn [2011 5589] at Causey Arch near Causey. Thick-bedded, locally cross-bedded, medium- to coarse-grained sandstone overlying Durham Low Main and Top Brass Thill coals.

2537 Greymare Hill. General view [0530 5225] from near Shotleyfield.

2541 Snow's Green Burn [0926 5338] upstream from Snow's Greenburn Bridge. Sandstone forming part of the 'Third Grit' overlying sandy shales, bright coal and seatearth.

QUATERNARY

1727 Dipton Burn [NY 9848 6021] south-west of Riding Mill. Glacial drainage channel excavated through 'First Grit' and underlying strata.

1728–30 Styford Gravel Pit [0140 6315; 0120 6350], south-east of Corbridge. Ill-sorted, cross-bedded, interbedded sands and gravels.

1731–32 Eddy's Bridge Farm [0405 5073; 0413 5100], Derwent Valley near Edmundbyers. Alluvium and river terrace deposits exposed during 'cut and cover' excavations for part of the Transfer Works (Kielder Water Scheme) at the Derwent Valley crossing.

1734 Eddy's Bridge Farm [0415 5152], Derwent Valley near Edmundbyers. Terrace deposits on north side of valley; scarps of Stainmore Group sandstones in background.

1743–46 Hown's Gill [0903 4941; 0953 4916; 0958 4900], Consett. Various aspects of Hown's Gill glacial drainage channel including views taken both up and down channel from top of Hown's Gill Viaduct.

1757 Edmundbyers–Blanchland road [0086 5140] near Pow Hill. Morainic mounds of glacially transported material.

1758 Bashaw Bank [0401 4991] near Muggleswick. Mounds of clayey sand and gravel; also slipped mass of sandstone.

1763 Hamsterley–Rowlands Gill road [1473 5702] near Lintzford. Alluvium and terrace flats of Derwent Valley.

1764 Lintzford Lane [1500 5766] near Rowlands Gill. Glacial sand and gravel. Hummocky topography.

1765 View [1734 5975] over Derwent Valley near Rowlands Gill. Hummocky topography of glacial sand and gravel in foreground.

1786–94 Total panoramic view from Oxpasture Hill [1980 5520] north-east of Tanfield Lea showing heavily drift-covered ground including the sand and gravel deposits of the Beamish Burn and Causey areas.

2535 800 m south [0040 5188] of Hunter House on the south side of the Derwent Reservoir. Drumlins on south flank of Derwent Valley.

SCENERY AND GENERAL INTEREST

1735–37 Airy Holm Farm [0458 5424; 0460 5378] near Shotleyfield. Boulder clay mantling slopes underlain by sandstones and shales in the upper part of the Stainmore Group.

1740 View [0415 5600] looking north from Greymare Hill towards Tyne Valley; also shows route of Pumping Main from Riding Mill — part of the Transfer Works of the Kielder Water Scheme.

1751 Pemberton Road [0923 5102] near Consett. View of part of the old slag heaps associated with the former British Steel Corporation, Consett (now landscaped).

1752 The former British Steel Corporation site at Consett; view [0757 5049] from near Allensford.

1753 The former British Steel Corporation site at Consett; view [0779 4963] from near Castleside.

1766 Castleside Quarry [077 495] looking east towards former British Steel Corporation site at Consett.

1772–77 Various views [0464 4950 to 0518 5000] of incised meander and gorge of River Derwent near Muggleswick.

1778–79 General view [1448 5304] looking north from the Consett–Sunniside Road near Pontop towards Pontop Burn and the Derwent Valley.

1780–82 General view [1664 5516] looking between north-west and north-east from Mountsett Quarry near Tantobie.

1783 View [1840 5727] looking west from Byermoor towards Burnopfield.

1784–85 View [1643 5683] looking between north-west and north-east across Derwent Valley from High Friarside, Burnopfield.

2681 Causey Arch [2013 5588], Causey Burn, near Causey.

2970 View [NY 9840 5410] looking south-east from the Kilnpit Hill to Slaley road across Derwent Valley and Reservoir.

2971 Airy Holm Dam and Headpond [043 539], near Shotleyfield; a component part of the Transfer Works of the Kielder Water Scheme.

In addition to the above, the following form part of an older collection of black and white photographs taken in the late 1940's. They all belong to Series N and may be supplied as black and white prints or lantern slides. Some of these photographs are of considerable historical interest (indicated below by an asterisk*).

STAINMORE GROUP

12 Newton Hall Quarry near Newton. Grindstone (Newton) Limestone [c.037 653].

80 Bank of River Tyne, east of Styford Hall. Styford Limestone [c.029 621].

82 Roadside quarry, 1/4 mile south-west of Pow Hill on Edmundbyers-Blanchland road. Slipped mass of 'Second Grit' [c.008 514].

83 Berry Bank Quarry, north of Edmundbyers. Medium-grained sandstone with irony concretions in 'Second Grit' [c.013 507].

84 Burnhope Burn, 1/4 mile east of Edmundbyers. Marine beds in upper part of Stainmore Group. Exact location not known [NZ 05 SW].

101 Sandstone ('Second Grit') in gorge of the River Derwent near Shotley Bridge [c.088 522].

102 Crag consisting of cross-bedded sandstone ('First Grit') on the south bank of the River Derwent. Exposure partly landslipped. West of Allensford [c.076 500].

COAL MEASURES

13* Heddon-on-the Wall Quarry. Face of medium- to coarse-grained cross-bedded sandstone and referred to 'Third Grit'. Stone used in the construction of Newcastle Central Railway Station [c.130 668].

72 Road cutting at Houghton, $^{1}/_{2}$ mile west of Heddon-on-the-Wall. Sandstones near the base of Lower Coal Measures [c.126 668].

73 Quarry south of Corbridge road, west of Heddon-on-the-Wall. Cross-bedded sandstone near base of Lower Coal Measures [c.129 668].

78–79* Bearl Sandstone Quarry, near Ovington. 'Third Grit' [c.053 641].

91 Causey Arch and Causey (Houghwell) Burn in deep rock gorge near Causey [c.201 559].

94–96* Hown's Quarry, Hown's Gill, near Consett. Sandstone quarry in Lower Coal Measures showing evenly bedded sandstones and shales overlying massive sandstone, including galleries cut into the face [098 489].

QUATERNARY

71 Old sand and clay pits. Glacial sand and gravel overlying clay; north side of Tyne Valley, near Newburn [c.174 653].

85 West side of road just west of hamlet of Muggleswick. Mound of clayey sand and gravel and slipped mass of sandstone [c.040 499].

86 Moor south of Edmundbyers–Blanchland road. Morainic mounds and slipped material under 'Second Grit' escarpment [c.008 514].

87* Team Valley Trading Estate, Gateshead, c.1949.Flat-bottomed valley of the River Team. [c.244 598]. Alluvium and laminated clay.

88–90* Panoramic views from Oxpasture Hill, north-east of Tanfield Lea, across heavily drift-covered ground including sand and gravel [1980 5520].

92* Hown's Gill, near Consett; view from north-west [c.093 493].

97* Hown's Gill, near Consett; view down glacial overflow channel from top of Hown's Gill viaduct [096 491].

98–99* View across Derwent Valley from south of Ebchester looking towards sand and gravel deposits in vicinity of Broad Oak Farm [c.113 558].

100 Postglacial gorge of the Cong Burn near Waldridge, 1 mile west of Chester-le-Street [c.245 500].

103 Postglacial gorge of the Cong Burn, Waldridge Fell, $1^{1}/_{2}$ miles west of Chester-le-Street [c.240 497].

104 Glacial drainage channel, near Edmondsley; looking south towards Sacriston [c.237 487].

105 Glacial drainage channel, near Edmondsley; looking west up valley [c.237 487].

106* Glacial drainage channel (now filled in by spoil from opencast mining). Charlaw Fell, 1 mile west of Edmondsley. Exact site unknown (west side of NZ 24 NW).

107–109 Glacial overflow channels and panorama on the northern flanks of the Twizell Burn near West Pelton [c.225 526].

SCENERY AND GENERAL INTEREST

93* Consett Ironworks from near Castleside [c.090 495].

110 Topography of the Beamish Burn Valley, taken from near High Urpeth [c.234 540].

111–112 Coldwell Hill, north-east of West Wylam [c.115 641].

113 Bradley Park, west of Crawcrook [c.125 630].

AUTHOR CITATIONS FOR FOSSIL SPECIES

To satisfy the rules and recommendations of the international codes of botanical and zoological nomenclature, authors of cited species are listed below.

The British Geological Survey holds a large and varied collection of fossil fauna and flora from the Newcastle district, including samples from outcrops and numerous boreholes acquired over a period of many decades. This material is stored at its Keyworth, Nottingham, headquarters and is open to examination on request. Many of the samples remain unidentified or older identifications require modern scrutiny. Only a small part of the collection has been taken into account during the preparation of this memoir, sufficient only to support the stratigraphical classification and description of the rocks. Thus, this is not a comprehensive list of the Carboniferous fauna and flora of the district.

Chapter 3

INDEX

BRITISH GEOLOGICAL SURVEY

Keyworth, Nottingham NG12 5GG
0115 936 3100

Murchison House, West Mains Road, Edinburgh
EH9 3LA 0131-667 1000

London Information Office, Natural History Museum
Earth Galleries, Exhibition Road, London SW7 2DE
0171-589 4090

The full range of Survey publications is available through the Sales Desks at Keyworth and at Murchison House, Edinburgh, and in the BGS London Information Office in the Natural History Museum (Earth Galleries). The adjacent bookshop stocks the more popular books for sale over the counter. Most BGS books and reports can be bought from The Stationery Office and through Stationery Office agents and retailers. Maps are listed in the BGS Map Catalogue, and can be bought together with books and reports through BGS-approved stockists and agents as well as direct from BGS.

The British Geological Survey carries out the geological survey of Great Britain and Northern Ireland (the latter as an agency service for the government of Northern Ireland), and of the surrounding continental shelf, as well as its basic research projects. It also undertakes programmes of British technical aid in geology in developing countries as arranged by the Department for International Development and other agencies.

The British Geological Survey is a component body of the Natural Environment Research Council.

Published by The Stationery Office and available from:

The Publications Centre
(mail, telephone and fax orders only)
PO Box 276, London SW8 5DT
General enquiries 0171 873 0011
Telephone orders 0171 873 9090
Fax orders 0171 873 8200

The Stationery Office Bookshops
123 Kingsway, London WC2B 6PQ
0171 242 6393 Fax 0171 242 6394
68–69 Bull Street, Birmingham B4 6AD
0121 236 9696 Fax 0121 236 9699
33 Wine Street, Bristol BS1 2BQ
0117 9264306 Fax 0117 9294515
9–21 Princess Street, Manchester M60 8AS
0161 834 7201 Fax 0161 833 0634
16 Arthur Street, Belfast BT1 4GD
01232 238451 Fax 01232 235401
The Stationery Office Oriel Bookshop
The Friary, Cardiff CF1 4AA
01222 395548 Fax 01222 384347
71 Lothian Road, Edinburgh EH3 9AZ
0131 228 4181 Fax 0131 622 7017

The Stationery Office's Accredited Agents
(see Yellow Pages)

and through good booksellers